Estuaries of the World

Series editor

Jean-Paul Ducrotoy

For further volumes:
http://www.springer.com/series/11705

Mudflats in the Banc d'Arguin National Park, functional delta derived from a fossil estuary, Mauritania—Credit: JF Hellio et N Van Ingen/FIBA Special thanks to Mathieu Ducrocq, regional coordinator PACO/Programme Marin et Côtier (MACO)/Union Internationale pour la Conservation de la Nature (UICN)—Dakar—Sénégal, for providing this picture

Salif Diop • Jean-Paul Barusseau
Cyr Descamps
Editors

The Land/Ocean Interactions in the Coastal Zone of West and Central Africa

Editors
Salif Diop
Cheikh Anta Diop University
Dakar
Senegal

Jean-Paul Barusseau
University of Perpignan
Perpignan
France

Cyr Descamps
University of Perpignan
Perpignan
France

ISSN 2214-1553 ISSN 2214-1561 (electroninc)
ISBN 978-3-319-06387-4 ISBN 978-3-319-06388-1 (eBook)
DOI 10.1007/978-3-319-06388-1
Springer Cham Heidelberg New York Dordrecht London

Library of Congress Control Number: 2014940389

© Springer International Publishing Switzerland 2014
This work is subject to copyright. All rights are reserved by the Publisher, whether the whole or part of the material is concerned, specifically the rights of translation, reprinting, reuse of illustrations, recitation, broadcasting, reproduction on microfilms or in any other physical way, and transmission or information storage and retrieval, electronic adaptation, computer software, or by similar or dissimilar methodology now known or hereafter developed. Exempted from this legal reservation are brief excerpts in connection with reviews or scholarly analysis or material supplied specifically for the purpose of being entered and executed on a computer system, for exclusive use by the purchaser of the work. Duplication of this publication or parts thereof is permitted only under the provisions of the Copyright Law of the Publisher's location, in its current version, and permission for use must always be obtained from Springer. Permissions for use may be obtained through RightsLink at the Copyright Clearance Center. Violations are liable to prosecution under the respective Copyright Law.
The use of general descriptive names, registered names, trademarks, service marks, etc. in this publication does not imply, even in the absence of a specific statement, that such names are exempt from the relevant protective laws and regulations and therefore free for general use.
While the advice and information in this book are believed to be true and accurate at the date of publication, neither the authors nor the editors nor the publisher can accept any legal responsibility for any errors or omissions that may be made. The publisher makes no warranty, express or implied, with respect to the material contained herein.

Printed on acid-free paper

Springer is part of Springer Science+Business Media (www.springer.com)

Mandela

This book is dedicated to Nelson Rolihlahla Mandela "Madiba" (1918–2013), The First post-apartheid President of South Africa. As Calestous Juma, Professor at Harvard University; USA, used to say: "Mandela will be remembered as one of the greatest leaders of all time. One of the best ways to live up to his loftiest aspirations for Africa are to give future generations science and technology education that gives them the skills to expand their economic opportunity"

Mandela's direct involvement in science may be seen as linked to his firm belief about the power of education in building democracy and development. He said, "Education is the most powerful weapon which you can use to change the world". Motivated by this concern, Mandela lent his name to the creation of a new generation of African Institutes of Science and Technology, seen as the beginning of a new generation of African research universities. Two have already been established, in Tanzania and Nigeria

To quote Dr. Ismail Serageldin, Director of the Bibliotheca Alexandrina in Egypt "Nelson Mandela was undoubtedly one of those immortal leaders. He now belongs to history, but we are fortunate to have lived in his time and to have been witness to his magic allure, his saintly demeanor, his twinkling mischievous eyes, his humor and his wisdom. We have witnessed his mind and his heart at work, and admired his unique combination of political genius and human warmth, his vision of the Rainbow Nation and how to make it a reality"

May Madiba rest in Peace, and may his legacy live forever

Foreword

The coastline of western and central Africa is made up of diverse marine and coastal ecosystems, such as estuaries, mangrove swamps and forests and offshore cold water springs. Although the focus of this publication is on the estuaries, as part of the "Estuaries of the World" series, its scope goes well beyond this particular coastal feature. Indeed, the estuary can only be considered as part of the life cycle of the entire river and the marine area it feeds into: an area particularly subject to human and natural pressures. Land degradation upstream, sea level rise downstream, salinization and drought, overexploitation of fisheries and mangroves, are all issues whose impact is felt at the river's mouth.

The vital role these estuaries play in the ecosystem of the region has been recognised in the creation of a number of protected areas, natural parks and reserves. Specific habitats such as mangroves, sea grass beds and sand banks provide refuge to many endangered species, and cover the flight path of most of the migratory birds of West and Central Africa.

The main estuaries and deltas of this region provide a variety of goods and services to its coastal population. The most important of them are related to critical fish habitat, wood and charcoal from mangroves, as well as space for agriculture, aquaculture, urban development, tourism and transport.

Mangroves, in particular, play a significant role in terms of flood control, groundwater replenishment, coastline stabilisation and protection against storms. They also retain sediments and nutrients, purify water and provide critical carbon storage. Such hydrological and ecological functions explain the focus on serving mangrove ecosystems and the nearby communities, which draw significant income from fishing, rice production, tourism, salt extraction and other activities such as harvesting honey and medicinal plants.

However, in recent decades, population growth, environmental degradation and climate change have led to an erosion of the biodiversity in these ecosystems. Resources, as a result, are becoming scarce, and the pressure upon local communities is increasing.

There is a need to focus and to prioritise research and data to help manage and protect sustainable estuaries in the region, for the benefit of future generations. A number of international and regional programmes have been undertaken to address the critical issues and to find appropriate solutions. Among the principal programmes involved are the UNEP Regional Seas and the Intergovernmental Oceanographic Commissions, and other IGO and NGO programmes.

There is still much to be done to achieve the goal of restoring and protecting the resource-rich estuarine and other coastal water ecosystems in West and Central Africa. This new publication constitutes a first step towards that goal, bringing together new and updated information, including maps, models, new data and knowledge on recent changes and evolution, and their implications in the management of coastal waters in the region.

The close cooperation between Prof. Salif Diop, Prof. Jean-Paul Barusseau, Prof. Cyr Descamps and Prof. Jean-Paul Ducrotoy has been the driving force behind this publication. The Editors of this book would like to thank all the authors who contributed their time, resources and expertise.

It is my firm belief that this book will provide important and up-to-date information essential for the public at large but more specifically for scientists, researchers, managers, decision makers all working together in order to safeguard, protect and sustainably manage estuaries, deltas and lagoons, and the coastal and ocean waters of Western and Central Africa.

Achim Steiner

Preface

Why are West African estuaries so important in land/sea interactions?

One of the major challenges that humans face today is the management of estuaries so that future generations can also enjoy the remarkable visual, cultural and food products that they provide. The book series "Estuaries of the World" (EOTW) by Springer uses a multidisciplinary approach in presenting the science of estuaries. Such an approach presupposes that all users of the environment can share views and are able to communicate effectively on the basis of robust science.

Estuaries are vulnerable because they are exposed to multiple human activities such as fish and shrimp farming, industrial and domestic pollution, dredging, land reclamation and agriculture in the watershed. The threat to coastal ecosystems posed by human activities is well recognised and documented, yet the mitigation of human impact remains a major challenge due to a lack of understanding of the scale and rate of observed changes. Mangroves, for instance, are subject to clear-cutting and overlogging and such disturbances increase the variability of natural systems. The variability of natural systems is difficult to include in any political agenda due to the certainty of information required for decision making. It is possible, however, to better understand how humans change the way in which ecosystems function using a combination of different approaches aimed at combining functional ecology studies and a pressure/risk assessment approach (both on ecological and socio-economic aspects). In this way, it is possible to integrate the novel and interdisciplinary scientific evidence of multiple research disciplines. Such a dynamic interplay between theory and empirical study forms the basis for the transdisciplinary approach of the EOTW series.

With this perspective in mind, it is important to assess the capacity of ecosystems in fulfilling their role within the biosphere. Integration can be seen as one of the tools or methodologies for realising this goal by encompassing all aspects of an issue through a collaborative approach between natural sciences and economic, socio-cultural, legal and institutional disciplines. Integrated Coastal Zone Management (ICZM) is still a relatively new and evolving concept and there is no consensus regarding issues such as the fundamental nature and structure of the coastal zone, the most appropriate timescales for the application of ICZM policies, or the key criteria for defining sustainability in coastal zone development. Integration needs to be established between disciplines, sectors and in governance across the land–water interface. Through improving the scientific understanding of the performance of coastal ecosystems in terms of fluxes of energy and matter in relation to human impacts, ICZM should be able to predict the effects of measures taken and find responses to the fast evolving demands from society. The EOTW series offers a framework for facilitating such integration.

The notion of ecosystem services is useful in that it provides insight into the resilience of ecosystems and how changes affect them. The reduction in marine biodiversity and productivity is multifactorial, especially in coastal waters. Direct habitat destruction through the erection of engineering and drainage works, which disturb the physical integrity of coastal and marine systems is the most drastic, as the habitat itself is changed to a point where the ecosystem loses its identity and assumes a different function. Poor fisheries management, including the uncontrolled exploitation of corals and molluscs and the by-catch of large numbers of non-target species in fisheries, is another pertinent example of detrimental marine resource exploitation. An integrated approach to coastal zone management of fisheries is predicted to prevent impoverished functioning of such ecosystems. The consequence of unchecked exploitation is that the productivity of fisheries and important ecosystems, such as mangroves and coral reefs, reduces which in turn causes suffering for the affected local communities.

In general, estuaries and salt marshes, mangrove forests and sea grass beds near cities and towns are severely degraded worldwide with many species now threatened to become extinct in the near future. Found in tropical and subtropical regions, mangroves are especially vulnerable. These salt-tolerant forested wetlands at the sea–land interface form the link between the terrestrial landscapes and the marine environment. Rapid changes in anthropogenic activities in coastal zones impact on the structure of organism populations, which will in turn affect the geochemical cycles of the ecosystem, to a point where such cycles might become dysfunctional. Changes in costal ecosystems can lead to an imbalance in fluxes of energy and minerals at the interface between land and sea. These localised changes have the propensity to reach a global level. The dynamics of such systems are complex and conservation should address all aspects of this complexity and not solely focus on fixing the coastline to its physical limits, or preventing erosion and sea level rise. Because costal systems are alive, they are able to cope with a multitude of changes. The critical determinant of an ecosystem's capacity to cope with change, however, is the rate of change, and it is the rapidity of change inflicted by humans to natural systems, which makes the anthropocene unique.

This volume in the EOTW series offers case studies in West and Central Africa and demonstrates that mangrove ecosystems are extremely valuable in mitigating effects from deleterious human activities, providing ecosystem services like carbon sequestration, protection from storms, floods and erosion, processing of waste and nutrient pollution, aquaculture and agriculture support and a refuge for aquatic and terrestrial species.

In order to discriminate between global and local influences, it is essential to acquire an in-depth knowledge of natural processes, as well as understand relevant institutional, cultural, economic, social and political frameworks based on a robust scientific approach. Suitable studies have been developed and used to analyse causal linkages within West African coastal ecosystems, forecasting the effects of acute or chronic interference on resource use, and to address wider, management-related issues such as the restoration of damaged habitats and the potential for aquaculture. The context of natural resource management in West Africa is complex. If the elements of ecosystems are interconnected and interdependent, those of regional environmental systems are even more so. Thus, the work as presented in this volume of the EOTW series contributes to improve the understanding of the dynamics and functioning of coastal ecosystems and habitats, including mangrove forests that constitute the most apparent features along western and central African coasts. Considering the highly threatened nature of marine and coastal ecosystems in this part of Africa and bearing in mind that the major drivers of change, degradation and loss of marine and coastal ecosystems and services are mainly anthropogenic, the question will be what types of options exist to respond to such challenges? By all means, addressing uncertainties and elaborating trade-offs could provide useful mechanisms for operational responses and this should be undertaken through established ecosystem-based approaches and improving the capacity of scientists to predict the consequences of the change of drivers in marine and coastal ecosystems. In this regard, long-term ecological processes and further research are needed in a number of areas in order to improve sustainable management policies of coastal and marine ecosystems of West and Central Africa.

The complex problems caused by human–environment interactions occur within the intricate structure of ecosystems, which are in a natural state of constant flux and change. This book explores the complex problems caused by human–environment interactions within the naturally and artificially fluctuating and changing coastal ecosystems of West and Central Africa. The authors have shared their knowledge and experience on ecological, social and cultural aspects simultaneously. This interdisciplinary approach makes the discovery of this fascinating region even more enriching.

Dakar, Senegal   Salif Diop
Hull, UK   Jean-Paul Ducrotoy

Acknowledgments

The authors would like to thank those experts involved in numerous marine and coastal programmes from various parts of the world and who have peer reviewed this overall publication, namely: François Blasco, Joan Fabres, Hartwig Kremer, Jacqueline Alder, Koranteng Kwame, Lorna Inniss, Marc Steyaert, Alan Simcock, Eric Wolanski, Beatrice Padovani Ferreira, Rice Jake, Johnson U. Kitheka, Enrique marschoff, Peyman Eghtesadi Araghi, Luiz Drude de Lacerda, Kalifa Goïta and Peter Scheren.

A special note of thanks should be dedicated to M. Ibrahim Thiaw, Deputy-Executive Director of the United Nations Environment Programme for his constant support during the preparation of this book, to Peter Saunders from UK who has proceeded for the English pre-editing of all articles prior to sending them to peer reviewers and to Walter Rast, Professor Emeritus and Director, Texas State University, USA for his extensive peer review of important chapters of the book. I would like to thank as well M. Taibou Ba, from the "Centre de Suivi Ecologique—Dakar—Senegal", for redesigning some of the maps contained in this publication, Joana Akrofi and Matthew Billot, from Scientific Assessment Branch, Saly Sambou and Birane Cisse, students at Doctoral Level in University Cheikh Anta Diop/CAD of Dakar—Senegal and Awa Niang, Senior Lecturer at the same University CAD/Dakar. Finally, the authors would also like to thank UNEP and especially colleagues from the Division of Early Warning Assessment (DEWA) and the Division of Environmental Policy and Implementation (DEPI) for their contribution during the preparation process of this volume dedicated to the African continent.

Contents

The Western and Central Africa Land–Sea Interface: A Vulnerable, Threatened, and Important Coastal Zone Within a Changing Environment.... 1
S. Diop, J. Fabres, R. Pravettoni, J.-P. Barusseau, C. Descamps, and J.-P. Ducrotoy

West African Coastal Area: Challenges and Outlook 9
Jean-Jacques Goussard and Mathieu Ducrocq

Morphological and Hydrodynamic Changes in the Lower Estuary of the Senegal River: Effects on the Environment of the Breach of the 'Langue De Barbarie' Sand Spit in 2003 23
Awa Niang and Alioune Kane

Management of a Tropical River: Impacts on the Resilience of the Senegal River Estuary 41
Coura Kane, Alioune Kane, and Joël Humbert

Combined Uses of Supervised Classification and Normalized Difference Vegetation Index Techniques to Monitor Land Degradation in the Saloum Saline Estuary System 49
Ndeye Maguette Dieng, Joel Dinis, Serigne Faye, Marçia Gonçalves, and Mário Caetano

Studies and Transactions on Pollution Assessment of the Lagos Lagoon System, Nigeria. 65
Babajide Alo, Kehinde Olayinka, Aderonke Oyeyiola, Temilola Oluseyi, Rose Alani, and Akeem Abayomi

Estuarine and Ocean Circulation Dynamics in the Niger Delta, Nigeria: Implications for Oil Spill and Pollution Management 77
Larry Awosika and Regina Folorunsho

Morphological Characteristics of the Bonny and Cross River (Calabar) Estuaries in Nigeria: Implications for Navigation and Environmental Hazards 87
Regina Folorunsho and Larry Awosika

Status of Large Marine Flagship Faunal Diversity Within Cameroon Estuaries of Central African Coast 97
Isidore Ayissi, Gordon N. Ajonina, and Hyacinthe Angoni

Morphology Analysis of Niger Delta Shoreline and Estuaries for Ecotourism Potential in Nigeria 109
O. Adeaga

Importance of Mangrove Litter Production in the Protection of Atlantic Coastal Forest of Cameroon and Ghana 123
Sylvie Carole Ondo Ntyam, A. Kojo Armah, Gordon N. Ajonina,
Wiafe George, J. K. Adomako, Nyarko Elvis, and Benjamin O. Obiang

Carbon Budget as a Tool for Assessing Mangrove Forests Degradation in the Western, Coastal Wetlands Complex (Ramsar Site 1017) of Southern Benin, West Africa 139
Gordon N. Ajonina, Expedit Evariste Ago, Gautier Amoussou,
Eugene Diyouke Mibog, Is Deen Akambi, and Eunice Dossa

Mangrove Conditions as Indicator for Potential Payment for Ecosystem Services in Some Estuaries of Western Region of Ghana, West Africa 151
Gordon N. Ajonina, Tundi Agardy, Winnie Lau, Kofi Agbogah,
and Balertey Gormey

Plantation Agriculture as a Driver of Deforestation and Degradation of Central African Coastal Estuarine Forest Landscape of South-Western Cameroon 167
Patience U. Ajonina, Francis A. Adesina, and Oluwagbenga O. I. Orimoogunje

Assessment of Mangrove Carbon Stocks in Cameroon, Gabon, the Republic of Congo (RoC) and the Democratic Republic of Congo (DRC) Including their Potential for Reducing Emissions from Deforestation and Forest Degradation (REDD+) 177
Gordon N. Ajonina, James Kairo, Gabriel Grimsditch, Thomas Sembres,
George Chuyong, and Eugene Diyouke

Governing Through Networks: Working Toward a Sustainable Management of West Africa's Coastal Mangrove Ecosystems 191
Dominique Duval-Diop, Ahmed Senhoury, and Pierre Campredon

The Importance of Scientific Knowledge as Support to Protection, Conservation and Management of West and Central African Estuaries 207
S. Diop, J.-P. Barusseau, and C. Descamps

Index 209

Contributors

Akeem Abayomi Analytical and Environmental Research Group, Department of Chemistry, University of Lagos, Lagos, Nigeria

O. Adeaga Department of Geography, University of Lagos, Lagos, Nigeria

Francis A. Adesina Department of Geography, Obafemi Awolowo University, Ile-Ife, Nigeria

J. K. Adomako Department of Botany, University of Ghana, Legon, Ghana

Tundi Agardy The Marine Ecosystem Service (MARES) Programme, Washington, DC, USA

Kofi Agbogah Coastal Resources Center, Takoradi, Ghana

Expedit Evariste Ago Unit of Biosystem Physics, University of Liege Gembloux Agro-Bio Tech (GxABT), Gembloux, Belgium

Gordon N. Ajonina CWCS Coastal Forests and Mangrove Programme, Mouanko, Littoral Region, Cameroon; Institute of Fisheries and Aquatic Sciences, University of Douala (Yabassi), Douala, Cameroon

Patience U. Ajonina Department of Geography, University of Buea, Buea, Cameroon

Is Deen Akambi Benin Ecotourism Concern (Eco-Benin), Zogbadjè, Rue début Clôture IITA, Jéricho, Benin

Rose Alani Analytical and Environmental Research Group, Department of Chemistry, University of Lagos, Lagos, Nigeria

Babajide Alo Analytical and Environmental Research Group, Department of Chemistry, University of Lagos, Lagos, Nigeria

Gautier Amoussou Benin Ecotourism Concern (Eco-Benin), Zogbadjè, Rue début Clôture IITA, Jéricho, Benin

Hyacinthe Angoni Faculty of Science, Department of Plant Biology, University of Yaounde I, Yaounde, Cameroon

Larry Awosika Nigerian Institute for Oceanography and Marine Research, Victoria Island, Lagos, Nigeria

Isidore Ayissi Cameroon Marine Biology Association (CMBA), Ayos, Cameroon; Specialized Research Center for Marine Ecosystems (CERECOMA), Institute of Agricultural Research for Development, Kribi, Cameroon; Institute of Fisheries and Aquatic Sciences, University of Douala, Yabassi, Cameroon

J.-P. Barusseau CEFREM, Universite de Perpignan, Perpignan, France

Mário Caetano Higher Institute of Statistics and Information Management, New University of Lisbon, Lisbon, Portugal

Pierre Campredon International Union for Conservation of Nature (UICN), Bissau, Republic of Guinea-Bissau

George Chuyong University of Buea, Buea, Cameroon

C. Descamps Maitre de Conférences Emeritus, Institut Fondamental d'Afrique Noire—Cheikh Anta Diop, Dakar, Senegal

Ndeye Maguette Dieng Geology Department, Faculty of Sciences and Techniques University Cheikh Anta Diop, Dakar, Senegal

Joel Dinis Higher Institute of Statistics and Information Management, New University of Lisbon, Lisbon, Portugal

S. Diop Universite Cheikh Anta Diop de Dakar, Dakar-Fann, Senegal

Eunice Dossa Benin Ecotourism Concern (Eco-Benin), Zogbadjè, Rue début Clôture IITA, Jéricho, Benin

Mathieu Ducrocq Coastal Ecosystems Group of the Commission on Ecosystem Management, IUCN, Gland, Switzerland

Dominique Duval-Diop West African Regional Network of Marine Protected Areas (RAMPAO), Dakar, Senegal

Nyarko Elvis Department of Marine and Fisheries Sciences, University of Ghana, Legon, Ghana

J. Fabres GRID-Arendal, Arendal, Norway

Serigne Faye Geology Department, Faculty of Sciences and Techniques University Cheikh Anta Diop, Dakar, Senegal

Regina Folorunsho Nigerian Institute for Oceanography and Marine Research, Victoria Island, Lagos, Nigeria

Wiafe George Department of Marine and Fisheries Sciences, University of Ghana, Legon, Ghana

Marçia Gonçalves Remote Sensing Unit of the Portuguese Geographic Institute, Lisbon, Portugal

Balertey Gormey Coastal Resources Center, Takoradi, Ghana

Jean-Jacques Goussard Coastal Ecosystems Group of the Commission on Ecosystem Management, IUCN, Gland, Switzerland

Gabriel Grimsditch UNEP, Nairobi, Kenya

Joël Humbert Laboratoire Image, Ville, Environnement—UMR 7362, CNRS, Université de Strasbourg, Strasbourg cedex, France

James Kairo Kenya Marine and Fisheries Research Institute, Mombasa, Kenya

Alioune Kane Ecole Doctorale «Eau, Qualité et Usages de l'Eau» (EDEQUE), Université Cheikh Anta Diop de Dakar, Dakar, Sénégal; Laboratoire LINUS «Littoraux: Interface Natures-Sociétés», Unité Mixte Internationale (Joint International Unit) 236 Résiliences, Institut de Recherche pour le Dévelopement (Research Institute for Development) Campus International de Dakar-Hann, Dakar, Sénégal

A. Kojo Armah Department of Marine and Fisheries Sciences, University of Ghana, Legon, Ghana

Winnie Lau The Marine Ecosystem Service (MARES) Programme, Washington, DC, USA

Eugene Diyouke Mibog CWCS Coastal Forests and Mangrove Programme, Mouanko, Cameroon

Awa Niang Ecole Doctorale «Eau, Qualité et Usages de l'Eau» (EDEQUE), Université Cheikh Anta Diop de Dakar, Dakar, Senegal; Laboratoire LINUS «Littoraux: Interface Natures-Sociétés», Unité Mixte Internationale (Joint International Unit) 236 Résiliences, Institut de Recherche pour le Dévelopement (Research Institute for Development) Campus International de Dakar-Hann, Dakar, Senegal

Sylvie Carole Ondo Ntyam Department of Marine and Fisheries Sciences, University of Ghana, Legon, Ghana; Centre for Coastal and Marine Research (CERECOMA/IRAD), Kribi, Cameroon

Benjamin O. Obiang CEPFILD Circle of Forest Promotion and Local Initiatives Development, Kribi, Cameroon

Kehinde Olayinka Analytical and Environmental Research Group, Department of Chemistry, University of Lagos, Lagos, Nigeria

Temilola Oluseyi Analytical and Environmental Research Group, Department of Chemistry, University of Lagos, Lagos, Nigeria

Oluwagbenga O. I. Orimoogunje Department of Geography, Obafemi Awolowo University, Ile-Ife, Nigeria

Aderonke Oyeyiola Analytical and Environmental Research Group, Department of Chemistry, University of Lagos, Lagos, Nigeria

R. Pravettoni GRID-Arendal, Arendal, Norway

Thomas Sembres UNEP, Nairobi, Kenya

Ahmed Senhoury Mobilization and Coordination Unit Regional Partnership for the Conservation of the West African Coastal and Marine Zone (PRCM), Nouakchott, Mauritania

About the Editors

Prof. S. Diop, who is a water expert with a number of referred related publications and broad experience in various aspects of scientific assessment of freshwater, coastal and marine resources as well as in areas related to the management and sustainable development of the environment. So far, he has pursued the coordination of important international programmes on freshwater, marine and coastal processes including scientific assessment of marine and coastal waters, assessment of groundwater resources in Africa and in other regions of the world, evaluation processes and scientific assessment of the world oceans and coasts; geosphere–biosphere interactions; processes on integrated management of marine areas and coastal interfaces land/sea/water and atmosphere, assessment of water resources as a key factor for sustainable development, the development of modules for the evaluation of freshwater resources, wetlands, marine waters and coastal areas, including the relations and impacts of climate change.

Prof. Salif Diop, Professor of University, UCAD
Member National Academy of Sciences and Techniques of Senegal (ANSTS) and African Academy of Sciences (AAS)
The World Academy of Sciences for the Advancement of Sciences in Developing Countries (TWAS)

Dr. Jean-Paul Barusseau, is attached to the centre of education, training and research on marine environments (CEFREM). He is a geologist sedimentologist, specialised in studies of the coastal zone. His interest is directed towards the history of these environments during the Late Quaternary, especially the second half of the Holocene, and their dynamic evolution in the Recent with a particular interest in the issue of erosion of sandy coasts. His work has been conducted in France (Bay of Biscay, Gulf of Lions) and Africa, particularly in Senegal and Mauritania. In Senegal, his work has concerned the Petite Côte, the post-dam evolution of Senegal River Valley and the Saloum river delta; in 2007–2009, he participated in the revision of the geological map, such as 1/200000 sheets of the river valley and 1/50000 sheets of the Petite Côte (PASMI program). In Mauritania, he participated in the study of the Banc d'Arguin area in order to present a land–

sea continuum of the coastal sedimentary unit evolution, from the coastal plain to the marine basin during the Late Holocene (PACOBA program). His work provided 156 publications. He was the director of 12 Ph.D. theses.

> Prof. Emeritus Jean-Paul Barusseau
> CEFREM—University Via Domitia—Perpignan (France)

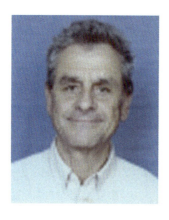

Prof. Cyr Descamps, lecturer in retirement, is attached to the France Centre for Historical Research on Mediterranean Societies (CRHiSM) at the University of Perpignan Via Domitia, and Senegal to the Fundamental Institute of Africa South of Sahara (IFAN) Ch. A. Diop University of Dakar. He participated in a series of pre-historical investigations in West African coastal regions including in estuaries, deltas and coastal lagoons, especially in the Senegal–Mauritania sub-region. One of his other focuses is the historical and pre-historical underwater investigations Prof. Descamps has undertaken in the Mediterranean. Prof. Descamps's key interests are people living in estuarine and deltaic areas who collect shellfish with significant accumulations on sandy shores, muddy and rocky. Ethno-archaeological approach and consideration of environmental parameters has allowed him to better understand the dynamics of these stands. He also participated in multidisciplinary work on changes of such historical and pre-historical sites at different timescales, bathymetry and coastal shorelines in tropical regions.

> Prof. Cyr Descamps Centre for Historical Research on Mediterranean Societies (CRHiSM)—University Via Domitia—Perpignan (France) and Fundamental Institute of Africa South of Sahara (IFAN) Ch. A. Diop
> University of Dakar—Senegal.

The Western and Central Africa Land–Sea Interface: A Vulnerable, Threatened, and Important Coastal Zone Within a Changing Environment

S. Diop, J. Fabres, R. Pravettoni, J.-P. Barusseau, C. Descamps, and J.-P. Ducrotoy

Abstract

The primary objective of this book, focusing on Western and Central African coastal areas, is to provide up-to-date scientific information and discuss quantitative data about selected estuaries and coastal ecosystems. As a volume of the series "Estuaries of the World" (EOTW), it is aiming at offering a better understanding of Land–Ocean Interactions in the Coastal Zone (LOICZ) in this region, including processes, functioning, and impacts in a changing environment. Although it focuses on estuarine and related coastal ecosystems, it includes consideration of the freshwater systems that feed these coastal environments. Indeed, the importance of the Western and Central African coastal areas cannot be overemphasized, with regard to both human and ecosystem needs, and watersheds have to be considered as influencing them directly and indirectly. Moreover, this up-to-date information and data discuss estuaries, deltas, and coastal lagoons, as well as their recent changes and evolution, and the implications of these changes in regard to managing these coastal waters. The content of this book also contributes to the global research analysis and synthesis agenda of the LOICZ especially on its primary focus related to river mouth systems, including estuaries and deltas. This book illustrates several themes organized around the concept of

S. Diop (✉)
Universite Cheikh Anta Diop de Dakar, B.P. 5346
Dakar-Fann, Senegal
e-mail: sal-fatd@orange.sn esalifdiop@gmail.com

J. Fabres · R. Pravettoni
GRID-Arendal, Postboks 183, 4802 Arendal, Norway
e-mail: joan.fabres@grida.no

R. Pravettoni
e-mail: riccardo.pravettoni@grida.no

J.-P. Barusseau
CEFREM, University Via Domitia—52, Avenue Paul-Alduy,
66860, Perpignan, France
e-mail: brs@univ-perp.fr

C. Descamps
Maitre de Conférences Emeritus, Institut Fondamental d'Afrique
Noire—Cheikh Anta Diop, B.P. 206, Dakar, Senegal
e-mail: cyrdescamps@yahoo.fr

J.-P. Ducrotoy
Institute of Estuarine and Coastal Studies, The University of Hull,
Cottingham Road, Hull, HU6 7RX, UK
e-mail: jean-paul.ducrotoy@hull.ac.uk

S. Diop et al. (eds.), *The Land/Ocean Interactions in the Coastal Zone of West and Central Africa*, Estuaries of the World, DOI: 10.1007/978-3-319-06388-1_1,
© Springer International Publishing Switzerland 2014

ecosystem services and benefits within estuaries, coastal, and marine areas in West and Central Africa, with six main objectives: (1) Introducing the framework of current studies, including factors related to oceanographic, geologic, geomorphologic, physicochemical and biogeochemical components of these ecosystems; (2) Emphasizing the need to study natural and human-induced impacts on the functioning and sustainability of estuaries and other coastal systems, with possible options for managing them for sustainable use; (3) Facilitating recognition by coastal scientists and managers of the unique features of estuaries and other relevant coastal environments in Africa, while also enhancing existing knowledge regarding the associated ecosystem services; (4) Encouraging young researchers, scientists, and advanced students to undertake holistic, integrated studies on estuaries and coastal areas in their regions, using an ecosystems assessment approach and ecosystem-based management; and (5) Providing professionals, students, and the general public with readily accessible and understandable scientific articles and papers on the economic and ecological importance of estuaries and other related coastal ecosystems in Africa.

Keywords

Ecosystem goods and services • Estuaries and coastal ecosystems • West and Central Africa • Ecosystem approach • Oceanographic • Geologic • Geomorphologic • Physicochemical • Biogeochemical • Socioeconomic factors

Introduction

The West and Central African coastline extends over around 6,000 km, from the shores of the sandy desert of Mauritania in the north to the lagoon areas and coastal belts of the Gulf of Guinea, including deeply cut coastlines of islands (for example, Guinea-Bissau and the Bissagos islands) and estuaries up to the Republic of Congo (Fig. 1). The immense Niger and Cross River delta and Congo River mouth are located at the eastern and southern end. The West African "ecomarine" region (more or less corresponding to the Canary Current Large Marine Ecosystem[1]) stretches along 3,500 km of coastline and covers six countries: Mauritania, Senegal, Gambia, Cape Verde, Guinea-Bissau, and Guinea (Diop 1990). The coastline exhibits a great variety of habitats, ranging from enormous extensions of seaweed prairies in the north, to rocky cliffs and long sandy beaches, to mangroves and well-developed estuaries in the south (UNEP-WCMC 2007). Beyond this ecoregion, the entire Guinea Current Large Marine Ecosystem is located to the south of Guinea, stretching as far as Nigeria, and well beyond Gabon, offshore of the two Congo's, Sao Tomé and Principe, and Equatorial Guinea (see chapter "Morphological and Hydrodynamic Changes in the Lower Estuary of the Senegal River: Effects on the Environment of the Breach of the 'Langue de Barbarie' Sand Spit in 2003").

The ecological importance of the Western and Central African coastal region and the adjacent uplands areas (from the continental margins to the offshore island nations), which contain a wide and varied range of important habitats and associated biota that exhibit a high biological diversity, cannot be overemphasized. The continental shelf along the coast is in general narrow, with widths between 20 and 50 km (but reaching up to 150 km in certain places like offshore Guinea-Bissau, Sierra Leone, and Ghana). One of the main characteristics of this African marine and coastal region is the occurrence of seasonal upwelling, which explains the abundance of substantial commercial stocks of demersal and pelagic fish in its coastal waters (World Bank Africa 1994). Indeed, the economy of most of the countries discussed in this report is highly dependent on their coasts and their marine environment.

Natural Conditions and Processes

Regional Morphology and River Basin Drainages

Four narrow coastal sedimentary basins, containing a few volcanic intrusions and other rocky outcrops at the major capes, have developed on the edges of the coastline, including the Senegalese-Mauritanian basin, the Cote d'Ivoire basin, the Niger basin (including Niger Delta), and the coastal basins from Gabon to Congo (UNEP 1999).

All four coastal sedimentary environments are strongly influenced by their river basin drainages. There are five major

---

[1] Large Marine Ecosystems encompass waters from river basins and estuaries to the seaward boundaries of continental shelves and margins of coastal currents and water masses (Sherman 1994).

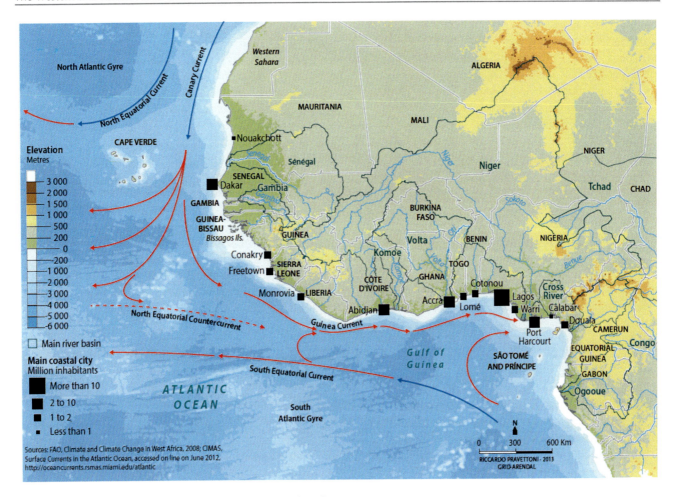

Fig. 1 Map of the coast of West Africa with main rivers and marine currents

river systems draining the entire coastline from Senegal to Congo. The most important rivers include the Niger, which drains an area of over 1 million km^2; the Volta, with a drainage basin of 390,000 km^2 (World Bank Africa 1994); and the Congo river, with the second largest mean annual runoff and catchment area in the world, freshwater and sediment discharge estimated to be 30–80 tons/km^2. For purposes of energy production, irrigation, and flood control, however, most of these rivers have been dammed, consequently significantly altering their hydrology and sediment flows, thereby causing inevitable downstream impacts and accelerating coastal erosion processes (see chapter "Management of a Tropical River: Impacts on the Resilience of the Senegal River Estuary"). The coastal basins, particularly along the Niger Delta, are gradually subsiding because of the geology of the area, as well as such human activities as oil mining and natural gas exploitation. On the other hand, the existing agrochemical and agricultural runoff within the Niger River basin, as well as the sediment load and urban and industrial wastewaters, have caused notable groundwater contamination and water quality degradation. These pollutant discharges directly affect the coastal ecosystems of the countries located along the coast. Regarding coastal concerns, the potential sea level rise and its impacts also are important, including shoreline retreat and coastal erosion, an increased frequency of coastal wetland submergence and saltwater intrusion into estuaries and coastal lagoons and aquifers.

General Oceanography, Coastal Morphology, and Processes

Four distinct and relatively persistent oceanic current systems are of importance off the shores of Western and Central African coasts (Fig. 1) in regard to the transport of substances, water temperature, meteorology, and biological conditions. They are as follows:

(a) The cold Canary Current, flowing southwestward along the coast in the northern part of the Western and Central African region (Mauritania, Senegal, Gambia, Guinea, etc.). It feeds the North Equatorial Countercurrent and the Guinea Current (Fig. 1);

(b) The North Equatorial Countercurrent flowing eastward from the central Atlantic;

(c) The Guinea Current, fed by the North Equatorial Countercurrent, involving warm waters flowing eastward and southeastward along the coast of the Gulf of Guinea (Sherman and Hempel 2008); and

(d) The South Equatorial Current, which flows at some distance from the coast, between 10°S and the Equator.

The Canary Current itself transports cool waters toward the Equator and has current speeds of approximately 20 cm/s. It is an essentially wind-driven current linked to the same regional wind systems responsible for the upwelling phenomenon that dominates the coastal waters up to several tens of kilometers offshore. The cool and richer upwelling waters prevail along the northwestern part from November to April/May and along limited parts of the northern parts of the Gulf of Guinea (Sherman and Hempel 2008).

High precipitation and numerous rivers on the central West African coast generate large masses of warm (above 24 °C) and low-salinity (less than 35‰) waters, the so-called Guinea waters.

In terms of coastal morphology, the succession consists of the following:

(a) Sandy arid coastal and plains bordered by eolian dunes (Mauritania and North coasts of Senegal);

(b) More-or-less sandy alluvial marshes with estuaries and deltas, colonized by mangrove vegetation (South of Senegal, Guinea-Bissau and Guinea, and Sierra Leone);

(c) Rocky scarps and sandy beaches with barrier islands, alternating with mangrove vegetation (Sierra Leone, Liberia, and eastern Nigeria to Gabon);

(d) Low sandy coastal plains, alternating with coastal lagoons along the Gulf of Guinea (Côte d'Ivoire, Ghana, Togo, Benin, and Congo estuary); and

(e) Huge marshy areas formed by the Niger Delta, with mangroves and rapidly growing Nypa Palms, indented by fluvial channels subject to tidal influences.

Further, a number of islands and archipelagos are located in the Atlantic Ocean offshore of the West Africa coasts (Canary and the Cape Verde Islands; Bissagos archipelago) and in the eastern part of the Gulf of Guinea (Sao Tome and Principe and Annabon in Equatorial Guinea, among others).

The tidal ranges along these west and central coasts are wide, exceeding 5 m in places, with the average for the whole coastal area being studied in the order of 1 m (See chapter "West African Coastal Area: Challenges and Outlook"). The highest tidal ranges recorded in the region are in Guinea-Bissau, Guinea, and Sierra Leone (from 2.8–4.7 m to 2.8 m; see Fig 2; Table 1).

Coastal Processes, Physical Alterations, and Habitat Modifications

Coastal erosion clearly constitutes the most serious coastal process problem in many West African countries. The rate of the coastal retreat can average several meters per year (e.g., in Fajara, Serekunda in the Gambia; in Keta, Ghana; and in the south coast of Dakar, Senegal). Although the coastline is very sensitive to natural erosion and sedimentation processes attributable to high wave energy, strong littoral drift transport, etc., human activities have significantly intensified coastal erosion, notably through sand mining, disturbance of the hydrological cycles, river damming, port construction, dredging, and mangrove deforestation, to cite a few examples. These examples are particularly relevant for the western part of Africa and mainly for the coastal countries in the Gulf of Guinea (Benin, Côte d'Ivoire, Ghana, Nigeria, and Togo).

Ecosystem Goods and Services and Species Diversity

A large variety of ecosystems and habitats exist along the western and central coasts of Africa, including the following:

(a) Wetland habitats, particularly those containing mangrove swamps and forests in a series of deltaic and estuarine formations, are the most apparent features. They extend more than 15,000 km^2 from Senegal to Congo, with the areas of highest mangrove concentration being located along the coasts south of Senegal, Guinea and Guinea-Bissau, Sierra Leone, and mainly in the Niger Delta (with more than 6,500 km^2). Although these mangrove forests are less diverse in terms of species than those in East Africa, they cover large surfaces and constitute the most extensive mangrove forests in Africa (see chapters "Combined Uses of Supervised Classification and Normalized Difference Vegetation Index Techniques to Monitor Land Degradation in the Saloum Saline Estuary System", "Importance of Mangrove Litter Production in the Protection of Atlantic Coastal Forest of Cameroon and Ghana", "Carbon Budget in Mangrove Forests of Varying Degradation Regimes in the Western Coastal Wetlands Complex (Ramsar Site 1017) of Southern Benin, West Africa", "Rapid Assessment of Mangrove Conditions for Potential Payment for Ecosystem Services in Some Estuaries of Western Region of Ghana, West Africa", "Assesment of Mangrove Carbon Stocks in Cameroon,

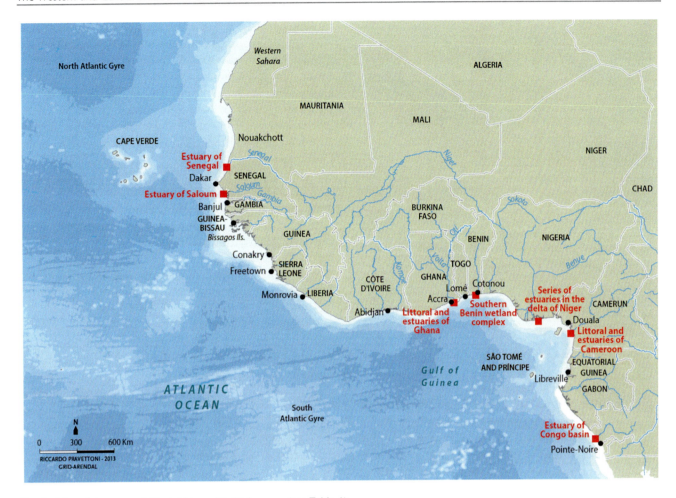

Fig. 2 Map of the coast of West Africa with tidal ranges (see Table 1)

Gabon, the Republic of Congo (RoC) and the Democratic Republic of Congo (DRC) Including Their Potential for Reducing Emissions from Deforestation and Forest Degradation (REDD+)"). Most of the coastal wetlands provide unique ecological conditions and habitats for migratory birds. They also function as nurseries for valuable fish and shellfish species, but remain unprotected in regard to impacts from natural and human influences and exploitation (UNEP-WCMC 2007);

(b) Coastal lagoons are found mainly in the Gulf of Guinea, from Côte d'Ivoire to east of Nigeria. They are associated with freshwater rivers, deltas, and estuaries and include a wide range of tidal swamps and seasonal marshlands;

(c) Sea grass beds are not very well developed in West Africa, although they exist off the shorelines of some estuaries and deltas (i.e., Saloum, Casamance in Senegal; Cacheu and Geba in Guinea-Bissau, etc.). There are no true coral reefs along the West African coast, mainly because of the cool waters of the Canary Currents;

(d) Sandy beaches, particularly in the Western African coast, along Mauritania and north of Senegal, and also in Guinea-Bissau, Sierra Leone, and Liberia and along the Gulf of Guinea, are considered important nesting ecosystems, particularly for sea turtles. Their exposure to strong currents and swells make the beaches extremely dangerous, often also being subject to marine debris and detritus accumulation in these areas.

In fact, the most important factor characterizing the open ocean waters off the shores of Gambia, Mauritania, and Senegal is the nearly permanent presence of upwelling, which is highly influenced in this region by the Canary Current. This is a reason why this region is well known for its rich fish production. Indeed, a large variety and diversity of marine resources species characterize the coastal marine waters of West and Central Africa. The wealth of estuaries, deltas, coastal lagoons, and the nutrient-rich upwelling of cold waters make a major contribution to the diversity of fish life. Nevertheless, these marine and coastal areas, including upstream freshwater regions, are presently affected by human activities, including over-exploitation and impacts from land-based settlements and pollutants from related industrial, agricultural, urban, and domestic sewage runoff and other mining activities (e.g., oil and gas), particularly off

Table 1 Average tidal range summary in various coastal locations along Western and Central Africa November 2013

Countries	Coastal stations	Low tide (m)	High tide (m)
Mauritania	Nouakchott	0.5	1.7
Senegal	Dakar	0.6	1.7
Gambia	Banjul	0.4	1.7
Guinea-Bissau	Bissau	1.0	4.7
Guinea	Conakry	1.0	3.2
Sierra Leone	Freetown	0.9	2.8
Liberia	Monrovia	0.3	1.5
Côte d'Ivoire	Abidjan	0.3	1.2
Ghana	Accra	0.4	1.7
Togo	Lome	0.4	1.8
Benin	Cotonou	0.2	1.5
Nigeria	Lagos	0.1	1.1
Cameroon	Douala	0.5	2.5
Equatorial Guinea	Bata	0.3	1.7
Gabon	Libreville	0.5	2.0
Congo	Pointe Noire	0.3	1.6

Source http://www.mareespeche.com

the shorelines of Angola, Gabon, and Nigeria (see chapters "Studies and Transactions on Pollution Assessment of the Lagos Lagoon System, Nigeria", "Estuarine and Ocean Circulation Dynamics in the Niger Delta, Nigeria: Implications for Oil Spill and Pollution Management", "Morphological Characteristics of the Bonny and Cross River (Calabar) Estuaries in Nigeria: Implications for Navigation and Environmental Hazards").

At the same time, however, the presence of invertebrates such as intertidal mollusks (*Senilia sp., Crassostrea sp.*, etc.), reptiles (turtles and crocodiles), marine mammals such as the West African manatee (*Trichechus senegalensis*), and some shark species, often threatened by hunting and trapping, demonstrates the variety of the species in the western and central part of Africa (World Bank Africa 1994; chapter "Status of Large Marine Flagship Faunal Diversity Within Cameroon Estuaries of Central African Coast"). The most remarkable collection of millions of migratory birds that seasonally visit the West African coast and mainland regions highlights the importance of preserving and maintaining the existing wetlands in this part of Africa. Large concentrations of seabirds are found seasonally in Mauritania, Senegal, Gambia, and Guinea-Bissau, including *Larus genei, Gechelidon nilotica,* and *Thalasseus maximus albididorsalis,* as well as the regionally large populations of great white pelican, white-breasted cormorant, and Caspian tern. Many of the islands contain large seabird nesting sites, an example being the Cape Verde Islands. The Gulf of Guinea Islands, near Principe and Sao Tome, also contain sizeable colonies of terns, noddies, and boobies. This species diversity and fauna richness is facilitating conservation and preservation policies being undertaken by some Western and Central African countries through creation and implementation of marine and coastal protected areas (Banc d'Arguin and Djawling in Mauritania, Djoudj in Senegal). Plans for the near future include implementation of regional-wide protected coastal and marine along the coastal zone in West and Central Africa.

Anthropogenic Impacts: Demography and Socioeconomic Activities, Including Cultural Heritage

Socioeconomic Implications of Anthropogenic Alterations

An estimated 100 million inhabitants occupy a narrow coastal margin some 60 km wide along the Atlantic coast between Mauritania and Namibia. This region exhibits a potential doubling time of 20–25 years, noting its present annual population growth rate of about 3 %. The highest population density centers are located in some key cities along the coast, including Accra-Tema, Abidjan, Cotonou, Dakar, Douala, Lagos, and Port Harcourt. In fact, most of the highly and densely populated coastal regions depend on the biological resources of the marine and coastal areas. Offshore and inshore waters, including estuaries, lagoons, and rivers, serve as major animal protein sources, in the form of fish and shellfish. Coastal fisheries also represent significant income sources (NOAA 2003).

Further, in spite of a low level of industrial development in West and Central Africa, the rate of industrialization continues to increase along the coastal areas.

Both the increasing impacts of urban population growth and the associated industries have created negative synergies in terms of human and environmental impact along the coastal regions. A variety of pollutants can be found in the region, including sewage, industrial, and solid waste disposal. Oil spills from shipping operations also are found increasingly in the region. As a result, water quality degradation is one of the most important components of environmental degradation within the coastal, marine, and freshwater areas in the in West and Central Africa region (UNEP 1999, i.e., microbiological and bacteriological contamination in certain confine coastal regions such as the Bay of Hann, near the city of Dakar; in Ebrie and Lagos lagoons, around Abidjan and Lagos, etc., Re: chapters "Studies and Transactions on Pollution Assessment of the Lagos Lagoon System, Nigeria", "Estuarine and Ocean Circulation Dynamics in the Niger Delta, Nigeria: Implications for Oil Spill and Pollution Management").

Agricultural runoff attributable to irrigation in the river valleys and floodplains (i.e., Senegal River delta, interior Niger Delta, Volta delta), including elevated concentrations of nutrients and pesticides, also contributes to increased eutrophication in the estuaries, deltas, coastal, and freshwater environments in West and Central Africa. River inputs also carry significant amounts of sediment resulting from soil erosion and deforestation, which contribute significantly to the siltation of coastal habitats and declining productivity. Combined with the pollutant inputs, it illustrates the problems currently encountered with significant seasonal invasive aquatic weeds in most freshwater aquatic areas, including Côte d'Ivoire, Nigeria, and Benin coastal lagoons. The same degradation is now occurring in the northern part of West Africa, particularly in the Lac de Guiers and the Senegal River delta.

Although tourism constitutes an important industry in many West and Central African coastal countries, including Senegal, Côte d'Ivoire, Gambia, Ghana, Guinea, Guinea-Bissau, Nigeria, and Gabon (see chapters" Morphology Analysis of Niger Delta Shoreline and Estuaries for Ecotourism Potential in Nigeria, Plantation Agriculture as a Driver of Deforestation and Degradation of Central African Coastal Estuarine Forest Landscape of South Western Cameroon"), it can nevertheless result in severe impacts on the coastal areas from Dakar to Douala. The constructions of hotels and other recreational facilities directly on the shore have been responsible for clearing of coastal vegetation, filling of wetlands, and increasing sewage and solid waste loads. At the same time, such constructions have contributed to accelerate coastal erosion, e.g., Southern coast of Senegal. Overall, such negative situation has been exacerbated over past decades by inadequate maintenance infrastructures.

Over-exploitation of marine resources and degradation of nurseries and habitats is currently a serious and accelerating phenomenon in this region. Because of over-fishing in the highly productive West African offshore and coastal waters, for example, the region is facing gradually and significantly depleted fish stocks. Accordingly, fishing activities, an important source of foreign currency for the economy of several West African countries, must be carefully managed, including appropriate quotas for the exploitation of different species, use of appropriate fishing equipment and fishing methods, improved management, legislation and regulatory measures, and reduction in and/or elimination of coastal water pollution from both land- and marine-based sources.

Human Impacts on Cultural Heritage Sites

In regard to cultural heritage sites, West and Central African coastal zones have been areas of human settlements for a long time. Evidence of early human occupation of the rich mangrove fishing areas was confirmed through the discovery of shell middens and pottery in many estuaries and on deltaic islands (Diop 1990; Diop et al. 2002). Additional evidence of pre-colonial occupation of the West African coastal regions exists at Grand Popo and Ouidah on the coast of Benin. Further, colonial buildings of considerable architectural interest can also be found in Porto Novo, Benin, and Grand Bassam in Côte d'Ivoire. Indeed, archeological and historical sites are located in all these African coastal countries, particularly Equatorial Guinea, Gambia, Ghana, and Senegal. Those cultural heritage sites must be preserved and included as integral parts in the coastal and marine protected areas. They must be protected from degradation or damage attributable to unplanned urbanization, pollution, and intensive industrialization activities (Diop et al. 2011).

Conclusions

It is clear that human-based land-use activities in coastal and marine areas of West and Central Africa, in combination with natural degradation processes, can induce significant impacts on the coastal environment. These impacts include loss of habitats, productivity and biodiversity, degraded water quality, and changes in natural coastal and marine environment equilibrium including increasing harmful effects.

Against this background, major environmental concerns regarding the associated socioeconomic impacts requiring closer monitoring and clear options for sustainable management include the following:

(a) Water quality deterioration, mainly around urban areas, including eutrophication and its associated impacts on environment and public health;
(b) Increasing pollution of coastal and associated freshwater environments from industrial and agricultural activities;
(c) Physical alterations and degradation, coastal erosion, and habitat modifications; and
(d) Loss of fishery resources and marine biodiversity, including associated ecosystem services.

In recognizing and utilizing the economic and ecological importance of estuaries and coastal ecosystems, coastal countries have the responsibility to provide the means to manage and protect the sustainability of such vital coastal features and their resources for the benefit of future generations.

This book is meant to provide a synthesis of some of the scientific knowledge of the region. To accurately determine the nature and severity of the problems occurring in the coastal Western and Central Africa region, in-depth scientific assessments, including socioeconomic aspects, had to be undertaken, particularly with regard to food security, poverty alleviation, public and ecosystem health, sustainable coastal and marine resources, biological diversity, and ecosystem services, including their socioeconomic benefits and uses. To this end, the various chapters in this book constitute a contribution to such scientific studies and research for managing the estuaries and the coastal waters of West and Central Africa for sustainable human and ecosystem use.

The content of this book also contributes to the global research analysis and synthesis agenda of the Land–Ocean Interactions in the Coastal Zone (LOICZ) especially on its core program of Earth System Science Project convened under the International Geosphere-Biosphere Programme (IGBP) and the International Human Dimensions Programme on Global Environmental Change (IHDP). The primary focus of LOICZ concerns related to river mouth systems, including estuaries and deltas (see www.loiczs.org).

References

Diop ES (ed) (1990) La côte Ouest-Africaine: du Saloum (Sénégal) à la Mellacorée (Rep de Guinée). In: Coll. Etudes et Thèses—Editions de l'ORSTOM – Paris, 2 vols + illustrations and map plates, 379p

Diop ES, Gordon C, Semesi AK et al (2002) Mangroves of Africa. In: de Lacerda LD (ed) Mangroves ecosystems: functions and management; and 55 tables. Springer Verlag Environmental Science Series, Berlin, , pp 61–121

Diop S, Arthurton R, Scheren P, Kitheka J, Koranteng K, Payet R (2011) The coastal and marine environment of Western and Eastern Africa: challenges to sustainable management and socio-economic development. In: Wolanski E, McLusky DS (eds) Treatise on estuarine and coastal science, vol 11. Academic Press, Waltham, pp 315–335

NOAA (2003) Guinea current large marine ecosystem, LM E No. 28. http://na.nefsc.noaa.gov/lme/text/lme28.htm

Sherman K (1994) Sustainability, biomass yields and health of coastal ecosystems: an ecological perspective. Mar Ecol Prog Ser 112:277–301

Sherman K, Hempel G (eds) (2008) The UNEP large marine ecosystem report: a perspective on changing conditions in LMEs of the World's Regional Seas. UNEP Regional Seas Reports and Studies No. 182. UNEP, Nairobi

UNEP (1999) Regional overview of land-based sources and activities affecting the coastal and associated freshwater environment in West and Central African region. In: Diop S (Report) UNEP/GPA Co-ordination Office and UNEP Regional Seas Programme, Nairobi, 110 p

UNEP-WCMC (2007) Mangroves of Western and Central Africa—UNEP-Regional Seas Programme and UN EP-WCMC report, p 88

World Bank Africa (1994) A framework for integrated coastal zone management. Land, water and natural division. Africa Environmentally Sustainable Division, Washington DC, 139 p

West African Coastal Area: Challenges and Outlook

Jean-Jacques Goussard and Mathieu Ducrocq

Abstract

The Economic and Monetary Union of West Africa—WAEMU, instructed by its Conference of Environment Ministers, with the assistance of the International Union for the Conservation of Nature—IUCN, the consultancy firm EOS.D2C and the Coastal Ecosystems Group of the Commission on Ecosystem Management, undertook a vast diagnostic and prospective study on coastal risk study including the formulation of proposals for rethinking the development of the West African coastal strip, from Mauritania to Benin (SDLAO, UEMOA–IUCN 2011). This study highlights the general trends that will characterise these coastal systems by 2030 and 2050. These trends are based on the fragility of coastal systems, urban and industrial developments, and uncertainties related to climate change in a context where the sedimentary deficits are compounded by large dams and growing demand for construction materials. The importance given to green infrastructure and soft, natural solutions, assuming their conservation, and sometimes restoration, was emphasised in the conclusions of the study, as was the necessity of building capacity in terms of observation and anticipation, in order to steer development decisions on different scales, from regional to State to local authority. The complexity of the mosaic of estuarine habitats determines their sensitivity to any changes in the environmental conditions and positions the estuarine zones, which are generally populated, as sentries for the marine environment, but also for the management of the upstream river basins. The subtle geography of the estuaries and, more generally, of fluvio-marine systems (lagoons, deltas, etc.) should also teach us how to better employ the various dimensions of the rich notion of coastal area and lay the foundations of a kind of development and planning that is integrated into the natural land matrix, buoyed by it and respectful of it.

Keywords

Coastal spatial planning • Coastal erosion • West Africa • Ecological services • Coastal risk management

---

The myth of a predominantly continental Africa "with its back turned to the sea (Pelissier 1990)" would appear to have worn rather thin. The development, in colonial times, of major urban centres, practically all on the coast, subsequently relayed by post-independence developments, and the migrations consecutive to the droughts in the 1970s and 1980s have given the West African coastline all the appearances of a veritable pioneering front.

J.-J. Goussard (✉) · M. Ducrocq
Coastal Ecosystems Group of the Commission on Ecosystem Management, IUCN, Gland, Switzerland
e-mail: jj.goussard@eco-evaluation.org

M. Ducrocq
e-mail: mathieuducrocq@gmail.com

S. Diop et al. (eds.), *The Land/Ocean Interactions in the Coastal Zone of West and Central Africa*, Estuaries of the World, DOI: 10.1007/978-3-319-06388-1_2,
© Springer International Publishing Switzerland 2014

While certain segments of the West African sea front have long been settled by traditional maritime peoples (the Balante in Guinea Bissau, the Lebou in Senegal, the Imraguen in Mauritania…), this is where colonial history left its mark, first of all through the trading posts, motivated by the mining of the natural and geological resources of the hinterland.

Until today, this exploitation of natural resources and customary usage values have been the main drivers of the development and use of West African coastal areas. Today, the need to protect people and goods, in a concept of security and social progress, imposes to revisit the relation between African societies and their coastal lands; this is true in particular for estuarine areas, interfaces between maritime trade and continental resources, which were historically the pioneer centres of coastal settlements and of the investments of the major economic sectors.

The acceleration of the building of new facilities, the urban extensions and densification with associated environmental deterioration, arouse the fear that part of the development potential associated with coastal ecosystem services will deteriorate over the coming decades. Furthermore, against a background of climate change which no longer leaves any doubt as to the eventuality of a gradual rise in sea level, the question of the risks of natural catastrophes in coastal areas becomes more strident.

The Economic and Monetary Union of West Africa—WAEMU, instructed by its Conference of Environment Ministers, with the assistance of the International Union for the Conservation of Nature—IUCN, the consultancy firm EOS.D2C and the Coastal Ecosystems Group of the Commission on Ecosystem Management, undertook a vast diagnostic study of the situation including the formulation of proposals for rethinking the development of the West African coastal strip, from Mauritania to Benin (SDLAO, UEMOA–IUCN 2011).[1] The importance given to green infrastructure and soft, natural solutions, assuming their conservation, and sometimes restoration, was emphasised in the conclusions of the study, as was the necessity of building capacity in terms of observation and anticipation, in order to steer development decisions on different scales, from regional to State to local authority.

---

[1] http://www.iucn.org/fr/propos/union/secretariat/bureaux/paco/programmes/programme_marin_et_cotier__maco/projets/thematique__amenagement_integre_du_littoral_/erosion_cotiere_et_schema_damenagement_du_littoral_ouest_africain/

Human Land Use Rapidly Becoming More Dense in the Coastal Areas

Sub-Saharan Africa is the last region in the world to undergo demographic transition. This process implies a population multiplied by a factor of almost ten between 1950 (approximately 180 million) and 2050 (more than 1.7 billion according to United Nations forecasts). The total population growth rates rose from 2.3 % in 1950 to 2.6 % in 2000. Forecasts predict a rate of 2.2 % in 2025 followed by a decrease to 1.7 % in 2050. This tenfold increase in the population of Sub-Saharan Africa will be differential and heterogeneous, with some desert areas or areas already densely populated to the point of saturation being evidently less concerned. Human land use in the coastal areas expresses the diversity of the living systems and systems of production (Fig. 1).

The current human footprint on these coastal areas appears to be dominated by the concentration of population and economic stakes related to the (i) urbanisation and its forerunners (communication routes, alleviation from isolation, electrification, recent changes in artisanal fishing strategies, etc.); and (ii) rapid development of tourism and residential areas, often on the periphery of urban areas. Access to water in dry areas also constitutes a key factor in organisation and distribution and in the growth of human settlements. The acceleration of the often anarchic and spontaneous use of coastal land is all the more pronounced as land ownership control often remains unclear, given that such areas were still rural a short time ago, where legal pluralism prevails in terms of land ownership (customary law and modern law). This human land use of the West African coastal areas is expressed in different ways:

Development of Built-up Areas and Urban Areas

A remarkable fact is that the coastal zone (arbitrarily defined here at a width of 25 km inwards) concentrates slightly more than half of the total urban population of the coastal countries in around one-twentieth of the total surface area of these countries. This proportion seems to be very gradually declining, from 57 % in the 1960s to 53 % in 2010. As an order of magnitude for West Africa, the average standard is 150 m^2 urbanised space per urban inhabitant (excluding parks, water features, land where building is not authorised, or which is not yet developed or inhabited). However, the footprint of the agglomerations is greater than built-up land area alone. According to the AFRICAPOLIS study, the total surface area occupied by agglomerations in 2000 was in the order of 200–300 m^2 per capita, and an average 210 m^2 per capita for the coastal agglomerations identified from Mauritania to Benin, if

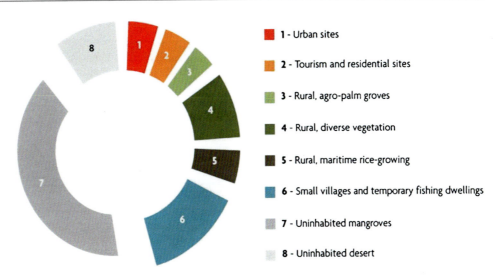

Fig. 1 Distribution (percentage of the coastline) of different forms of human land use in the first kilometre from the coastal strip inwards—the high percentage of mangroves is related to the high fractal dimension of the shoreline along these seaboards (SDLAO, UEMOA–IUCN 2011)

Abidjan, which is said to have an abnormally low rate of space consumption, is not included.

> The AFRICAPOLIS report states: "The average density of agglomerations did not increase from 1950 to 2000 as far as we can estimate in the current state of our work based on a sample of 97 towns for which we have the surface area in 1960. This sample accounts for 1/9th of the total agglomerations with populations of more than 10,000 in West Africa but 44 % of the urbanised land. This primarily concerns the largest agglomerations... Between 1950 and 2000, the urbanised area of our sample increased from 766 to 6,381 km², the average annual extension of urbanised land was therefore 5.1 % compared to 4.3 for the population."

Spread: The first thing that stands out about the growth of agglomerations is the horizontal spread of built-up areas, with the evident consequence of considerable land use, rising cost of facilities (roads, power, sanitation, etc.), accentuated by the often low-lying, flat topography of the littoral areas occupied by coastal agglomerations. These are often situated on the edge of a lagoon, in situations that complicate the collection and evacuation of waste waters and rainwater... Spread is often also responsible for the "exiling" of population groups in a precarious economic situation to peripheral areas far from the centre. Note that the historical centres of the largest agglomerations are typically located in proximity to the sea front.

Corridorisation: Agglomerations usually spread in corridors following the busiest communication routes. This corridor development can take on considerable dimensions in some cases, evolving into a long conurbation, as is the case between Keta (in Ghana) and the border with Togo, or between Lomé and Cotonou, in both cases along the coastal interstate road. In the case of Ghana, the model differs somewhat, with a mesh of "micro-centres", hub crossroads and small agglomerations in satellite positions around the major cities. These growth centres have a tendency to join up in the long term. In Senegal, there is an intermediate situation, where corridor concentration is sometimes weighted (except in the case of the Dakar peninsula) by a regular and relatively balanced road grid with tentacles stretching towards the different expanding secondary towns.

Apart from the situation of the consolidated districts (historical, partly) of urban centres, the dynamics of extension to peri-urban areas or areas in proximity to the sea is organised succinctly around for basic types:

- Extensions related to industrial activities, with, in particular, the attractiveness of harbours, which constitute business and investment centres.
- Precarious (or random) residential districts close to the centres and often located in areas highly exposed to risks and often historically unoccupied.
- Peri-urban extensions for residential purposes (seaside residences) or tourism, often along the main coastal roads, on north sides of the main agglomeration, the rapid development of which often follows speculative dynamics.
- The fishermen's districts located very close to the shore and canoe landing areas, in more or less precarious settlements.

These urban developments also imply the mobilisation of building materials leading to extractions and quarries on natural sites. The extraction sites are logically located as close as possible to the sectors being extended and concern fragile coastal formations, such as dune rims, for the extraction of sand. In other cases, rocky materials are employed either in blocks, or crushed, as is the case in Togo for the beach rock freed by erosion, an effective natural protection for a coast that is under threat, today subject to exploitation. Generally, regulatory measures eventually ban the practice, but either these come late or they are difficult to apply as long as viable economic and environmental alternatives have not been identified and possibly accompanied by public action.

To summarise:

- The coastal zone today concentrates 31 % of the total population and 51 % of the urban population of the coastal States.
- The total urban population of the coastal area in the 11 countries may well double, from 18 to 36 million between 2000 and 2020, while the rural population is expected to grow by half.
- From 2020 to 2050, the urban population of the coast could increase from 36 to over 80 million under the business-as-usual scenario and to 74 million under the moderate "controlling disparities" scenario.
- Almost all the administrative and/or economic capitals are situated there.
- The level of urbanisation on the coast is twice as high as in the hinterland.
- The current average population density is 260 per km^2, with maxima of 1,000 per km^2 in Togo and Benin and zones with less than 10 per km^2 in Liberia or Guinea Bissau. These densities could exceed 2,000 people per km^2 in certain countries in the Gulf of Guinea, such as Benin, for example. Certain areas remain unoccupied in the desert regions or large mangroves.
- The coastal fringe today accounts for approximately 56 % of the GDP of the coastal countries.

Growing Demand for Coastline

The generalisation of a global residential and leisure model confers particular appeal on the coast, which is expressed through (i) urban sprawl on coastal areas, followed by (ii) the densification of building in peri-urban and/or interurban coastal areas. The most attractive segments of coast (not isolated, with the right exposure, possessing a heritage of attractive beaches preferably not too far from the urban centres) are experiencing rapid growth related to the expansion of a tourist clientele mainly international initially, but increasingly regional and national as various more or less wealthy middle classes emerge. The stability of the coastal areas "under tourism" in this way is threatened:

- By the facilities built on the rim and on the backshore, depriving it of its sediment reserves and the exchanges that ensure its equilibrium.
- By disrupting the coastal drift, either through the building of leisure facilities affecting the beach, or even the intertidal zone, or even more so by individual or "spontaneous" anti-erosion structures put in place by owners anxious to preserve an heritage of beach exposure, when it is already deteriorating.

A further consequence of the intensification of tourism resides in the specialisation of the functions of the coast, breeching its multi-functional equilibrium, with the emergence of conflicts over use on the beaches which are gradually contracting because of erosion, in particular with the activities related to fishing (canoe landing, fish processing, smoking, etc.) (SDLAO, UEMOA–IUCN 2011).

The multiplication of spontaneous individual coastal defence and protection actions turns out to be completely counter productive and aggravates the sediment deficits observed globally. This is also a factor that reinforces inequalities, for the attempts at localised solutions implemented by owners with sufficient means triggers an acceleration of erosion downstream of coastal drift currents, affecting the more modest properties of poorer population groups or even traditional villages, when these alternate with major hotel infrastructure, as on the Petite Côte in Senegal. This "privatisation of the sediment heritage" is only one of the signs of the accelerated privatisation of the coast in sites that are "under tourism", which also leads to the public being denied access to the beaches in front of the seaside hotels and residences, or by the closure (walls and building) of sea views from public tracks and areas. The future demand for seaside leisure resorts, in particular in the major metropolises, and the respect of the landscape identity of all the coastal sites, should as an imperative be anticipated through a "back to basics" approach regarding the inalienable nature of the public maritime domain.

Economic and Infrastructure Growth

The prospective study that was carried out in the frame of the SDLAO (UEMOA–IUCN 2011) also forecasts a probable acceleration of economic growth in West Africa, with rates exceeding 5 % over the long term. This growth will support the pace of urbanisation throughout the region, which will see a reinforcement of the concentration of economic activity along the coast, with the building of heavy industrial plant and the development of agro-industrial production. Pressure on the raw materials market is also already being expressed through different projects to build ore ports, mainly oriented towards the estuaries, as in Guinea Bissau and Guinea. We were able to observe how in Maritime Guinea, the port of Kamsar caused a village to grow to the second largest city in the country in under 20 years.

In this favourable growth context in the sub-region, most African ports have already begun to undertake work to

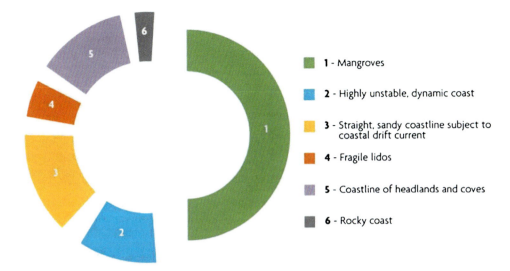

Fig. 2 Proportion of length of shoreline according to the different coastal facies (SDLAO, UEMOA–IUCN 2011). *Note* the high proportion of mangrove coasts, also related to the highly fractal dimension that characterises the shoreline in these milieus

extend their capacity, or will do so in the near future. The increasing penetration of the private sector in the management and even building of ports (for ore ports) should act as an incentive to the States to be vigilant in taking into account the environmental and coastal impacts of these new facilities.

A Fragile, Dynamic Coastline

On the coasts constituted of sedimentary accumulation, which are by far the most common in West Africa, the mobility of the shoreline largely depends on the local balance of supply and removal in the sediment budget. Removal operates under the action of natural agents (coastal drift, ocean waves, wind, etc.), which are also partly responsible for sediment supply. Removal may also be the result of human activity, either directly (extraction from the beaches of raw materials for building activities, for instance), or indirectly (the creation of surfaces that reflect wave energy or installations that disrupt the operation and the exchanges between the different sediment compartments of the beaches or that disturb the coastal drift parallel to the shore). The dams situated on the catchment areas also constitute traps for continental sediment which no longer reaches the coast, increasing the sediment deficit, particularly at the level of the estuaries and mouths of rivers.

Of the estimated less than 6,000 km of coastline (at a scale of 1:75,000) from Mauritania to Benin, rocky coasts represent fewer than 3 % of the coast line. These coasts are made of rock that is often altered and fractured, subject to landslides and erosion. There are, however, a few rock outcrops that structure this coast in headlands that are less soft but often fractured and fragile, and especially few in number:

- Basalts and other rocky formations on the Cape Verde Peninsula (Senegal).
- Rock outcrops at Cap Verga and the Conakry peninsula (Guinea).
- Breakwater at Freetown (Sierra Leone).
- Relict of sandstone or hardpan spared by erosion (sandstone on the Senegalese Petite Côte, the Bijagos and around the periphery of Accra).
- Granites and metamorphic rocks presents on all of Liberia, Western Côte d'Ivoire and the central part of the coast of Ghana.

The remainder of the coast line is composed of:

Unstable and/or very dynamic coasts

- Sand banks, estuaries, river mouths, spits and islets by nature also very unstable and dynamic (12 %).
- Mangroves, continuously evolving (48 %).
- Sandy formations of lidos, thin sandy rim between a lagoon and the sea shore, also unstable and highly changing (7 %).

Less dynamic coasts, but still subject to natural episodes of erosion and accretion outside of human intervention.

- More or less straight sandy coasts, fashioned by the coastal drift currents, relatively stable but subject to cyclical phases of erosion and accretion, also very sensitive to nay disruptions of the coastal drift (16 %).
- Stepped coasts or headlands and coves, where the coves are compartments more or less separated by rock outcrops or less soft. Their stability strongly depends on the orientation in relation to the ocean waves and currents (14 %). The sediment stocks here are often very limited (Fig. 2).

The whole of this coastal system is first of all conditioned by the sediment legacies dating from the last transgressions and remobilised by the morphogenic agents (currents, winds and ocean waves). Continental fluxes, whether aeolian or fluviatile, only partially contribute to maintaining the legacy stocks. This is nonetheless a hypothesis which has not yet been confirmed.

Fig. 3 Continental shelf from Mauritania to Benin (blue shading from 0 to 300 m)

A Narrow Continental Shelf

The landform is, on the whole, not very rugged. The continental shelf is narrow in the main, around 30 km on average, except from Guinea Bissau to the Sherbro islands in Sierra Leone, where it widens considerably to 200 km. This continental shelf is marked by some major deep features: the Khayar canyons in Senegal to the North of the Cape Verde Peninsula, and the deep canyon ("Trou sans fond" (Bottomless pit)) that cuts through the shelf perpendicular to Abidjan in Côte d'Ivoire. For certain authors, these bathymetric features contribute to trapping the sediment transported by the coastal drift current parallel to the coast (Fig. 3).

Circulation and Redistribution of Sediment

The propagation of ocean waves affects the whole of the two major sea fronts along this coastline, west and south, with an orientation that is generally oblique, which contributes to the generation of a significant coastal drift current more or less parallel to the coast.

The circulation and redistribution of sediment is governed on the major part of the coast by this coastal drift current, which is subject to annual variations, but the resultants of which are globally north–south all along the western sea front (from Mauritania to Guinea Bissau) and west–easterly along the Gulf of Guinea. In certain cases, these variations are considerable: the drift is reversed seasonally on the Grande Côte of Senegal, and the resultant observed through the physiography of river mouths and estuaries indicates an east–west orientation for a large part of the coast of Liberia and Sierra Leone.

In the portion of coast between Guinea Bissau, Guinea, and the north of Sierra Leone, called the Southern Rivers region, sediment circulation and redistribution is primarily governed by tidal removal currents, combined with river spates in these regions with high seasonal rainfall. The role of the mangrove estuaries in trapping sediment and building sediment accumulations, which are partially expelled in periods of spate, contributes strongly to modelling the facies of this portion of the West African coast. It should be noted that the tidal ranges are very wide in this zone, exceeding 5 m in places, while the average for the whole coastal area studied is in the order of 1 m. In these regions where rainfall determines important annual spates, the dams built across rivers can reduce these spates and restrict the expulsion of the mud plugs, an important source of sediment supply usually put into circulation in the coastal waters during these episodes.

Five Major Coastal Profiles

There are five distinct major coastal profiles from north to south:

- *The straight coastal regions from Mauritania to the Cape Verde peninsula* composed for the most part of sandy formations subject to the direct action of the coastal drift. In the immediate proximity of and behind the ridge/sandbar, there are vast expanses of low-lying salt marshes situated below sea level in places.
- *A coastal region with headlands and softened coves from Cape Verde peninsula to Basse Casamance* structured by

the major estuaries of Sine Saloum and the Gambia. This coast is structured by rocky outcrops of sandstone and badly deteriorated, fragile ferruginous cuirass.

- *The mangrove coastlines associated with the estuaries in the Southern Rivers region from Sine Saloum in Senegal to the Sherbro islands in Sierra Leone.*
- *A coastal region highly structured into rocky headlands and sandy coves from Liberia to the West of Coast d'Ivoire.* This same profile is also to be found in the central part of Ghana.
- *From the west of Côte d'Ivoire to Benin* stretches two large sediment basins of soft coastline (Côte d'Ivoire and Dahomey basins) also characterised by important lagoon and channel systems parallel to the coast and situated behind a sandbank that is very narrow in places (lidos). These two large sediment basins are separated by the Three Points Cape in Ghana and a few adjacent formations that are more or less rocky (sandstone) or in headlands, right to the mouth of the Volta.

> The major part of the coast in West Africa has a high sensitivity to coastal erosion related to (i) the nature of the materials (mobile sandy sediment or highly altered and fractured rocks; (ii) the circulating sediment fluxes which remain limited either due to continental or river mouth trapping, or due to the coastal sediment partitioning that can be observed on coasts that are more predominantly structured in headlands and coves.
>
> The developments and infrastructures that disturb a coastal drift that is typically parallel to the shore create observable direct impacts: siltation upstream and erosion downstream of the portions of the coastal region that have undergone human artificialisation, in particular through structures placed perpendicular to the shore and the coastal drift current (ports of Nouakchott, Lomé, Cotonou, etc.).
>
> In the stepped coasts, or in headlands and coves, the diversity of the situations in relation to the predominant ocean waves and the coastal drift current requires a case-by-case analysis of the sensitivity to erosion, which is largely conditioned by the local configuration.

Biodiversity and Ecological Services

The biodiversity of the coastal ecosystems of West Africa is directly related to the variety of types of coast and to the steep bioclimatic gradient characteristic of the region, covering the Saharan, Sahelian, Sudanese and Guinea-Congolese zones (White 1983). The extensive wetlands, corresponding to the morphology of flat, low-lying topography of the major part of the coast and to the interpenetration of fluvio-marine influences, in particular in the estuarine areas, constitute the striking characteristics of this maritime façade.

Major Natural Areas

There are four major, extensive, more or less protected natural areas along this coastline: the Banc d'Arguin National Park in Mauritania, the Delta of the River Senegal, the Bijagos archipelago and the Sherbro-Robertsport complex between Sierra Leone and Liberia.

Between these major units, some of which have already been subject to conservation measures for some time (Arguin, Bijagos), is an interspersed network of natural areas that are still relatively preserved, some of which are subject to local protection measures (RAMSAR sites, marine protected areas in Senegal, the national network of marine and coastal protected areas in Guinea Bissau, etc.).

Dense Guinea-Congolese forest

The last relics of the Guinean coastal forests (Guinea Bissau, Guinea) are today largely deteriorated, or have simply disappeared. However, a few dense forest areas remain in places, in particular in Liberia, with little data available on the actual status and distribution of these formations. Secondary forests from the recolonising of plantation areas that were previously artificialised are better represented from Liberia to Ghana. Note that the dense, evergreen Guinea-Congolese forests extend from Cap Palmas to Cape Coast in Ghana, at some distance from the coast. These formations extend to Nigeria after an interruption (Dahomey gap), due to the bioclimatic reasons from Keta in Ghana to Benin inclusive. There is practically no forest remaining on the actual edge of the coastal area.

These hinterland forest facies vary depending on the edaphic conditions (rock outcrops and cuirasses, wetland depressions, leached sand on coastal terraces), orographic conditions (reliefs), bioclimatic conditions (duration of the dry season) and the intensity of human intervention (fires, conversions and secondary regrowth after plantation or slash-and-burn).

This coastal evergreen Guinea-Congo rainforest is recognised as exceptionally rich with a diversified flora including a notable proportion of endemic species. Like the dense coastal forests of Guinea, these forest entities are also highly threatened.

Mangroves

The mangroves of West Africa are completely different in their composition from those in East Africa. The seven main species they comprise (*Rhizophora mangle, R.*

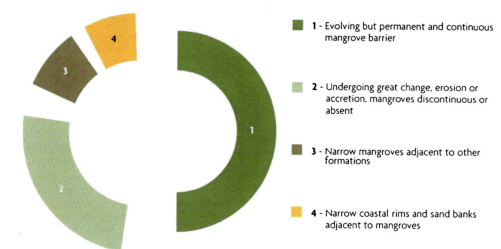

Fig. 4 Proportion of coastline according to physiography of mangrove stands (the whole of the study area) (SDLAO)

harrisonnii, R. racemosa, Avicennia germinans, and *Laguncularia racemosa, Acrostichum aureum, Conocarpus erectus*) are also distributed on the eastern coasts of tropical America. These mangroves grow in the intertidal zone. They cover approximately 14,000 km^2 in the zone under study and are subject to the influence of various factors: oceanographic, sedimentary, geomorphologic, but also and increasingly, anthropic.

Not very diversified from a floristic point of view, these mangrove communities nonetheless play a vital role for the coastal ecosystems as a whole, in particular by the high net production that characterises them, which is exported to marine milieus and enables a rich and diverse piscifauna to be maintained. Their physiographic characteristics (shallow gradients, the cryptic nature of milieus that are crowded with the roots of the mangrove trees) also make them essential reproduction zones for a high proportion of fish species that make up the region's fishing stocks. The small areas of estuarine mangroves in the Gulf of Guinea, particularly in countries like Togo and Benin, are extremely threatened and specific protection measures are required. The pressure on the mangroves and on these coastal ecosystems is increasing today. A distinction should be made between:

- *Biomass removal*: wood for energy (supplying the nearby urban concentrations for smoking fish or producing salt in Guinea), for services, fisheries and the gathering of attached organisms, molluscs and crustaceans, removal of bark and of various species in traditional pharmacopeia.
- *Conversions and clearance*: artificialisation for rice production, salt production, or shrimp farming, which is expected to expand in the future, in particular in Guinea. The surface area occupied by mangroves in the region is thought to have decreased by a quarter between 1980 and 2006 (Corcoran 2009)

The changes in the surface areas of mangroves (contraction or expansion) are, however, also governed by continental drought cycles or, on the contrary, abundant rainfall. These systems are particularly dynamic and sensitive to changes in mud banks, the physiognomy and topography of which are under the influence of coastal hydro-sediment forcing.

Preserving these original systems also conditions that of the veritable ethno-ecosystem that characterise these areas and their population which was largely dependent on a daily basis on resources valorised locally through complex production systems adapted to these particular milieus. The contribution of these ecosystems to the subsistence strategies of certain coastal societies (in Sine Saloum, Casamance, the Gambia, Guinea Bissau, Guinea and Sierra Leone) is fundamental, as much at the level of food, food security and pharmacopeia as from a cultural point of view (Figs. 4 and 5).

> Five West African governments (Mauritania, the Gambia, Guinea, Guinea Bissau and Serra Leone) have ratified a charter that commits them to cooperate for the protection of the mangrove in the sub-region. This mangrove protection charter also comprises detailed plans of action specific to each country.

Estuaries

Estuaries constitute areas of particular importance for the interpenetration of marine and fluvial environments, favouring different ecological processes and the exchange of nutrients, which, depending on the season, are used by a wide variety of fishes and crustaceans species, some at specific stages in their life cycle. The estuaries of West Africa are home to the major expanses of mangroves, which are reputed for their biological productiveness and their role

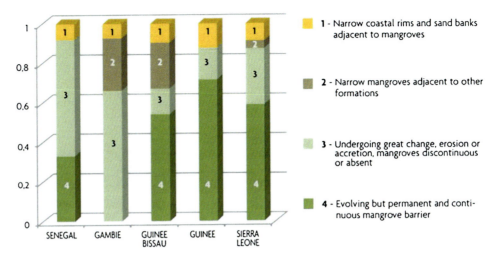

Fig. 5 Proportion of coastline according to the physiography of mangrove stands (by country) (SDLAO, UEMOA–IUCN 2011)

of being nurseries and feeding areas for large numbers of marine species and water birds.

Beyond the mangrove area associated with the major estuaries and the archipelago estuary of the Bijagos, the string of small estuaries from Sierra Leone to Côte d'Ivoire also represents a network of extremely rich ecosystems, sheltering a fluvio-marine and brackish water fauna the diversity of which remains relatively unexplored and unknown. As far as fish species are concerned, the estuaries are home to relatively diverse communities that are uniform on a regional scale, with low seasonal variations compared to adjacent marine waters.

The biological diversity and ecological processes associated with the estuaries relates to the multiple gradients (salinity, temperature, bathymetry and topography of the banks and coastal lagoons, conditions and sediment dynamics and local current systems) which characterise these milieus, whose conservation certainly constitutes a priority on a regional scale, in particular given their richness, their sensitivity to pollution and the risks associated with the development dynamics of harbour and urban infrastructure.

The dramatic effects of dams on coastal areas:

Hydrologic constraints have led to the building of dams on the majority of large rivers, often for hydroelectric power (50 % of dams), but also for agricultural purposes. In certain cases, there are several purposes, the Senegal River Basin Development Organization for example tries to reconcile agricultural production goals with the production of hydroelectric power and navigation. There are approximately 150 dams in West Africa, with several more scheduled. This number is relatively low when compared to Southern Africa, however, which has the majority of dams (there is a total of 1,300 dams in Africa and 45,000 worldwide). The two largest dams in West Africa are the Akosombo on the Volta in Ghana, built in 1964, which stands 134 m high (4th highest in Africa) and has a capacity of 150 billion m^3 (3rd in Africa) and the Kossou on the Bandama in Côte d'Ivoire, which has a capacity of 28 billion m^3 (6th in Africa).

The consequences of these developments are multiple, in particular in terms of conserving biodiversity, but also in reducing sediment load and the speed of flows particularly during flood peaks. The consequences in the coastal zone and the deltas are often major: salinisation of soils and surface waters, erosion provoked by sediment deficits, accretion and delta formation related to energy reduction when annual expulsion of silt plugs should normally occur.

With the exception of special cases such as the Gambia and Senegal, which have regional river development organisations (OMVG, OMVS), the majority of these dams were designed at national level, and therefore often without taking into account in depth the remote impacts of the developments, which should be considered on a sub-regional scale.

Coastal Conservation and Natural Infrastructure

The natural coastal milieus in West Africa, and in particular the estuary systems and fluvio-marine connections, contribute directly to producing ecological services that are useful or even indispensable to the coastal societies, perhaps even more so in the context of climate change on the agenda today. These ecological services procure identifiable benefits on every scale, including global: carbon sequestration by the mangroves, sea grass beds and coastal marshlands, the importance of which is recognised.

- *Self-maintenance services*: constitution of habitats and of the milieu. Maintaining of energy flows and nutritional cycles through primary production, inter- and intra-eco-system services and functions, reproduction, nourishment, etc.
- *Provisioning services*: fisheries (artisanal, staple, and commercial), agriculture, firewood, ligneous and non-ligneous gathered food products, aquaculture, crafts, building (materials and service wood), pharmacopeia, genetic resources, etc.
- *Regulation services*: climatic (carbon sequestration), sediment trapping and coastal protection against marine erosion and extreme marine weather events, treatment and recycling terrigenous and effluent input from human activities, waste water purification, protection against floods from continental waters, stabilisation of mobile dunes, etc.
- *Cultural services*: landscape appeal and environmental quality (formation of beaches, islands and coastal landscapes), leisure activities (urban beaches for example), research and education, cultural and religious heritage (sacred sites, customs, traditional ways of life, artistic expression), etc.

Not all of these ecosystem services, which strongly contribute to the development potential of the coastal countries, are subject to systematic economic valorisation to date, except for a few sectors such as fishing. This also implies that these services are globally still functional. Nonetheless, the concerns related to coastal erosion show that functional deficits in these natural systems can have a considerable economic impact. The numerous instances of deterioration observed along the West African coastal zone are generally attributable to inappropriate development practices implemented with no concern for the anticipation of how the natural systems will respond (SDLAO, UE-MOA–IUCN 2011).

Fishing: A major sector of activity, weakened by fish stocks depletion

With an EEZ of more than two million km^2, and the existence of upwellings (essentially in Mauritania, Senegal and the Gambia) rendering the waters highly productive, fishing livelihoods are an essential component of the development strategies of the coastal States of West Africa, not only in building their GNP, but also in the struggle to attenuate poverty and malnutrition. Fishing is an important sector for employment (approximately 600,000 jobs in Senegal, more than 500,000 in Ghana).

The competition for access to these fishing resources is intensified in a global context of increasing demand, and access to pelagic resources (but also to a lesser extent, demersal resources) is coveted by foreign fleets—European fleets through fishing agreements, but also Asian (Korea, China) and Eastern European fleets. The way these foreign fleets respect the access conditions is sometimes relative, and the fiscalisation of the activity remains very unequal, depending on the states and their foreign partners.

While total catches have increased regularly since 1950, this growth should also be compared to the regular increase in fishing efforts and the efficiency of fishing units. The observed depletion of certain stocks of demersal species, associated with the reduction in the diversity of the communities, the sensitivity and fluctuations recorded in certain specific fisheries, such as cephalopods, closely dependent on the conditions of the milieu (in particular of the upwelling), certainly attests to a deterioration in the composition, structure and organisation of marine biological communities.

Climate: Facing the Possible Futures

The uncertainty that characterises the future of these coastal systems in a context is subject to climate change, which will very probably have significant impacts on the state of the coastal sea and the coast area.

West African Climate Models

There are systematic biases in the simulation of the African climate by most of the climate models that contributed to the 4th report of the IPCC (Intergovernmental Panel on Climate Change). A total of 90 % of these models overestimate the precipitations on a large part of the continent (Christensen et al. 2007). The temperatures simulated also show bias, but this is not significant enough to call into question the credibility of the projections. The intertropical convergence zone simulated is moved towards the equator in most of these models. The surface sea temperatures are overestimated by 1 to 2° on the Gulf of Guinea. A large part of these models have no monsoon, as they cannot properly reproduce the northward movement of precipitations on the continent. Only 4 of the 18 global ocean-atmosphere models in the 4th IPCC report examined by Cook and Vizy (2006) are able to produce quite realistically the interannual variability of surface water temperatures in the Gulf of Guinea and the dipolar structure of precipitations between the Sahel region and the Guinean coast.

Future Projections to 2050

Temperatures: The projections for Africa show temperature rises that will very probably be well above the average global rise, with an accentuation on the arid zones (AC-MAD, in SDLAO, UEMOA–IUCN 2011).

Precipitations: The global ocean-atmosphere models have more difficulty simulating precipitations than temperatures. In several regions of the world, these models agree on the rise or fall in precipitations, but they diverge greatly in their projection of precipitations in West Africa and the signal for variations in precipitations on the Sahel and Guinean coast remain uncertain. The overall average of the various models presents a downward trend in precipitations in JJA (June July August) on the West African coast to the north of the 10th degree of latitude, which is approximately the domain of the maritime trade wind and the north of the Liberian–Guinean domain. This decline would be accompanied by an increase in the intensity of precipitations and a decrease in the number of rainfall events (Tebaldi et al. 2006). In the south, on the domain of the permanent Atlantic monsoon, the models do not agree on the signal of change, even though the average presents a slight upward trend.

> The increase in the intensity of precipitations and the reduction in the return periods of certain extreme events could cause the flooding of coastal zones and aggravate erosion phenomena (which could in certain specific cases lead to increased siltation). The global reduction of rainfall in the course of the twenty-first century would cause a decrease in the flow rates of the major rivers such as the Senegal and the Volta, which would be accompanied by a sediment deficit and an aggravation of coastal erosion. To this should be added the influence of works such as dams on these watercourses, which only aggravate the trend/phenomenon.

Frequencies of Extreme Events

Among the most important extreme events affecting the coasts of West Africa, the episodes of intense precipitations, depressions and tropical storms can cause considerable damage. There is disagreement between the different studies on the projected the frequency of extreme events (including cyclones) as a result of global warming. There seems to be more of an agreement on their increase in intensity because of a perceptible rise in the temperature of marine surface waters. Furthermore, storm surges depend greatly on local conditions, in particular bathymetric and related to tidal regimes. This means studies of storm surge statistics are specific to each region and cannot be generalised.

Significant Wave Height

There are a limited number of studies on wave climatology projections (Weisse and von Storch 2010). These studies nonetheless allow for a considerable increase in the significant height of waves in the North Atlantic, consistent with the deviation of storm paths towards the poles. These studies do not predict an upward trend in low latitudes. For West Africa, the change will therefore come especially from the increase in the frequency and duration of tidal wave events, in particular related to extreme marine weather events.

Higher Sea Level and Storm Surges

The fact that sea level is rising seems to have been largely confirmed. *The historical tide gauges show that in the course of the past 100 years, the level of the sea has risen by an average 20 cm.* The current estimations are between 20 and 50 cm by the end of the century. Much more dramatic estimates evoke (on a conservative hypothesis) a rise of 3.3 m according to a number of possible scenarios, such as, for example, the complete disintegration of the West Antarctic ice sheet (Bamber 2009).

The spatial distribution of the sea rise signal is nonetheless far from uniform. First of all, the surface of the oceans is not regular and for example, in the subtropical Atlantic, we note a convex area of approximately 1 m in elevation. This spatial distribution also depends on climate variability and the hazards of marine circulation. These spatial disparities were already observed in the data for the decade 1993–2003.

At regional level, this rise can significantly deviate from the global average due to little known local factors such as land subsidence, change in atmospheric circulation and wind regime, the redistribution of atmospheric pressure or the unequal distribution of thermal expansion. Our current state of knowledge does not allow more accurate estimations.

According to the IPCC's 2007 report, in 2090–2099 average sea level will have risen by around 18–59 cm compared to 1980–1999. By 2050, the rise will be in the order of 10–20 cm. This rise does not take into account the probable acceleration of ice melt, which could add a further 10–20 cm. There is a lot of uncertainty surrounding these values, which could be exceeded (Meehl et al. 2007).

Alarming Conclusions

The erosion and flooding (submersion) of coastal areas which largely contributes to the receding shoreline will be aggravated in the course of the twenty-first century following an increase in average sea level.

Africa is one of the regions in the world whose coastal zones, estuaries and deltas are the most exposed to risks of flooding related to the rise in mean sea level (Nicholls and Tol 2006). This rise in sea level, combined with increased intensity or frequency of extreme events, will have serious consequences for the development of the coastal zone. Many coastal or island areas will be submerged or subject to increasingly frequent flooding, causing considerable damage.

In West Africa, although this rise cannot be estimated accurately, a rise greater than the global average is expected. There could be dramatic consequences for certain areas, such as around Nouakchott, an area which is already below sea level. Major conurbations are greatly at risk. *The destructive effect of this rise in water level will lead to an increase in the frequency of storm surges and their submersion potential, particularly in the estuaries and river deltas. There will be more frequent intrusions of saline waters which may gradually make aquifers unfit for consumption and agriculture (the advancement of the saltwater wedge, alteration of freshwater lenses and surface aquifers).*

The consequences remain, however, extremely difficult to evaluate and should only be envisaged through *a detailed study of local situations*. The hypothesis of a rise in sea level of 1 m would cause a considerable aggravation in coastal hazards, first of all, reaching the low sandy coasts and mangroves, as well as coastal zones composed of easily erodible sandstone or marno-limestone cliffs. The major lagoon systems will also be affected. The lowest-lying sectors will be subject to increased erosion or temporary or permanent submersion.

To complete these hypotheses, it should be added that natural coastal systems are not in fact passive with regard to the rise in sea level, and there are numerous threshold effects, for these systems also react and adapt to the new configurations. For example, in the case of submersion hazard, coastal plant formations can trap sediment, river flow rates can be modified by the variability of continental precipitations, lagoon or estuary outlet streams can be partially closed by the advance of sand spits, etc. Any evaluation of the impacts of the rise in sea level should therefore remain cautious and avoid swinging into simplistic, reductionist or "mechanical" calculations or representations, in particular in the field of economics. The submersion hazard, when the stakes justify this, can only be properly considered through a detailed local hydraulic approach.

- *Sandy coastline*: increasing erosion of sandy systems is expected, aggravating the risks of submersion. The receding of the shoreline already observed should accentuate.
- *Dune ridges and lidos*: lidos and dune ridges will migrate inland, at least for the narrowest lidos. Certain lidos will become fragmented. A tracking programme should enable local identification of the lidos able to migrate and those likely to fragment.
- *Lagoon systems*: the hydrology of lagoons comprises exchanges with continental waters, but also with the sea. In addition, these are located at a height close to sea level. The ecology of lagoons is based on two main parameters: depth and salinity. While the former is not expected to change much,[2] the salinity is expected to change (i) following the rise in sea level; (ii) by the salinisation of aquifers; (iii) by a possible decrease in the freshwater supply consecutive to a reduction in rainfall and therefore in flood peaks. On the other hand, the tendencies for lagoons to fill in by terrigenous supply could be partially counteracted. Note also that the multiplication of dams, by removing the very small spates, may contribute to reducing the salinity of certain brackish waters.
- *Closing of lagoon outlets*: the closure or strangling of lagoon outlets due to the development of spits and local accretions leads to the eutrophication of the aquatic milieus concerned. The filling in of these outlets also implies flooding in periods of spate.
- *Estuaries*: the way estuary systems work could also be profoundly modified, in particular in relation to changes in sediment supply, but also to the distribution of salinity, the surface temperatures of the waters, etc., all parameters which strongly determine the biological potential of the gradients that characterise these very specific milieus. The impact of the decisions made in the management of the major dams will also be largely as important as trends cause by climate change.

> These different elements of climate forecast must, however, be balanced by the recognition of the non-linear and chaotic nature of the dynamics under

---

[2] The migration of lagoons, if this is possible, should conserve the initial depth gradient in most cases, even if the depth of the lagoons increases slightly.

consideration, and the threshold effects that characterise the different manifestations of climate change. The combination of permanent shoreline monitoring with the monitoring of changes in climate conditions should enable the production of scenarios to be updated regularly, in order to reduce the uncertainty that today besets any forecasting in terms of climate.

environment, but also for the management of the upstream river basins.

The subtle geography of the estuaries and, more generally, of fluvio-marine systems (lagoons, deltas, etc.), should also teach us how to better employ the various dimensions of the rich notion of coastal area and lay the foundations of a kind of development and planning that is integrated into the natural land matrix, buoyed by it and respectful of it.

Estuarine Sentries

In West Africa, the concentration of social and economic stakes on the coastal strip, along with the multiple rapid changes already underway, calls for the implementation of forward-thinking, deliberate land planning policies, in order to preserve the fragile coastal interface on which the future depends—the future, not only of the coastal States, but also, partially, of the Sahelian States (population mobility; economic exchanges).

Interconnecting the inland areas with the coastal sea, the estuaries aggregate these stakes. Historically and geographically, estuaries are structural components of the West African seaboard, and are today in the front line of exposure to the general trends affecting the entire coastal complex.

The complexity of the mosaic of estuarine habitats determines their sensitivity to any changes in the environmental conditions and positions the estuarine zones, which are generally populated, as sentries for the marine

References

Bamber JL (2009) Reassessment of the potential sea-level rise from a collapse of the West Antarctic Ice Sheet. Science 324:901–903

Cook KH, Vizy EK (2006) Coupled model simulations of the West African monsoon system: twentieth-and twenty-first-century simulations. J Clim 19:3681–3703

Corcoran E et al (2009) Les mangroves de l'Afrique de l'Ouest et centrale. PNUE—Programme des Mers Régionales

Meehl GA et al (2007) Global Climate Projections

Pelissier P (1990) Post-scriptum à Rivages. L'Afrique tourne-t-elle le dos à la mer? Cahiers d'Etudes Africaines 117(XXX-1):7–15

Tebaldi C, Hayhoe K, Arblaster J, Meehl G (2006) Going to the extremes. Clim Change 79(3):185–211

UEMOA–IUCN (2011) Schéma Directeur du Littoral d'Afrique de l'Ouest—SDLAO

Weisse R, von Storch H (2010) Marine climate change: ocean waves, storms and surges in the perspective of climate change. Springer, Berlin

White F (1983) The vegetation of Africa, UNESCO

Christensen JH et al (2007) Regional climate projections

Nicholls R, Tol R (2006) Impacts and responses to sea-level rise: a global analysis of the SRES scenarios over the twenty-first century. Philoso Trans A 364(1841):1073

Morphological and Hydrodynamic Changes in the Lower Estuary of the Senegal River: Effects on the Environment of the Breach of the 'Langue De Barbarie' Sand Spit in 2003

Awa Niang and Alioune Kane

Abstract

The Senegal River estuary and its coastal interface, the 'Langue de Barbarie', a long sandy spit shaped by the littoral dynamics, are located in Sahelian zone. Their instability results in considerable risk to their hydrological, climatic and ecological balances. The ecosystems there have suffered severely from the effects of drought and reduced freshwater inflows. Dams have partly dealt with the problem of water availability, especially in the upper basin. But the effects of their management on the environment have often been criticized. The breaching of the Langue de Barbarie sand spit on 4 October 2003 was justified by the imminent flooding of the city of St. Louis. It allowed the rapid escape of the flood waters of the Senegal River and thus saved St. Louis. The initial channel of 4 m width is now 2 km wide, with significant changes to the environment. This rapid evolution of the breach was accompanied by major impacts on the environment. Today, the lower estuary of the Senegal River is at a critical stage of its history with the accumulation of vulnerability factors such as the development of a marine dynamic, the over-salinization of water and lands and the rapid morphological change of the Langue de Barbarie sand spit caused by severe coastal erosion. In socioeconomic terms, despite the attempts of the communities to adapt through the development of activities such as salt extraction or the move of market-gardening activities towards less disadvantaged areas, the situation remains alarming, in view of the impoverishment of the local communities.

Keywords

Senegal river • Estuary • Sand spit • Langue de Barbarie • Hydrodynamics • Morphology of the estuary • Salinization • Environmental change

Introduction

The lower estuary of the Senegal River, like the rest of the basin, has, since the beginning of the great drought of 1970, seen a series of developments and actions in line with significant changes in its environment. This is one of the

A. Niang (✉) · A. Kane
Ecole Doctorale «Eau, Qualité et Usages de l'Eau» (EDEQUE), Université Cheikh Anta Diop de Dakar, Dakar-Fann, BP 5005, Dakar, Senegal
e-mail: awa10.fall@ucad.edu.sn

A. Niang · A. Kane
Laboratoire LINUS «Littoraux: Interface Natures-Sociétés», Unité Mixte Internationale (Joint International Unit) 236 Résiliences, Institut de Recherche pour le Dévelopement (Research Institute for Development) Campus International de Dakar-Hann, Dakar, Senegal

S. Diop et al. (eds.), *The Land/Ocean Interactions in the Coastal Zone of West and Central Africa*, Estuaries of the World, DOI: 10.1007/978-3-319-06388-1_3,
© Springer International Publishing Switzerland 2014

reasons for the programme of large hydro-agricultural installations of the OMVS.[1]

Since the early years of operation of large dams on the Senegal River, many research studies have been devoted to the consequences of these structures on the hydrological cycle, water quality and environmental changes on the local to regional scale (Diakhate 1988; Kane 1997; Equesen 1993). A near unanimity has emerged about their appropriateness and their effects, some of which are considered highly beneficial.

In 2003, the issue of water in the Senegal River basin has raised new developments with the opening of a new channel. The creation of water reservoirs behind the dams offers now the opportunity to develop the agriculture through irrigation during a longer dry period. The return to better rainfall conditions in 2003, particularly in the upper basin, resulted in threats to the viability of the hydraulic installations and especially the city of St. Louis, in the lower part of the Senegal River. To avoid flooding, the Government of Senegal decided to open a channel, 7 km south of St. Louis, on the night of 3-4 October 2003.

Designed specifically to allow a release of hydraulic energy and preserve the city from flooding, this breach now functions as the real mouth of the Senegal River. From an initial width of 4 m, the opening today is more than 2 km wide and the changes continue, since the baseline scenario was that the gap would close very quickly.

However, this breach, the new mouth of the Senegal River, is now considered as a real option for flood control of the city of St. Louis, despite all the criticisms raised by environmentalists. This explains the decision to pay great attention to increasing knowledge of the Langue de Barbarie and the evolution of the Senegal River's mouth. Indeed, at present, the river mouth is the centre of wide-reaching projects whose results are likely to be predictable and harmful.

The purpose of this study, in addition to understanding the dynamics of the breach, is to determine its major effects on the functioning of the estuary at the hydrodynamic and morphological level. It is based on an analysis of the evolution of factors such as tidal levels, salinization of land and water, and a cartographic analysis of morphological changes in the estuary.

The opening of the breach, a decision of the Government of Senegal, has met expectations well; however, the absence of accompanying measures and monitoring plans has resulted in an environmental and socioeconomic disaster unprecedented throughout the Senegal River estuary, which is now severely threatened. It is therefore necessary to identify research directions in the short and medium term and also to consider aspects related to forecasting the future evolution of the lower Senegal River estuary.

The Study Area

The Langue de Barbarie sand spit and the mouth are central focus in the management of the Senegal River estuary. The social environmental system is highly vulnerable to change due to decade of various threats impacting the Senegal River estuary (Kane 2010). Vulnerability is related to the damage which an ecosystem is likely to suffer following natural or anthropogenic catastrophes (D'Ercole 1994). Vulnerability is a concept related to resilience which translates the capacity and the mechanisms of response of a community to a shock that is comparable to natural disasters related to water (Berkes 2007). To understand the current configuration of the estuary, it is necessary to review recent trends and especially the mechanisms that led to the current situation.

The Senegal River Estuary

The Senegal River estuary, the lower part of the basin (Fig. 1), with landscapes strongly marked by salinity, includes the Langue de Barbarie and its hinterland, also named Gandiolais. The notion of an estuary of the Senegal River has been widely discussed by experts on coastal geomorphology such as Tricart (1961) who defended the thesis of a pseudo-delta. Kane (1985 and 1997) talk of an estuary despite the discharge of river water through a single channel, a phenomenon that can be explained by the hydrological regime and coastal dynamics in the lower part of the Senegal River Basin.

This estuary was formed between 6800 and 4200 B.C. by the formation of a lagoon behind coastal sand, the Langue de Barbarie (Kane 1997). The late Holocene evolution has been studied by Elouard et al. (1977), Monteillet (1977 and 1988), Kane (1985) and Diouf (1989), mainly on the basis of an analysis of the sea level variations.

The city of St. Louis was built in the estuary on flat land and semi-fixed dunes with alternating mudflats, creeks and tributaries of the Senegal River. It is a city only slightly above sea level, almost everywhere it is less than 2.5 m NGI (Durand et al. 2010), which exposes it to frequent flooding.

The Senegal River estuary is composed of:

- Two water bodies on the right bank, immediately downstream of the Diama dam (the Nthialakh and Gueyeloubé) draining lagoons located in Mauritania;
- The Djeuss, crossed at Dakar-Bango by a bridge–dam and whose upstream part constitutes the water reservoir

---

[1] Organisation pour la Mise en Valeur du fleuve Sénégal (Organization for the Development of the Senegal River).

Morphological and Hydrodynamic Changes

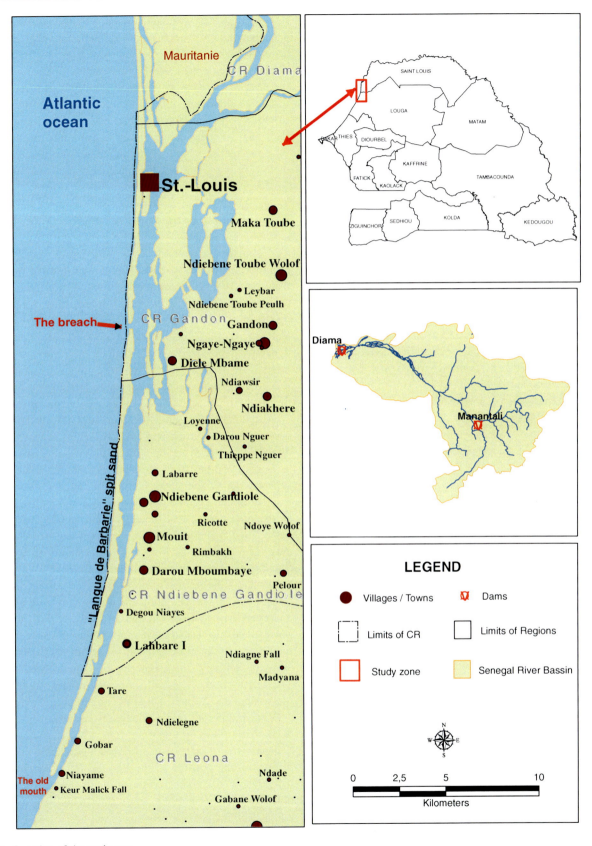

Fig. 1 Location of the study area

of St. Louis and the downstream part, mainly open to the estuary, is bordered by a modest mangrove swamp;

- Backwaters Khor and Marméal located upstream of St. Louis and which join the river on the south,
- The short arm of the river between the Langue de Barbarie and the island of St. Louis, which is in communication with the main river,
- the lagoon complex on the left bank, between St. Louis and Gandiole, which includes Leybar and Ndel backwaters, shallow channels, temporarily flooded by the tides, partially lined with mangroves.

The estuary also includes the National Park of the Langue de Barbarie (PNLB) located about 12 km downstream of St. Louis covering an area of about 2,000 ha. The PNLB extends, partly, onto the coastal zone with the presence of some brackish lagoons and mangrove relic. Nesting areas for thousands of birds, but also spawning fish, and for shrimp including the park's environment are strongly threatened by sewage from St. Louis.

Before constructing the Diama dam, saline intrusion was a major determinant of all socioeconomic activities in the Senegal River Delta. In normal hydrological conditions, saline water reached Podor, about 300 km from the mouth. During extreme years, sea water intrusion reached Dioudé Diabé, 150 km upstream of Podor. In the estuary, the river flooded many areas occupied by mangrove mudflats and halophytic steppes. Between Gandiole and St. Louis, the landscape is composed, in the area between the road and the river, of *Rhizophora*, *Laguncularia racemosa* and *Avicennia africana*. The hinterland is mainly occupied by clumps of vegetation of different varieties. In sandy areas, we note the dominance of *Sporobolus spicatus*.

The beginning of the major hydraulic structures of OMVS with the Diama dam in 1985 resulted in a new balance in the evolution of the estuary of the Senegal River. From the first months of operation of the dam, the hydrodynamic conditions were highly modified leading to a continuation of morphological, sedimentological and ecological changes.

Since 1985, the estuary of the Senegal River has been confined to a very narrow zone between the Diama and the Langue de Barbarie sand spit. The periodic opening of the dam to prevent hyper-salinization of water due to evaporation also led to a decline in mangrove mudflats, which have been converted mainly into residential areas. Vulnerability to flooding increased, particularly in low-lying areas surrounding the city of St. Louis.

The change in the hydrological regime of the Senegal River related to the dams of Diama (downstream) and Manantali (upstream) has had a strong impact on the morphological evolution of the region. Designed in the context of drought, these two dams were essential to ensure the sustainability of water resources but did not take into account factors such as the sedimentology and ecology of these environments. Indeed, the reduction of sediment yield has transformed the estuarine system and introduced disturbances that have altered and/or perpetuated the natural instability of the estuary, particularly the mouth. The hydrological functioning of the estuary has changed significantly in recent years from the effects of these dams and their very important transformations of the environment.

Langue de Barbarie and the Senegal River Mouth

The Langue de Barbarie is a long sandy spit of about 30 km, oriented NNW–SSE; it separates the Senegal River from the Atlantic Ocean and is divided into three sectors (Sall 1982): an external or shoreline maritime sector swept by the swell; a dune area under the influence of the wind; and a fluvial sector or internal shoreline subjected to fluvial dynamics. The Gandiolais, which is the hinterland of the Langue de Barbarie, is an area of dunes punctuated by mangroves, mudflats and saline areas. It also corresponds to the northern part of the Niayes region (Niang 2002).

At the maximum, the Quaternarian Nouakchottian transgression between 5500 and 4000 B.C., the sea invaded the Senegal River delta and built an estuary which then extended to Boghé, 250 km from the coast (Monteillet et al. 1981). Between 4000 and 1800 B.C., a significant littoral drift resulted in the formation of sandy offshore bars from east to west along the coast of Senegalo-Mauritanian sedimentary basin. The Langue de Barbarie, whose construction continued until the sub-actual period, was the result of North–South–South currents resulting from the reflection of northwest swells along the coast. These currents, very strong, cause an intense marine sedimentation and force the river to erode the coast and then elongate the Langue de Barbarie. The strong power of the swells and the currents makes the Senegal River mouth one of the most agitated in the world (Coleman and Wright 1975): the power of the swell there, $112.42.10^7$ ergs.s^{-1}, is 550 times that at the mouth of the Mississippi and 5 times that at the mouth of the Nile.

Very fragile and morphologically unstable, the Langue de Barbarie plays an important role in the dynamics of the Senegal River. Indeed, its end has determined, since its formation, the position of the Senegal River mouth which has always been subject to high spatial and temporal mobility (Gac et al. 1982; Kane 1997; Dia 2000; Lamagat 2000). The location of the mouth south of St. Louis probably dates back from the mid-seventeenth century. Between the Senegal River and the Atlantic Ocean, the Langue de

Fig. 2 Wavelet analysis of zonal wind fields from the ERA40 data (Courtesy to Massei N., M2C Rouen)

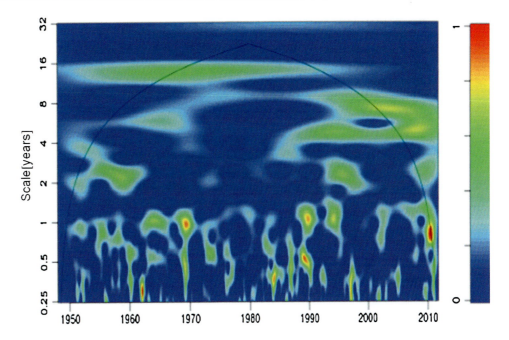

Barbarie is home to about 24 % of the population of the city of St. Louis.

Maps and historical documents have been used to trace the evolution of the mouth since 1658 and set a periodicity of 14 years for breaks around the Langue de Barbarie and the maximum penetration of the tide into the estuary (Rochette 1974). This periodicity seems perfectly correlated with the results of an analysis of zonal wind fields from the ERA40 climate reanalysis data, which also show a cycle of between 12 and 16 years, observable until the early 1990s (Fig. 2), coinciding with the start-up of the Diama and Manantali dams. This change in the 1990s was also recorded in the Maghreb, including Morocco, Tunisia and Algeria (Massei et al. 2007; Laignel et al. 2010; Louamri et al. 2011).

So since 1973, no significant breakdown of the Langue de Barbarie has been detected; the Senegal River mouth has steadily migrated south (Fig. 3), carrying with it the littoral spit which has neither been widened nor grown since its formation (Diouf and Kane 2002). The continuation of this southern trajectory implies the mobilization and deposition of sediments on the northern edge and the erosion of the southern edge of the spit. Indeed, the Langue de Barbarie is a sedimentary transit area of North–South direction. The estimation of sedimentary transport on the external shore, carried out according to various methods and by several authors, gives values between 1.5 million m^3 $year^{-1}$ and 365,000 m^3 $year^{-1}$ (Minot 1934; Riou 1936; Surveyer, Nenninger & Chenevert 1972; Pinson-Mouillot 1980; Sall 1982; BBL/SW et al. 1985). Based on topographical surveys and annual average lengthening southward, the estimates made by Kane (1985 and 1997) reported an annual average volume of approximately 600,000 m^3 of sediment transported annually to the forefront of the Langue de Barbarie. More detailed estimates by Barusseau (1980) give values between 223,000 and 495,000 m^3 $year^{-1}$ for mobilized materials of dimension varying between 0.1 and 0.5 mm.

On the internal shore, corresponding to the river sector, suspended sediments were estimated, in particular by Kane (1985 and 1997) and the EQUESEN programme (1993). Flows of continental origin transiting to the river mouth were 725,000 tonnes for the period 1989–1990. This discharge of continental sediments is in the form of a turbid plume, very characteristic of the Senegal River mouth (Fig. 3). According to Michel et al. (1993), the sediment load varies between 10^6 tonnes in low flood to 2.8 10^6 tonnes in very high flood. These suspended sediments are composed mainly of clay (92 %), fine silt (6 %) and coarse silt (2 %).

The Breach of the Langue de Barbarie

The first years of the Diama dam resulted in profound changes, both morphological and hydro-chemical, in the lower estuary, the operation of which was strongly altered. The dam became the artificial boundary upstream of the estuary while the limit downstream remained the mouth, a unique outlet where sea water can go back into the river (Kane 1997). The Manantali dam completion in 1988 contributed to the complexity of the system. In addition to

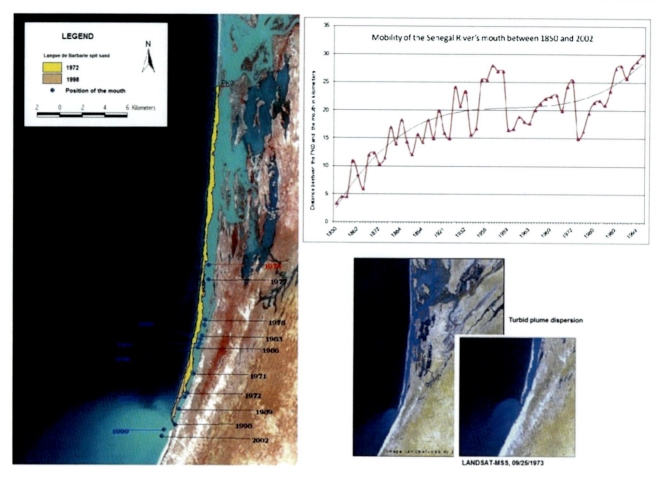

Fig. 3 Spatial and temporal mobility of the Langue de Barbarie and the Senegal River mouth from 1970 to 2002 and turbid plume dispersion (Niang 2002)

rainfall, the hydrological functioning of the system was found to be highly dependent on the management options of the dams.

The chronic instability of the Senegal River mouth (Gac et al. 1981; Kane 1985; Nakamura et al. 2002) grew blurred, giving way to a regular progression of the Langue de Barbarie towards the south. This southward progression of the spit constitutes indeed a characteristic of the Langue de Barbarie.

In the case of extreme events, the flood risks are also multiplied. The vulnerability of the city of St. Louis to floods has increased significantly due to the double influence of maritime and fluvial waters on the estuary.

The context of the emergency that led to the opening of the breach makes it possible to understand, partly, the impacts that it caused on the environment of the estuary. Monitoring the dynamics of the breach gives us hope for the definition, in the long term, of a model of evolution of the Senegal River estuary.

Context of Opening

The establishment of dams on the Senegal River, especially the Diama dam, initiated a new operating pattern in the lower part of the basin. Sediment retention at Diama is one of the first elements that influenced the evolution of the lower estuary. Indeed, under the predominant effect of marine dynamics, the sandy spit is subjected to intense degradations. The immediate consequence is a significant reduction in the depth and the width of the estuarine area which thus prevents a normal discharge of floodwaters to the sea, causing flooding in St. Louis and its environs (Dia 2000).

A city that is hardly 7 m above sea level, St. Louis, because of its configuration and its urban dynamic space, is highly vulnerable to flooding from rain and rivers. Storm floods are primarily related to the coverage of wetlands, about 900 ha, which were previously able to absorb these waters. Low parts of the city are flooded when daily rainfall

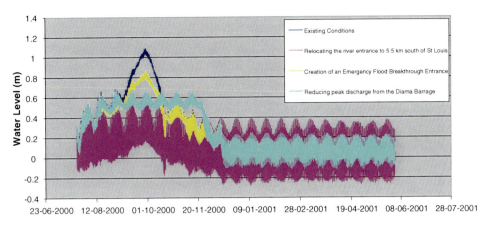

Fig. 4 Evolution of water levels in St. Louis under different scenarios of management of the Senegal River estuary (*Source* DHI 2002)

Photo 1 Aerial views of St. Louis suburb of Sor and during the 1950 flood (Courtesy to NDIAYE Gora)

of more or less 90 mm is recorded (Kane 2010). River floods are related to the low topography of St. Louis Island which was flooded when the water level reached 1.5 m IGN; the ordinary high watermark of the river is 1.2 m IGN. As a result of hydraulic installations such as the jetty and rehabilitation of the docks, these values have been revised upwards. The alarm level is now set at 1.75 m for St. Louis while the peripheral districts such as Diaminar, Pikine and Khor are under threat from a level of 1.20 m IGN (Fig. 4).

Since the nineteenth century, the city of St. Louis has experienced floods, more or less severe depending on the quality of the interventions by the local authorities and/or management of the Senegal River Basin. Kane (1997) recorded more than ten exceptional floods between 1850 and 1950, all of which resulted in severe flooding of the city and its surroundings, in particular the suburb of Sor. More recently, several flood events were noted in the lower estuary with water levels higher than 1.80 m IGN. In 1999, a record height of 2.10 m IGN was reached in October, followed by particularly severe flooding, probably less severe than in 1950 (Photo 1) but more severe in relation to the spatial extension of the city.

In 2003, heavy rainfall in the upper basin resulted in the arrival of five successive flood waves at the Bakel station between August and October 2003. These affected the valley and the estuary which recorded early and significant water levels, generally above alarm levels. The alert level in St. Louis is 1.50 m IGN, a height reached on 19 August 2003. In mid-September, the water level has always been between 1 and 1.50 m in St. Louis and between 1.35 and 1.60 in Diama. The dam discharged more than 1,000 $m^3\ s^{-1}$ over a period of more than two months. The rising waters become alarming for St. Louis; the managers

Photo 2 Progression of the gap between October and November 2003 (*Photos* 1. Diop, 2003)

of the Diama dam were forced to allow releases to the lower estuary, contributing to raising the water level in the Senegal River.

The vulnerability of the city of St. Louis to floods is related to physical, human and organizational factors: (1) the city is built on a site with very low topographic heights, barely exceeding 3 m except in dune areas where altitudes can reach 10 m; (2) building on the flood plain, which is characterized by important deposits of muddy sediments, prevents proper drainage of large floods; (3) urban planning did not allow for the constraint represented by the river—the long-lasting drought since the 1970s had virtually removed flooding from the list of urban risks. Other factors such as lack of dredging of the river channel and from the river mouth to the city of St. Louis provide a better understanding the exposure of the lower estuary to the recurrent floods which it has experienced in recent years.

To limit this threat and protect St. Louis and its heritage, the government of Senegal decided to open a breach in the Langue de Barbarie on the night of 3–4 October 2003. The main objective of this action was to rapidly release the floodwaters that threatened the city. The gap thus formed lowered the water level from 1.95 to 1 m IGN, with discharges of 100 $m^3 s^{-1}$ at the opening.

This particularly severe action was decided and executed without preliminary study of the potential impacts or any accompanying measures to mitigate their effects in the longer term.

The idea of opening a breach in the Langue de Barbarie and/or dredging the mouth is not new; several authors and studies suggested it since the early nineteenth century. Studies of the seaworthiness of the Senegal River led by Cosec (2002) also recommended digging a channel for improving the navigation and security of crossing the bar. Modelling studies conducted within the framework of an Integrated Coastal Area and River Management Programme in 2001–2002 under the supervision of the Water Ministry proposed, among other solutions, the opening of a channel with floodgates at about 5.5 km south of St. Louis. The effectiveness of the channel was to be improved by dredging the river channel to allow faster drainage. These studies have clearly not been taken into account.

Spatial and Temporal Evolution of the Breach

The breach in the Langue de Barbarie was opened about 7 km south of St. Louis, on the night of 3–4 October 2003. Its initial dimensions were 4 m broad and 100 m length, with a depth of 1.5 m (Photo 2). The decision to open the breach was made by high-ranking government authorities.

From the early hours of its opening, the breach fully met expectations with the creation of a hydraulic head and a breaking of the sand spit whose effect was to rapidly drain the estuary and thus preserve the city of St. Louis from flooding. The water level in St. Louis fells from 1.95 to 1 m, nearly 50 %, in 26 h.

The rapid evolution of the breach, under the combined effects of the swell and longshore drift, is still a concern, justifying the various monitoring performed by the Regional Water Resources Services and the Department of Geography of Cheikh Anta Diop University between October 2003 and July 2008. With a width of 200 m, the discharge flow rate stood at 1,906 $m^3 s^{-1}$ on 6 October 2003. On October 23, the breach was 329 m wide and the discharge flow rate was 1,968 $m^3 s^{-1}$. In December 2003, the width was 490 m.

The initial width of 4 m increased rapidly to 1,300 m in one year and then to 1,700 m in July 2008, according to bathymetric measurements carried out regularly by the Regional Services Hydraulics St. Louis (Kane et al. 2011). In addition to widening rapidly, the gap also quickly

Fig. 5 Enlargement of the breach and elongation of the north spit between October 2003 and October 2012 as measured on a series of LANDSAT images (www.usgs.gov)

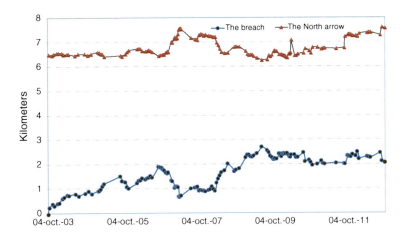

deepened its initial depth of 1.5 m on 4 October and increased to 6 m by 26 October.

These data are confirmed by measurements from a series of more than 130 LANDSAT images between October 2003 and October 2012 (Fig. 5). In contrast to the field data, which showed an exponential growth of the breach, the measures using LANDSAT imagery show that the width of the breach showed high intra-seasonal and inter-annual variations, probably related to oceanic forcing (tides, sea level pressure) and winds (zonal winds and southern winds to 800 HPA, winds to 10 m). This assumption needs to be confirmed with different correlations between the width of the breach and the evolution of these different elements.

After an almost exponential trend between 2003 and 2006, with a width increasing from 4 m to about 2 km in September 2006, the breach experienced a period of decline, with widths ranging between 700 and 900 m. A new phase of exponential growth then began, with increasing widths reaching a record 2.7 km in July 2009. The width measured on the image of 20 September 2012 was 2.38 km. Starting from September 2010, with the appearance of a small island which resulted in the creation of a second passage way, the breach behaved as a real mouth similar to the old mouth, moving slowly but surely towards the south.

The speed of progression of the sandy spit seems synchronized with that of the widening of the breach (Fig. 5). The length of the northern portion of the spit has rarely exceeded 7.5 km since 2003. On 6 October 2012, the length of the northern portion of the Langue de Barbarie was 7.3 km.

The gap functions as a real mouth, wide open to the sea, very deep and subjected to a strong agitation. Since 2003, the city of St. Louis, in particular in its island component, has been protected from river flooding. When a flood arrives, water is quickly discharged to the sea. Indeed, the river flow dynamics have been reinforced by the currents and marine dynamics, and this new situation has raised serious concerns about the nature of the flood hazard in the region which could now come from the sea (Durand et al. 2010). More particularly, much of the country around St. Louis is now below sea level (−0 m IGN). The island of Doun Baba Dieye, located to the right of the breach, which threatened the early months of operation of the new mouth, is now almost engulfed because of the tides and the high penetration of marine waters into the estuary. This possible disappearance of the Doun Baba Dieye Island has already been studied by Dieng (2010) and resulted in increased vulnerability of the population of the island, which forced them to relocate on the mainland.

Today with two new openings south of the channel of 2003, the entire lower estuary is threatened with destruction. The establishment of a monitoring mechanism for the Langue de Barbarie is a necessity in the context of the development of the lower estuary, in particular with a view to safeguarding human settlements. The challenge of managing this environment revolves around the evolution in the short and medium term of these three openings which have tripled the contacts with the Atlantic Ocean (Box 1).

The breach of the Langue de Barbarie in October 2003 established the new position of the Senegal River mouth. In addition to shortening the transit time of water from the Senegal River to the sea, this has also resulted in a change in the flow conditions upstream and downstream of Diama.

At the scientific level, the first reports indicate a very alarming diagnosis of the morphology of the littoral spit, the hydrodynamics, the biological diversity and environmental changes, which in the long term may result in a disturbance of this region that is without precedent.

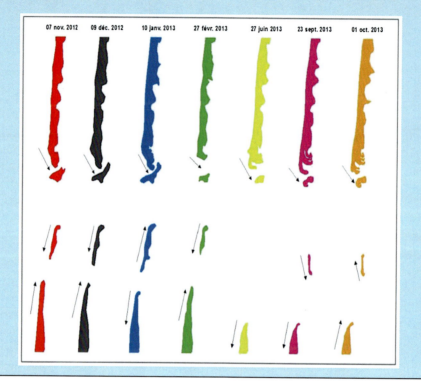

The breach on Langue de Barbarie, a real threat for the Senegal River estuary

The breach opened on Langue de Barbarie spit sand in october 2003 *cause* significant morphological changes in the Senegal River estuary. Based on LANDSAT imagery, the survey conducted on sandy spit from October 2003 to October 2013 demonstrates major environmental changes on the lower estuary. From November 2012 to October 2013, the width of the breach is increased from 4 to *around* 5.8 km. The rapidly changing environment is a strong concern for people whose vulnerability has tripled since 2003.

Impacts on the Estuary

The study of the hydrodynamic modifications, the water quality and the morphology of the spit will provide some answers about the projected risk of increased vulnerability of the environment of the lower estuary and local populations. A recent study (Kane 2010) mentioned the depletion of some fish species with high economic value. The market-gardening that was practiced in Gandiolais, already under limiting conditions before 2003, has now fallen into neglect. This means that all the difficulties related to the impacts of the breach on the environment of the Senegal River estuary, including with the hydrodynamics, the salinization of lands and waters of Gandiolais region and the morphology of the lower estuary, are worsening.

On the Hydrodynamics

The main objective that guided the opening of the breach was the creation of a hydraulic head and a breaking of the sand spit, enough to discharge all the water that threatened the city of St. Louis in early August 2003 (Fig. 6). Only a few hours after the opening of the channel, the decrease in the water level between St. Louis and Diama was nearly 50 cm. The effectiveness of this opening on flood management no longer needs to be demonstrated; indeed, no floods of river origin have been recorded in the lower estuary since 2003, except overflows of the river in the low zones around the jetties.

In the first months following the opening of the breach, the fishermen were delighted by the shortening of the river–sea distance, allowing substantial economies of gasoline. However, many accidents, nearly always fatal, around the breach

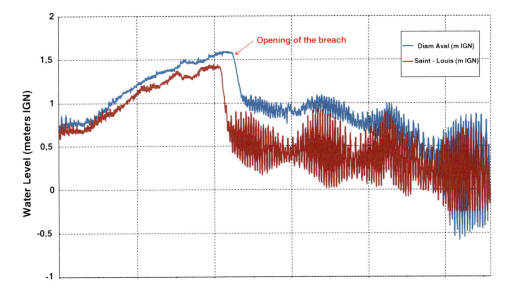

Fig. 6 Instantaneous data in Diama and St. Louis (Quay) before and after the opening of the breach, between August 1st and on 30 November 2003 (*source data* IRD)

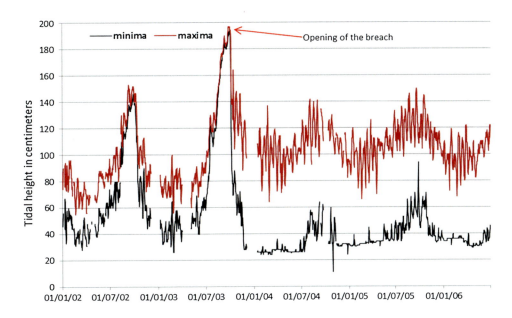

Fig. 7 Evolution of the tidal range in St. Louis between 2002 and 2006

are often reported by the press. These could be related to the strong turbulence of the water and the strong currents.

Since 1985, the hydrodynamics of the Senegal River estuary has been under the double influence of the tide during periods of low water and of the operations of floodgates at Diama dam in times of flood. The tide in the Senegal River estuary is semi-diurnal, with a period of 6 h.

Before the breach, the tidal range was attenuated by the distance between the mouth and the town of St. Louis (approximately 30 km); today an inversion of the situation is seen, depending on the distance to the sea (approximately 7 km).

Since October 2003, the modifications of the tidal signal in St. Louis have been one of the most remarkable consequences of the opening of the breach (Fig. 7). In the dry season, the river follows the fluctuations of the oceanic tides. In periods of low tide, the water level in St. Louis is close to − 0.44 m IGN; when the waters of the river are withdrawn and one can see the piles below Pont Faidherbe (Photo 3), a very unusual sight for the population of St. Louis.

Examination of the tidal measurements for 2002–2006 highlights important changes: water heights recorded as often for low as for high tides, indicate large tidal range downstream of Diama, three times the previous level, increasing from 0.30 m (average 2001–2002) to 0.93 m (average 2004–2005) according to Kane et al. (2010). This represents a major change in the hydrological functioning of the lower estuary which is now under the almost exclusive control of the tides.

The tides are now experienced all the year, even during high waters of the Senegal River, contrasting with what was

Photo 3 Withdrawal of the Senegal river at low tide at the Faidherbe bridge (Pictures from Kane A., December 2003)

traditionally known in this area. This new situation also affects the management of the Diama dam whose function has changed from an anti-salt barrier to a storage dam. Indeed, the average water level at Diama has increased from 1.50 m in 1992 to 1.75 m in 1995, 1.90 m in 1997, 2.0 m in 1999 and 2.10 m since 2002. The aim of this increase in water level is to satisfy the increasing demand for agricultural water (Duvail 2001). This also contributes to the increasing water levels in the estuary, which periodically receives freshwater discharged by the Diama dam.

The current estuarine hydrodynamics is strongly dependent on the constraints of management of Diama which requires periodic opening of the floodgates to regulate the levels upstream. Indeed, the operating rules of Diama require the energy of dissipation not to exceed 1,000 $m^3\ s^{-1}$ under a 1 m fall (Coyne and Bellier Sogreah 1987). But according to a new study, following the increase of the tidal estuary, the problems of energy dissipation no longer arise since this energy could be fifteen times greater without any consequence for the dam.

Salinization of Land and Water

Salinization of land and water has always represented a major challenge for socioeconomic development in the lower estuary of the Senegal River, especially the region of Gandiolais, in line with the natural region of Niayes.

The digging of the drainage channel brought the estuary to the sea, causing an increase in the salinization of groundwater and surface water. The effects of ocean dynamics, which now dominates the river, are now being felt throughout the lower estuary and almost throughout the year. Containment of the estuary between the breach and the Diama dam contributes to the worsening of the situation. Accordingly to the opinion of the Gandiolais local people, salinity is increasing exponentially and is now reaching key areas hitherto spared. The river salinity data in St. Louis show a sharp contrast between before and after the construction of the Diama dam and after the opening of the breach. Salinity levels are much lower than in the period immediately after the construction of the both Manantali

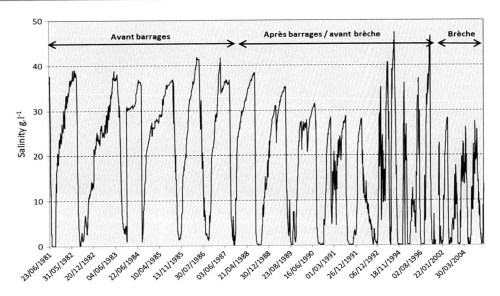

Fig. 8 Evolution of the salinity range in St. Louis from 1981 to 2006

Table 1 Rates of salinity in the Senegal River in St. Louis in 2003 and 200

Years	Maximum		Minimum		Number of days of zero salinity
	Salinity (g l^{-1})	Date	Salinity (g l^{-1})	Date	
2003	17.8	May 21	0.48	July 15	129
2004	26.3	May 31	0.208	July 30	52

Source St. Louis Regional Service of Hydraulics

(upstream) and Diama (downstream) dams (Fig. 8) but with the volumes of water much higher. However, the general trend of the change of surface salinity in St. Louis has been on the rise since the opening of the channel.

Salt is present in the estuary most of the year, due to the proximity of the sea and tidal effects which are dominant compared to freshwater from the Senegal River, the result being that the river is salty almost throughout the year (Table 1). Indeed, the number of days where the salinity is zero more than halved between 2003 and 2004, from 129 to 52 days.

Some measurements of water quality conducted in some wells in the Gandiolais region between 2005 and 2012 show increasing trends of salinity in groundwater, already affected by residual salt of geological origin. In villages located in the lower estuary, lower freshwater inflows have caused enormous difficulties in water supply for local populations, which are now seriously threatened by a drastic reduction in freshwater sources despite the programme of water purification units installed in several locations in the Senegal River estuary.

Exploitation of salt in the Gandiolais dates from the late nineteenth century, the salt resulting from marine transgressions during the Quaternary, in particular during Inchirian and Nouakchottian periods. During the withdrawal of the sea, the evacuation of the marine deposits was not entirely realized because of topographical and climatic factors. The marine deposits thus accumulated in the lower estuary and caused salinization of land.

Today, the boom in the exploitation of salt provides the best indicator of salinity in the estuary. Salt extraction is practiced mainly by women, in the Gandiolais region, particularly Tassinere, Mouit and Ndiebene Gandiole, around the basin of Ngaye–Ngaye, tributary of the basin of Gueumbeul, where the salinity rate generally exceeds 35 g l^{-1} at the beginning of winter (Corea 2006).

The volumes of salt produced have increased steadily over the past ten years that is since the opening of the breach and especially since the closure of the old mouth, which turned into a lagoon since 2005 and operates almost like a salty swamp. The size of salt farms gives a clear indication of the presence of salt in the estuary (Fig. 4). This salt mining can be seen as an adaptation attempt of local communities to degradation of environmental conditions. However, the development of this salinization now paralyses the market-gardening activities which are reducing and could even disappear. As an alternative, the population shifted their activities, when possible, to areas less affected by the salinization. Is this a sign of a developing resilience of these communities of Gandiolais? Only socioeconomic studies and surveys can provide an answer to this fundamental question (Photo 4).

Photo 4 Salt exploitation at Ndiebene Gandiole (LINUS, December 3, 2012)

Effects on the Morphology of the Lower Estuary

The Senegal River estuary and particularly the littoral spit of the Langue de Barbarie have always been subjected to a strong instability of their morphology, conditioned primarily by hydroclimatic and hydrodynamic factors.

The mobility of the furthest part of the Langue de Barbarie which always marked the Senegal River mouth is used as an indication of the morphological evolution of all the area. According to measurements carried out on the mobility of the spit since 1658, movement towards the south has generally been regular, punctuated of temporary movements back towards the north. After each breach, the spit returned to its movement towards south. It was this typical evolution which was observed until the commissioning of the Diama dam in 1985 and the opening of the breach in October 2003. The opening of this breach was a major and significant change of the environment of the Langue de Barbarie. Indeed, the Senegal River mouth was shifted by 30 km, to about 7 km south of St. Louis, with a similar shortening of the littoral spit (Table 2).

A multi-annual monitoring of the Langue de Barbarie has been made using LANDSAT imagery. Methodologically, the choice of LANDSAT imagery is justified primarily by its accessibility; the resolution of the images (30 m) and their scale (approximately 1/200,000e) constitute one of the major limits to their use. However, the selected images enabled us to trace, with acceptable accuracy, the limits of the Langue de Barbarie and its evolution from 2003 to the present. The mapping of the extent of the Langue de Barbarie is based on the use of the instantaneous shoreline as a reference line for all the years studied. This methodological choice may be debatable, given the heterogeneity of the images in terms of dates of acquisition, tidal conditions, etc. (Faye et al. 2008). The identification of the shoreline was carried out automatically by the vectorization module of the ArcGis 10 software. The use of this process is also justified by the significant number of images composing our working database. Moreover, the goal here was to highlight the evolution of the limits of the Langue de Barbarie between October 2003 and October 2012. As a reminder, measurement of the length of the Langue de Barbarie is based on a reference line located at the right of the Faidherbe Bridge; it is the same reference as that used in the historical documents (Gac et al. 1982).

There is evidence, according to analysis of the data for the width of the breach and length of the northern segment of the spit (Fig. 5), that there have been four major periods in the evolution of this environment. Generally, one notes strong erosion which results in important surface losses on the coastal side of the Langue de Barbarie and on the internal banks of Gandiolais. The most affected zone seems to be Doun Baba Dieye Island which, in one decade, lost more than three quarters of its surface area, resulting in the displacement of many families and the loss of their agricultural lands and thus their incomes.

From 2003 to 2006, there was a steady increase in the size of the breach and a relative stability of the northern portion of the sandy spit. The year 2005 was marked by the closure of the old mouth, thus creating a closed lagoon, suggesting significant ecological problems, because of the situation of the Langue de Barbarie National Park. Indeed, the disappearance of the small island which shelters the breeding of some migratory birds could affect tourism, one of the principal businesses of the St. Louis region.

Between 2007 and 2008, there was a reduction in the width of the breach which was accompanied by a regression of the spit. From 2009 to 2010, the breach widened steadily without resulting in a significant movement of the spit southward. In addition, this period was marked by the appearance of a small island to the south of the northern spit, in June 2010.

Table 2 Variations of the gap width and the length of the northern portion of the Langue de Barbarie

Date	Breach		Langue de Barbarie	
	Width (m)	Expansion/decline (m)	Length (m)	Extension/drop (m)
Oct.-04-2003	04	–	≈ 7,000	–
Oct.-14-2003	273	269	6,497	−503
Oct.-06-2006	1,920	1,467	6,471	−26
Oct.-06-2009	2,374	454	6,408	−63
Oct.-06-2012	2,168	−206	7,612	1,204
Oct. 2003-Oct 2012	–	2,164		612

(*Source images* www.usgs.gov)

From 2011, there has been a relative stability of the size of the breach which fluctuates between 2,500 and 2,700 m. The width of 3.1 km measured by the US Army (Barry and Kraus 2009) has not been found in our series of images. The channel consequently seems to reach a critical size which can cause weakening of the sandy spit. This threat is now proven with the appearance of two new openings south of the breach of 2003. Monitoring of these two openings is essential because of the challenges to the evolution of the lower estuary. What is now the probable evolution for the lower estuary of Senegal River and the Gandiolais area? Prospective studies should allow a better vision of the possible patterns of evolution of the area at the morphological level and especially the socioeconomic and socio environmental changes (Fig. 9).

Conclusion

The lower estuary of the Senegal River is a structurally fragile, unstable area, changing very rapidly depending on climatic, hydrodynamic and socioeconomic conditions. Its evolution, since geological times, shows its strong dependence on climatic and hydrodynamic factors. The change of configuration of the Senegal River basin resulting from the implementation of structural installations like the Diama and Manantali dams was not without environmental impacts on the estuary, whose cycles of evolution were consequently broken.

The opening of the breach on the Langue de Barbarie, in fact, corresponds to the displacement of the Senegal River mouth by 30–7 km downstream from St. Louis. Originally designed as a temporary structure intended to prevent flooding of the city of St. Louis, the breach became established and was accompanied by multiple environmental and socioeconomic problems. Thus, we observe salinization of surface water and groundwater, and of land, difficulties of access to drinking water for basically underprivileged populations, land loss due to erosion, hydrodynamic modifications, and the reduction of freshwater supplies. Any problems already present in the estuary were exacerbated by the opening of the breach. The socioeconomic consequences, such as income loss related to the impossibility of practicing market-gardening, exodus towards urban areas and extreme cases of clandestine immigration on board fishing boats, also constitute many obstacles to the sustainable development of the whole region.

The opening of the Langue of Barbarie sand spit, in an emergency situation, fulfilled its primary objective, which was to preserve St. Louis from flooding. However, no accompanying measures were taken to mitigate its predictable effects on the environment of the estuary. The negative impacts on hydrodynamics, salinization of water and land, and the morphology of the lower estuary are such as to naturally lead to questions on the appropriateness and relevance of this government action.

The vulnerability of the Gandiolais region has both natural and anthropogenic origins. The climate acts through reduction in precipitation, rise of temperatures and physical (gravitational water and groundwater) and physiological (plant species and animal) evaporation; while the estuarine location (at below sea level) supports the penetration of sea water into the depressions and backwater tributaries of the river.

The Diama dam, as a result of the way it is managed, especially in periods of low water (closing of the floodgates for approximately 7 months), upsets the balance of river–marine dynamics by storing water upstream, resulting in an exclusively marine dynamics downstream.

The breach, through the replacement and the closing of the previous river mouth, has accentuated the effects of marine dynamics such as erosion and the salinization of land and water. Today, all of these factors operate together in the lower estuary, causing an upheaval of the local environmental and socioeconomic situation.

Fig. 9 Multi-annual evolution of the Langue de Barbarie from 2003 to 2009

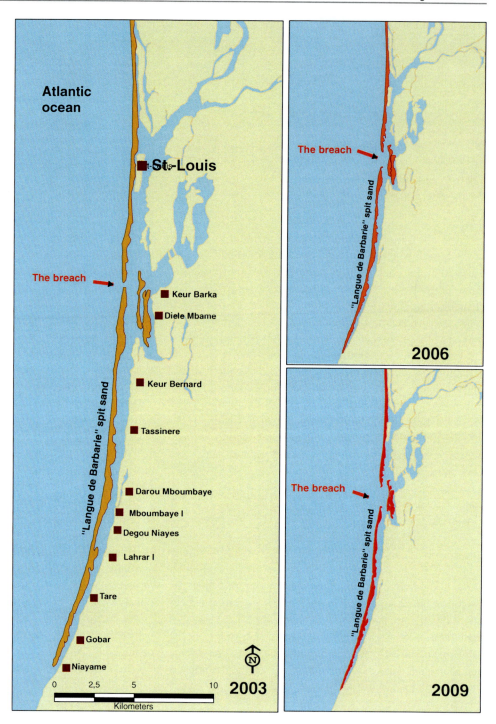

The vulnerability of the estuarine environment is thus an established fact but, in retrospect, one can question the resilience of this littoral ecosystem which is almost always rebuilt but at the cost of renewed shocks. Future studies will have to also reflect, as well as the questions of vulnerabilities, understanding of the mechanisms of rebuilding the estuary.

References

Barry KM, Kraus NC (2009) Stability of blocked river mouth on West Coast of Africa: inlet of the Senegal River Estuary. US army corps of engineers, coastal and hydraulics laboratory. Final report, 56 p

Barusseau JP (1980) Essai d'évaluation des transports littoraux sableux sous l'action des houles entre Saint-Louis et Joal. Bull. liaison ASEQUA Sénégal, n 58–59 pp 421–429

BBL/SW, OMVS, ACDI (1985) Etudes des ports et escales du fleuve Sénégal. Port de Saint-Louis. Etudes hydrographiques et hydrauliques. Rapport n 20, 149 p, annexes

Berkes F (2007) Understanding uncertainty and reducing vulnerability: lessons from resilience thinking. Nat Hazards 41:283–295

Coleman JM, Wright LD (1975) Modern river deltas: variability of process and sand bodies. In: Broussard ML (ed) Deltas. Houston Geological Society, Houston, pp 99–149

Corea M (2006) Analyse situationnelle des ressources en Eau dans l'estuaire du fleuve Sénégal: la dynamique de la salinisation dans le bief estuarien. *Mémoire de DEA Chaire UNESCO*, UCAD/Département de Géographie, 76 p

Cosec, Sogreah, Afid (2002) Étude de l'accessibilité et de l'implantation du port de Saint-Louis. Rapport SOGREAH-ELC-712064 CR, 18 p

Coyne et Bellier, Sogreah (1987) Consignes générales d'exploitation et d'entretien du barrage de Diama, 21 p

D'Ercole R (1994) Les vulnérabilités des sociétés et des espaces urbanisés: concepts, typologies, modes d'analyse. Revue de Géographie Alpine 4:87–96

Dia AM (2000) Ecoulements et inondations dans l'estuaire du fleuve Sénégal: le cas de la ville côtière de Saint-Louis. Mémoire DEA Géographie, Chaire UNESCO/UCAD 'Gestion intégrée et développement durable des régions côtières et des petites îles'. 65 p

Diakhate M, (1988) Ecodynamique des milieux et effets d'impact potentiels du barrage de Diama dans le delta du fleuve Sénégal. Thèse de doctorat de 3e cycle, Université de Lyon II, 450 p

Dieng D (2010) Apport de l'imagerie satellitaire au suivi de l'évolution environnementale de l'estuaire du fleuve Sénégal : cas du canal de délestage et de l'île de Doun Baba Dieye. RGLL N°08, pp 101–118

Diouf MB (1989) Sédimentologie, minéralogie et géochimie des grès carbonatés quaternaires du littoral sénégalo-mauritanien. Thèse Sciences, Océanologie, Univ, Perpignan 237 p

Diouf MB, Kane A (2002) Modifications morphologiques dans la zone de l'embouchure du fleuve Sénégal après la construction du barrage de Diama. Rapport GILIF, volet 1 bis, 33 p

Durand P, Anselme B, Thomas YF (2010) L'impact de l'ouverture de la brèche dans la langue de Barbarie à Saint-Louis du Sénégal en 2003: un changement de nature de l'aléa inondation ? Cybergeo: Eur J Geogr, Environ, Nature, Paysage, document 496, http://cybergeo.revues.org/index23017.html

Duvail S (2001) Scénarios hydrologiques et modèles de développement en aval d'un grand barrage. Les usages de l'eau et le partage des ressources dans le delta mauritanien du fleuve Sénégal. Thèse de doctorat en Géographie, Université de Strasbourg, 313 p

Elouard P, Faure H, Hebrard L (1977) Variations du niveau de la mer au cours des 15 000 dernières années autour de la presqu'île du Cap-Vert (Sénégal). Ass. Sénégal. Et. Quatern. Afr., Bulletin liaison n°50, pp 29–49

Equesen (1993) Environnement et qualité des eaux du fleuve Sénégal, Rapport final, 6 Tomes 12 chap., Projet CEE/ORSTOM/UCAD/ISRA, Dakar, 35 p. Doc. multigr

Faye IN, Hénaff A, Gourmelon F, Diaw AT (2008) Évolution du trait de côte à Nouakchott (Mauritanie) de 1954 à 2005 par photo-interprétation. Norois, n 208, 2008/3, pp 11–27

Gac JY, Faure H, Monteillet J (1981) Variations du niveau de la mer ou crues fluviales ? In: Proceedings of the symposium on variations in sea level in the last 15,000 years, magnitude and causes. Columbia (USA.), 6–10 avril

Gac JY, Kane A, Monteillet J (1982) Migrations de l'embouchure du fleuve Sénégal depuis 1950. Cahier ORSTOM, Série. Géol. XII 1:61–64

Kane A (1985)—Le bassin du Sénégal à l'embouchure. Flux continentaux dissous et particulaires. Invasion marine dans la vallée du fleuve. Contribution à l'hydrologie fluviale en milieu tropical humide et à la dynamique estuarienne en domaine sahélien. Thèse doctorat 3ème cycle, Univ. Nancy II, 205 p

Kane A (1997) L'après barrage dans la vallée du fleuve Sénégal : Modifications hydrologiques, morphologiques, géochimiques et sédimentologiques. Conséquences sur le milieu naturel et les aménagements hydro-agricoles. Thèse de doctorat d'Etat de Géographie physique, UCAD, 551 p

Kane C (2010) Vulnérabilité du système socio-environnemental en domaine sahélien : l'exemple de l'estuaire du fleuve Sénégal. De la perception à la gestion des risques naturels. Thèse de doctorat en Géographie, Université de Strasbourg/Université Cheikh Anta Diop, 317 p

Kane C, Humbert J, Kane A (2011) Modifications de l'embouchure du fleuve Sénégal : impacts sur la vulnérabilité des sociétés du bas estuaire. Natures tropicales. Enjeux actuels et perspectives, sous la direction de François BART. Espaces Tropicaux n 20, pp 316–326

Laignel B, Massei N, Rossi A, Mesquita J, Slimani S (2010) Water resources variability in the context of climatic fluctuations on both sides of the Atlantic Ocean. IAHS/Friends

Lamagat JP (2000) Inondation de la ville de Saint-Louis. Diagnostic. OMVS-IRD, 12 p

Louamri A, Mebarki A, Laignel B (2011) Variabilité interannuelle et intra-annuelle des transports solides de l'Oued Bouhamdane, à l'amont du barrage Hammam Debagh (Algérie orientale). Paper from the man, environment and sediment transport workshop, Tipaza, Algeria

Massei N, Durand A, Deloffre J, Dupont JP, Valdes D, Laignel B (2007) Investigating possible links between the North Atlantic oscillation and rainfall variability in northwestern France over the past 35 years. J Geophys Res 112:D09121. doi:10.1029/2005JD007000 10 p

Michel P, Barusseau JP, Richard JF, Sall M (1993) L'après barrages dans la vallée du Sénégal : modifications hydrodynamiques et sédimentologiques, conséquences sur le milieu et les aménagements hydro-agricoles. Projet Campus (1989–1992). UFR de Géographie Céróg (Strasbourg), Labo. De rech. En sédimentologie marine (Perpignan), Départ. de Géog., Départ. de Géol. de Dakar, coll. Etudes, Presses Universitaires de Perpignan 152 p

Minot A (1934) Contribution à l'étude du fleuve Sénégal. Bulletin du Comité d'Etudes Historiques et Scientifiques de l'AOF, t XVII, pp 385–416

Monteillet J (1977) Tourbes de l'Holocène inférieur (Tchadien) dans le nord du delta du Sénégal. Bull Asequa No 50:23–28

Monteillet J (1988) Environnements sédimentaires et paléoécologie du delta du Sénégal au Quaternaire. Thèse de doctorat 1986, Université de Perpignan, Labo. De Sédimentologie Marine (ed) Perpignan, 267 p

Monteillet J, Gac JY, Faure H (1981) Influence des oscillations du niveau de la mer et des variations des débits fluviatiles sur l'évolution des écosystèmes estuariens. In: Proceedings of the international symposium on 'Coastal Lagoons'. Bordeaux, 8–14 Sept 1981

Nakamura T, Hansen PB, Coly A, Niang A et al (2002) Vers une gestion intégrée du littoral et du bassin fluvial. Programme pilote du delta du fleuve Sénégal et de sa zone côtière. Rapport final, PNUE/UCC-Water/SGPRE, 114 p

Niang A (2002) Description des changements de paysage au niveau de la Langue de Barbarie, de l'estuaire et la zone côtière du Delta du fleuve Sénégal. Rapport GILIF, Phase II—Volet 2, 36 p. mult

Pinson-Mouillot J (1980) Les environnements sédimentaires actuels et quaternaires du plateau continental sénégalais (Nord de la presqu'île du Cap-Vert). Thèse 3e cycle, Univ. de Bordeaux I, 106 p

Riou C (1936) Rapport général. Proposition pour l'amélioration de la navigabilité du fleuve Sénégal. Aménagement de la vallée et utilisation de ses forces hydrauliques. Rapport MAS, Bull; 53, septembre 1936, Saint-Louis, Sénégal, 80 p. + tabl

Rochette C (1974) Le bassin du Fleuve Sénégal. Monographies hydrologiques. Orstom, 329 p. Doc. multigr

Sall M (1982) Dynamique et morphogenèse actuelles au Sénégal occidental. Thèse de Doctorat d'Etat. Université Louis Pasteur, Strasbourg 604 p

Surveyer, Nenniger, Chenevert (1972) Etude de la navigabilité des ports du fleuve Sénégal. Etudes portuaires à Saint-Louis, Kayes et Ambidédi. En collaboration avec C. Ostenfield et W. Jonson. Danish Institute of AppliedHydraulics. Projet Onu 51/71 Saint-Louis, Omvs

Tricart J (1961) Notice explicative de la carte géomorphologique du delta du Sénégal. Ed. Technip, Collection Mémoires du Bureau de recherches géologiques et minières, 139 p

Management of a Tropical River: Impacts on the Resilience of the Senegal River Estuary

Coura Kane, Alioune Kane, and Joël Humbert

Abstract

Tropical rivers are under constant pressure and are subjected to flood control policy and planning. The Senegal River has been, for more than thirty years, under multiple types of management and planning. That has become recently more important because of significant hydro-climatic variability that has occurred during recent years. Higher levels of winter rainfall have resulted in frequent occurrence of flooding which guided Senegalese authorities to create an artificial opening on the sand of spit so-called Langue de Barbarie. The major consequence of this has been the closing of the old mouth of the Senegal estuary 25 km downstream of the town of Saint-Louis. However, two new mouths have opened themselves recently within 150 m of the current mouth. The increasing vulnerability in the Senegal River estuary has been one of the main impacts of the watershed management of the Senegalese northern River accordingly. The developments of Senegal River management policy have been based on the concept of remedying the consequences of climatic events; then, the utilization of water resources presented risks that are far greater than expected.

Keywords

Watershed management • Resilience • Risk • Estuary • Senegal River

Introduction

Due to rapid growth of the population and lack of resources in developing countries, people tend to settle along coastal areas and develop activities in estuaries which receive constant natural and anthropogenic actions. Human beings have not been insensitive to the wide array of opportunities provided by the coasts and have been attracted to them, making the coasts the most favored locations to either live permanently, for leisure, recreational activities or tourism (Martinez et al. 2007; Culliton et al. 1990; Miller and Hadley 2005). Thus, Senegal River estuary has undergone significant alterations as a consequence of the hydrodynamic conditions that have prevailed there following human actions established in the area. With the construction of the Diama and Manantali Dams, in addition to the new opening on the mouth of the "Langue de Barbarie," the Senegal River has experienced major changes due to many water planning decisions (Kane et al. 2013). These decisions were made in the context of significant hydro-climatic variability. This resulted in the constant hydrological deficit recorded

---

C. Kane (✉) · A. Kane
Laboratoire de Géomorphologie et d'Hydrologie, Département de Géographie, Faculté des Lettres et Sciences Humaines, Université Cheikh Anta Diop, Dakar, Senegal
e-mail: courakane@yahoo.fr

A. Kane
e-mail: alioune.kane@ucad.edu.sn

J. Humbert
Laboratoire Image, Ville, Environnement—UMR 7362, CNRS, Université de Strasbourg, 3, rue de l'Argonne, 67083 Strasbourg cedex, France

S. Diop et al. (eds.), *The Land/Ocean Interactions in the Coastal Zone of West and Central Africa*, Estuaries of the World, DOI: 10.1007/978-3-319-06388-1_4,
© Springer International Publishing Switzerland 2014

during the 1970s, but also exploited the higher levels of rainfall, particularly prevalent during those years.

The Senegal River estuary is an extremely fragile environment; it is threatened by natural hazards such as the recurrent floods. This situation stems from the physical environmental conditions. The risk is thus permanent, which can be exacerbated by human presence and its multiple actions on the environment. By interventions was generally failed to seriously predict the effects induced, and they frequently destabilized environments characterized by precarious equilibriums. The new mouth of the Senegal River is causing significant damage to the estuary. This arrangement does not fail to change the hydrological regime of the estuary and especially led to changes in estuarine ecosystems and increased the vulnerability of the estuary. Anthropogenic actions lead, most often, uncertainty of the occurrence of catastrophic events. In this context, the domino effect can be triggered, resulting in, as a result, increasing vulnerability. The impacts associated with climate change will be much more catastrophic in Africa because of the weakness of the economy and the rapid growth of the coastal population (Nicholls and Cazenave 2010). Climate change and rising of sea levels will increase the risks to which the estuary of the Senegal River is already exposed, whereas the system has low capacity to absorb disturbance.

Enhancing disaster risk reduction before a disaster occurs, and also during the reconstruction process, requires enhanced knowledge regarding the most vulnerable groups, the areas at risk, and the driving forces that influence and generate vulnerability and risk (Birkmann 2007; Bogardi and Birkmann 2004). Better take into account of the physical environment in policy development is essential. Environmental assessment aims to contribute to environmental awareness among proponents and competent authorities and to environmental protection by requiring the ex ante evaluation of the environmental impacts of a wide range of public and private initiatives such as spatial plans (Runhaar et al. 2013). The vulnerability of the Senegal River estuary area was accentuated by the water management of the Senegal River with the opening of the new mouth that was made without prior impact assessment. To a certain extent, the resilience of the system is strongly found to disturb. For this reason, it is essential to carry out detailed studies of the impacts of development on the hydrological regime of the river and the consequences in the upstream and downstream portions. All positive and negative impacts should be considered before making a layout and fairly weigh the consequences of choices.

In this chapter, the consequences of the watershed management of Senegal River in the estuarine area are analyzed. This last appears as a space where a variety of risk factors which may cause significant damage. Its occupation makes these areas vulnerable for local riparian. The latter has a low capacity of resilience from watershed management issues.

Human Pressures and Functioning of the Senegal River Estuary

The presence of a risk linked to existence water resources in an area determines the way the population exploit it; that situation influences strongly the type of structure and management settled consequently (Beucher and Rode 2009). The situation of the Senegal River obeys to this concept. Somehow, it could be promoted some existing geo-engineering technologies when used properly have low or no significant negative impacts (Olson 2012).

Climatic variability is determining factors in risk management. In the 1970s, the Sahel region experienced a period of exceptional water deficit in terms of duration and intensity. This period of rainfall decreasing is still on going (Dai et al. 2004; L'Hôte et al. 2002, 2003) although less strong since the mid 1990s than during the 1980s. The average rainfall over the last 37 years has remained below the 1900–1970 average in this region with large-scale consequences (Hulme et al. 2005). For almost all rivers of West Africa has decreased (Mahé and Paturel 2009). More than 125,000 people died in the African French-speaking countries during this period, in addition to the successive food shortages of the 1972–1974 period (Dauphine 2003). A loss of 600,000 tonnes of the grain crops was also recorded, equivalent to about a 15 % reduction in average annual income. In livestock farming and during the same period of 1970s, there was an 80 % cattle loss (Guillaumie et al. 2005). As a consequence, the survival of the socio-ecological system was endangered, in addition to being subject to severe risks that resulted in destructive disasters. Water control appeared to be essential for reducing the system's vulnerability. Thus, the "taming" of nature became a significant challenge for riparian states and the construction of the Diama and Manantali dams on the Senegal River (Fig. 1) was the solution adopted to ensure sustainable development and food security.

The Diama Dam resulted in considerable modification of the hydrological regime of the Senegal River downstream of the structure. Freshwaters are stored in the upstream of the dam. They arrived in the estuary during the rainy season (Saint-Louis station), when dam gates are opened and when they are some releases from Diama Dam. In addition, the principle of functioning of the Diama dam is to close it in periods of low water to prevent the intrusion of saltwater and to open in period of floods to allow floodwaters to flow downstream (OMVS 2003). The role of the Diama Dam in controlling the Senegal River is strengthened by the Manantali Dam.

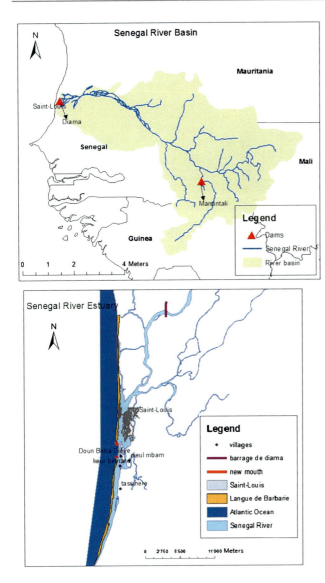

Fig. 1 Location of the Senegal River basin and estuary

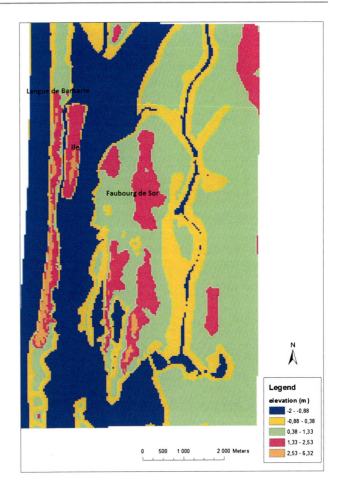

Fig. 2 Digital elevation model Saint-Louis

The Sahelian zone has in recent years recorded a return of heavy rainy seasons. In the tropical wet area of Guinea, high basin River, we note abundant precipitation creating an increase in the river water flow. The increase in the river water levels generates overflows which further increase the risks of floods. The town of Saint-Louis stuck between the Senegal River and Atlantic Ocean (Fig. 1) is built on an area dominated by deposits of mud and is particularized by the omnipresence of water (Kane et al. 2013). The digital elevation model (Fig. 2) shows the lower elevations of the city of Saint-Louis and identifies the most sensitive areas to flooding. Low areas represent 53 % of which 17 % have altitudes between −1 and 0 m and 36 % emerging barely above sea level is between 0 and 3 m. They correspond, in fact, the main bed of the river is located in "Faubourg de Sor." The slopes are also very low; they are between 0 and 9 %. In "Faubourg de Sor," the slopes are practically nil and they vary between 0 and 2 %. Also, at this point, changes in the water level will be low in case of flooding and lead to greater variation in the extent of water. Many cities are located in flood areas as they are likely to be places favorable for urban development (Tingsanchali 2012). The town of Saint-Louis is in a similar situation to many others throughout the world, for example, in Ethiopia where floods occur regularly in all parts of the country at times of rising water levels downstream of large basins and in low ground (Ayalew 2009). In Saint-Louis, following many years of drought, the local population occupied much of the river bed normally under water; consequently, many districts may be flooded in the event of a major rise in water level. This raises the important issue of population relocation during rainy seasons, as it is also a limitation to the smooth flow of people and goods. This weakening of the socio-economic system in the face of the threats of flooding prompted the Senegalese authorities to open a new mouth on the "Langue de Barbarie" in 2003 to enable a rapid escape of excess water toward the sea.

Thus, since July 1986, the Diama Dam has controlled most of the water influx into the estuary. An average annual of 255 m^3/s is released into the river below the dam during this period. Water flows through the dam since 1986

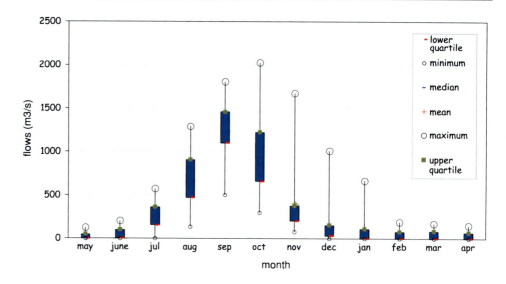

Fig. 3 Senegal River flows at Diama dam station (1986–2010) (OMVS Data) Organisation pour la Mise en Valeur du Fleuve Sénégal: Organization for the Development of the Senegal River

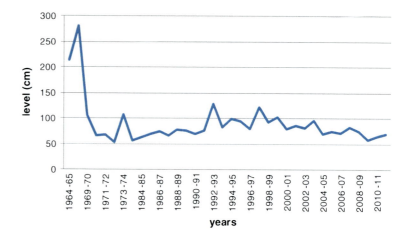

Fig. 4 Changes in water level at the Saint-Louis Station (1964–2012) (Data source: Division Régionale de l'Hydraulique de Saint-Louis/ Ministère de l'Hydraulique et de l'Assainissement)

(Fig. 3) enable determination of the estuary's monthly and annual flows, showing two periods:

- a 4-month period (August to November) during which the estuary is invaded by flooding water and
- a 8-month period during which the estuary is subject to marine influence.

The dam's highest water flow is in October. The highest flows are in August, September, October, and November, and the lowest usually in December, January, February, March, May, April, June, and July.

In order to ensure that instructions on the dam's proper management are followed, occasional releases can be carried out by managers to guarantee its safety. The openings of floodgates are related to the support of raising water levels at the Manantali dam, which at certain times requires substantial water releases to be taken from reserves intended for other needs, while higher than turbine output (Bader 1997). Moreover, with the operation of the hydroelectric station at the Manantali Dam since 2002, the floodgates of the Diama Dam are regularly opened to avoid overflow of the river banks.

Figure 4 shows changes in water level at the Saint-Louis station from May 1964 to April 2012, showing a succession of peaks which reflects the strong inter-annual variability of the flow. Water flows have reduced since the 1970s with a clear decline in water level, from a peak of 281 cm in 1965–1966 to an average of 54 cm in 1972–1973 reflecting changes in the climatic conditions in the Senegal River basin. Although insignificant, the river's water level seems to have experienced a slight increase in recent years, with an average of 129 cm in the 1992–1993 period. As a whole, there is a gradual fall from 1964 to 1971 and since about 1997.

The monthly water levels at the Saint-Louis station vary according to the totals discharged by the dam. They reach their maximum in October with, respectively, 1.32 m NGI[1]; 1.19 and 1.13 m NGI. There is nearly steady flow throughout the entire flooding period, becoming irregular after November. The minimum heights are recorded in May

---

[1] NGI measurements constitute the National Geographical Institute measurements of the various water levels.

Table 1 Rules of management of the Diama Dam before the opening of the new mouth

Period	Objective	Level of reserve
Low water level	Increased level in the reservoir	+2.10 to +2.20 m NGI (since the low water level of 2001–2002)
Beginning of flood waves	Maintenance of an increased level for as long as possible, then lowering preventive when flood waves arrives reach Bakel	+2.10 to +2.20 m NGI then lowering to +1.50 m NGI
Period of flood waves	Level lowered in reservoir	+1.50 m NGI as long as flows are such that it is necessary to entirely open close the floodgates
End of flood waves	Raising the level of the reservoir as soon as possible	+1.50 m NGI then rising to +2.10 to +2.20 m NGI

with, respectively, −0.3 m NGI; 0.01 and 0.05 m NGI. It appears that since 2004–2005, water level in the river has been decreasing constantly as a result of the opening of the new mouth. The hydrodynamic modifications generated by the opening have resulted in a disturbance of estuarine water circulation.

Difficulties in the Management of the Diama Dam Following the Opening of the New Mouth

The opening of the new mouth has lowered water level at the Saint-Louis station. The opening of the new mouth has changed the flow conditions downstream of the dam to a share by lowering the average level downstream side making it more sensitive to the effect level of the tide (OMVS, Coyne and Bellier 2005). Thus, at the Diama Dam, the tidal range has significant changes. Before the opening of the new mouth estuarine, tides were semi-diurnal with heights rarely exceeding the meter during whitewater and 0.5 m during neap tides. Currently, the maximum daily tidal range recorded downstream of Diama Dam has tripled from 0.30 to 0.93 m near is noticeable throughout the year, including during the rainy season, which makes it difficult handling the valve system. This had an impact on the management of the Diama Dam. Since tides are the determining factors that influence the final stretch of the river, their variation determines water levels. Water levels downstream of the dam were substantial due to flood waves, whereas the upstream part experienced a sharp decrease in water level. This difference in height may cause the dam's failure. It has also become difficult to maintain water levels upstream for irrigation needs while ensuring the dam's safety. Following OMVS et al. (2005), management rules prior to the opening of the new mouth are shown in Table 1.

The new context has required various studies (IRD[2]-OMVS 2004; OMVS et al. 2005) financed by the OMVS in order to revise the rules of management at the Diama Dam. These studies have enabled a more flexible management of the dam to enable levels of 1.50 m to be reached upstream.

These installations on the Senegal River estuary have strongly disturbed natural hydrological function and strikingly pose the issue concerning the risk of serious flooding in the case of a rise in seawater level, hypersalinization of water in the lower estuary, the development of halophytic plants, the degradation of ecosystems, etc. In conclusion, these installations, which aim at alleviating extreme events of climatic origin, actually pose significant risks.

Consequences of Water Management for the Senegal River Estuary

The changes introduced by the new mouth have clearly resulted in environmental impacts through disturbances in river water circulation. The new river mouth has resulted in a shorter residence time of freshwater since the part located downstream of the new mouth has evolved into a lagoon. The quality of water in the lower estuary remains close to that of the sea. According to Troussellier et al. (2004), salinity reached 35 ‰ more than 12 km south of the town of Saint-Louis. Since this section of the river has developed into a lagoon, exchange between marine and river water remains very limited. In fact, perturbations in the exchange between marine and river water and releases of salt from old deposits will certainly increase this salinization. The estuary is thus controlled by the Diama Dam and the marine dynamics by the opening of the new mouth. From the 2003 annual evaporation average (1,207 mm), we have calculated the volume of water evaporated from the lagoon, which corresponds to more than 13 million m^3. This strong evaporation explains the excessive salinization of the lagoon, which will certainly impact the groundwater recharge of the lower estuary. The average salinity level measured in 2007 in the lower estuary was 47 g/l, which is higher the concentration than in seawater.

---

[2] Institut de Recherche pour le Développement: Research Institute for Development.

In addition, groundwater changes related to the developments of rivers are seen as a major issue in many parts the world. In Morocco, for example, following the construction of the Mohamed V Dam, the coastal wetlands of Moulouya are no longer subject to flooding from rivers, which is responsible for the recharge of groundwater used for irrigation. Both water and soil have become increasingly saline as arable lands are abandoned; this has caused farm workers to emigrate (Arthurton et al. 2008). Groundwater contaminated by saltwater intrusion makes inappropriate irrigated crops. The population around the lower estuary is highly dependent on groundwater which constitutes their irrigation system for market gardening. The problem remains severe since the opening of the new mouth; the salinization of groundwater has reached critical levels, causing an extension of the salt marshes and a progressive loss of farmland.

Important changes were noted in the coastal hydrodynamics such as the tidal patterns. During high tides, there is a large increase in marine water with an important evacuation of the river flows during low tides. The acceleration of coastal erosion cause real threats to the area. The mangrove swamp has shown a progressive degradation following the opening of the new mouth. The area of the mangrove swamp has fallen by 636 ha (or 59 %), from 1086 ha in 1979 to 450 ha in 2003 (Kane 2010). The area of the island of Doun Baba Dièye decreased by 8 % between 2003 and 2007, from 257 to 236 ha and now has completely damaged since October 2012.

Many villages such as Dieule Mbame, Tassinere and Keur Bernard (Fig. 2) are under threat. Resettlement of populations constitutes indeed a real problem as rehousing in the lower estuary is a big issue relatively to the availability of land.

> Box 1: The opening of two new mouths
> Recently, in October 2012, it has been reported that the two natural openings which are distant of 150 m have accelerated the process of dismantling of the sandy spit of "Langue de Barbarie," and the disappearance of the coastal habitats. After having amalgamated with the breach opened in 2003, they reinforced the problem of the salinization of the grounds in the lower estuary.

Risks Linked to Water Management in the Senegal River Estuary

The major risk in this area is linked to population settlements and the impacts of the various changes to the ecosystem and the resources. The Senegal River estuary appears to be a risk-prone area, where the conjunction of several factors can produce incommensurable effects. Physical conditions (geomorphological, lower altitude) determine an area where risk is permanent and expose populations to hazard. Moreover, communities may aggravate the intensity or frequency of certain phenomena and the reach of their effects.

To prevent the occurrence of flooding requires the means to contain floods, such as the construction of dams or river dikes and the establishment of non-structural measures such as alarm systems for flood prevention, public participation, and institutional commitment (Tingsanchali 2012). However, the new opening of the mouth, which aims at reducing population vulnerability within the estuary, has been favored over precautionary principles. Also, questions concerning the human impact on risk aggravation are inevitably directed toward inquiring about negligence during establishment of these new developments, weaknesses in protection systems including alert and prevention, safety devices, and toward any negligence which may obstruct proper control and handling of these phenomena as well as reduction in the consequences.

Considering the impact of an extremely vulnerable area at the opening of "Langue de Barbarie," which has had the consequences described above, the occurrence of a natural disaster will only increase the fragility of the system. It is probable that rising of sea levels will intensify environmental pressures and undermine sustainable development. This will result in higher risks of flooding and greater ingress of seawater while the lower estuary is already experiencing groundwater salinization due to exchanges between river and marine water following the opening of the new mouth. Populations located in low and coastal areas as well as in deltas and which are estimated to make up approximately 10 % of the world's population will be particularly vulnerable to climate change (Nicholls et al. 2008; McGranahan et al. 2007). Simulations conducted by Durand et al. (2010), based on the potential height of water at the end of the rainy season and carried out for several scenarios for the rise in sea level, show that a 0.5-m sea-level rise would be enough to flood part of the town of Saint-Louis at times of annual high waters and that it would be almost completely submerged in the event of a 1-m rise.

Figure 5 represents the probable evolution of submergence of St-Louis based on increase in sea-level areas. It should be noted that the island and the sandy spit of "Langue de Barbarie" will be less affected than the "Faubourg de Sor." The surface losses are 3 and 126 ha or 4 % of the territory up to 1 m. However, as we mentioned earlier, the "Langue de Barbarie" currently experienced recurring phenomena of coastal erosion. This situation can be explained by the importance of altitudes of more than

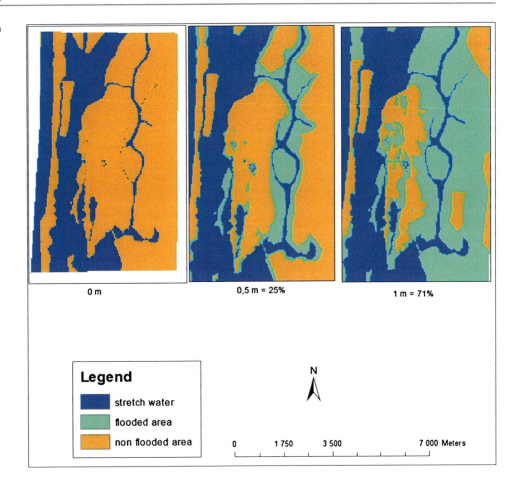

Fig. 5 Flooded areas in the town of Saint-Louis in case of sea-level rise

6 m. It is mostly in the "Faubourg de Sor" that flood risk will be greater due to lower altitudes. The occurrence of extreme events is increasingly determined by the coincidence of hydro-climatic events affecting a changed environment by human activities (Bogardi 2004). This area is experiencing unplanned occupation which only exacerbates the risk of flooding in the case of sea-level rise. This shows that the ignorance of risk is a source of vulnerability. At a height of 1 m of submersion, 71 % of the city will be under water. The entire area east of the city will be particularly affected.

The realization of large-scale new developments in sensitive areas may show a lack of perception by communities of the risks involved. Lefèvre and Schneider (2002) stress that extreme difficulty in risk recognition and management is due to their limited perception by human communities. The exposed population's lack of perception of their own vulnerability to natural disasters is itself a factor of vulnerability which will constrain the establishment of preventive measures (Leone and Vinet 2006). Thus, the estuary of the Senegal River appears vulnerable and subject to the shock of unforeseeable and uncontrollable devastating events.

Conclusion

Management of tropical rivers can increase risks and vulnerability. In the Senegal River, it has resulted in consequences that impacted seriously the functioning of the estuary. The sea-level rise and the rapid erosion of beaches are not always mastered. In spite of the estuary's particular nature, it has been subject to developments carried out without preliminary studies in regard to their impact. The resulting changes may be determining factors in the evolution of the local environment. In the Sahel, despite the large climatic variability which causes events of extreme nature, risk-related consequences have not been sufficiently considered. Preventive measures seem not to be related to knowledge of risk, but to knowledge of this perception.

The new mouth has created a dysfunctioning which, in the long term, will have much more significant repercussions. The consequences of these modifications generally erode resilience of the system, which has difficulty resisting, to adapt to a major change. The rise of sea level in the lower estuary would flood most low ground, causing damage to the socio-ecological system. Such a rise would constitute a serious threat to the socio-ecological system in addition to having serious consequences geomorphological and socio-economic. Nevertheless, knowledge and perception of risk play a fundamental role in the policies of prevention concerning risks. Even if the latter appear to have been controlled, there has not been sufficient consideration of the consequences associated with the prevalence of risk, which can generally lead to irreversible actions. Considerations of risk in the various management policies are of significant interest in a context characterized by an increase in disasters.

References

Arthurton R, Le Tissier M, SnoussI M et al (2008) AfriCat: LOICZ-global change assessment and synthesis of River catchment—Coastal Sea interactions and human dimensions in Africa, LOICZ Reports and studies n. 30, LOICZ IPO, Geesthacht, 122 p

Ayalew L (2009) Analyzing the effects of historical and recent floods on channel pattern and the environment in the Lower Omo basin of Ethiopia using satellite images and GIS. Environ Geol 58:1713–1726

Bader JC (1997) Le soutien de crue mobile dans le Fleuve Sénégal, à partir du barrage de Manantali. Hydrol Sci J des Sci Hydrol 42(6):815–831

Beucher S, Rode S (2009) L'aménagement des territoires face au risque d'inondation : regards croisés sur la Loire moyenne et le Val-de-Marne. http://mappemonde.mgm.fr/num22/articles/art09202htlm, 19 p

Birkmann J (2007) Risk and vulnerability indicators at different scales: applicability, usefulness and policy implications. Environ Hazards 7:20–31

Bogardi JJ (2004) Hazards, risks and vulnerabilities in a changing environment: the unexpected onslaught on human security? Glob Environ Change 14:361–365

Bogardi J, Birkmann J (2004) Vulnerability assessement: the first step towards sustainable risk reduction. In: Malzahn D, Plapp T (eds) Disaster and society—From hazard assessment to risk reduction. Logos Verlag, Berlin, pp 75–82

Culliton TJ, Warren MA, Goodspeed TR, Remer DG, Blackwell CM, MacDonough JJ (1990) 50 years of population change along the Nation's Coasts, 1960–2010. National Oceanic and Atmospheric Administration, Rockville 41 p

Dai AG, Lamb PJ, Trenberth KE, Hulme M, Jones P, Xie P (2004) The recent Sahel drought is real. Int J Climatol 24:1323–1331

Dauphine A (2003) Risques et catastrophes: observer, spatialiser, comprendre, gérer. Ed. Armand Colin, 288 p

Durand P, Anselme B, Thomas YF (2010) L'impact de l'ouverture de la brèche dans la langue de Barbarie à Saint-Louis du Sénégal en 2003 : un changement de nature de l'aléa inondation ? Cybergeo: European Journal of Geograph, Environnement, Nature, Paysage, document 496, http://cybergeo.revues.org/index23017.html

Guillaumie K, Hassoun C, Manero A et al (2005) La sécheresse au Sahel: un exemple de changement climatique. Atelier Changement ENPC-Département VET http://www.enpc.fr/fr/formations/ecole_virt/trav-eleves/cc/cc0405/sahel.pdf, 40 p

Hulme M, Doherty R, Ngara T (2005) Global warming and African climate change: a re-assessment. In: Low PS (ed) Climate change and Africa. Cambridge University Press, Cambridge, pp 29–40

IRD-OMVS (2004) Manuel de gestion du barrage de Diama. Version révisée pour les conditions d'écoulement prévalant depuis octobre 2003 (ouverture de la brèche dans la langue de barbarie), 52 p

Kane C, Humbert J, Kane A (2013) Responding to climate variability: the opening of an artificial mouth on the Senegal River. Reg Environ Change 13:125–136

Kane C (2010) Vulnérabilité du système socio-environnemental en domaine sahélien: l'exemple de l'estuaire du fleuve Sénégal. De la perception à la gestion des risques naturels. Thèse de doctorat en Géographie, Université de Strasbourg/Université Cheikh Anta Diop, 317 p

Lefèvre C, Schneider JL (2002) Les risques naturels majeurs. Ed. Contemporary Publishing International GB Science Publisher, 306 p

Leone F, Vinet F (2006) La vulnérabilité, un concept fondamental au cœur des méthodes d'évaluation des risques, pp 9–25. In«La vulnérabilité des sociétés et des territoires face aux menaces naturelles. Analyses géographiques». Sous la direction de Leone F., et Vinet F. Collection «Géorisques», n°1. Montpellier 3, 140 p

L'Hôte Y, Mahé G, Some B, Triboulet JP (2002) Analysis of a Sahelian annual rainfall index updated from 1896 to 2000; the drought still goes on. Hydrol Sci J 47:563–572

L'Hôte Y, Mahé G, Some B (2003) The 1990s rainfall in the Sahel: the third driest decade since the beginning of the century, Reply to discussion. Hydrol Sci J 48:493–496

Mahé G, Paturel JE (2009) 1896–2006 Sahelian annual rainfall variability and runoff increase of Sahelian Rivers. C R Geoscience 341:538–546

Martinez ML, Intralawan A, Vázquez G et al (2007) The coasts of our world: ecological, economic and social importance. Ecol Econ 63:254–272

McGranahan G, Balk D, Anderson B (2007) The rising tide: assessing the risks of climate change and human settlements in low elevation coastal zones. Environ Urbanization 19–1:17–37

Miller ML, Hadley NP (2005) Tourism and coastal development. In: Schwartz ML (ed) Encyclopedia of coastal science. Springer, Dordrecht, pp 1002–1008

Nicholls RJ, Hanson S, Herweijer C, Patmore N, Hallegatte S, Corfee-Morlot J, Château J, Muir-Wood R (2008) Ranking port cities with high exposure and vulnerability to climate change extremes exposure estimates. Organization for economic co-operation and development: environment working papers no 1 Publishing http://dx.doi.org/10.1787/011766488208

Nicholls R, Cazenave A (2010) Sea-level rise and its impact on coastal zones. Science 328:1517–1520

OMVS (2003) Etude de base pour la phase initiale de mise en place de l'observatoire de l'environnement, rapport technique, version finale provisoire V2.0, 295 p

OMVS Coyne Bellier (2005) Etude d'évaluation des impacts de l'ouverture d'une embouchure artificielle à l'aval de Saint-Louis sur le barrage de Diama. Rapport final, 39 p

Olson RL (2012) Soft geoengineering: a gentler approach to addressing climate change. Environ Sci Policy Sustain Dev 54(5):29–39

Runhaar H, Laerhoven FV, Driessen P, Arts J (2013) Environmental assessment in the Netherlands: effectively governing environmental protection? A discourse analysis. Environ Impact Assess Rev 39:13–25

Tingsanchali T (2012) Urban flood disaster management. Procedia Eng 32:25–37

Troussellier M, Got P, Bouvy M, M'Boup M, Arfi R et al (2004) Water quality and health status of the Senegal River estuary. Mar Pollut Bull 48:852–862

Combined Uses of Supervised Classification and Normalized Difference Vegetation Index Techniques to Monitor Land Degradation in the Saloum Saline Estuary System

Ndeye Maguette Dieng, Joel Dinis, Serigne Faye, Marçia Gonçalves, and Mário Caetano

Abstract

Saltwater contamination constitutes a serious problem in Saloum estuary, due to the intermittent and reverse tide flows of the Saloum River. This phenomenon is caused by the runoff deficit, which forces the advance of saltwater 60 km upstream, contaminating surface water and thus causing the degradation of biodiversity and large areas of agricultural soils in this region. The present study aims to evaluate the consequences of saltwater contamination in the last three decades in this estuary by assessing the land-cover dynamics. Thus, latter consists of tracking the landscape-changing process over time to identify land-cover transitions. These transitions are closely related to the ecosystem-setting condition and can be used to assess the combined impacts of both natural and human-induced phenomena over a given period of time. In this study, special attention was given to mangrove degradation and to temporal progression of the salty barren soils locally called "tan". The loss of mangrove areas to tan and the general increase in salty barren soil areas can reflect the increase in the level of salinization in the study area over the time period under consideration. To fulfill this objective, four Landsat satellite images from the same season in the years 1984, 1992, 1999, and 2010 were used to infer time series land-use and land-cover maps of the Saloum estuary area. In addition to satellite imagery, rainfall records were used to evaluate climatic variation in terms of high-to-low precipitation during the time span considered. Spectral analysis indicated that from 1984 to 2010, mangroves and savanna/rain-fed agriculture are converted to "tan" (denuded and salty soils). In addition, these results showed that significant changes in land use/land cover occur within the whole estuary system and reflecting therefore environmental degradation, such as land desertification and salinization, and vegetation degradation which reflect the advanced of salinity.

N. M. Dieng (✉) · S. Faye
Geology Department, Faculty of Sciences and Techniques
University Cheikh Anta Diop, P.O. Box 5005, Dakar, Senegal
e-mail: diengmaguette@yahoo.fr; ndeye81.dieng@ucad.edu.sn

S. Faye
e-mail: fayeserigne1@gmail.com

J. Dinis · M. Caetano
Higher Institute of Statistics and Information Management,
New University of Lisbon, Lisbon, Portugal

M. Gonçalves
Remote Sensing Unit of the Portuguese Geographic Institute,
Lisbon, Portugal

S. Diop et al. (eds.), *The Land/Ocean Interactions in the Coastal Zone of West and Central Africa*, Estuaries of the World, DOI: 10.1007/978-3-319-06388-1_5,
© Springer International Publishing Switzerland 2014

Keywords

Saloum inverse estuary • Salinization • Normalized difference vegetation index • Mangrove degradation • Change detection • Inverse estuary

Introduction

Evidence is mounting that we are in a period of climate changes brought about by increasing atmospheric concentrations of greenhouse gases. Due to increases in global temperature, sea level may rise from 0.5 m to more than 1.0 m above current mean sea level by the year 2100 (Church and White 2006; Overpeck et al. 2006; IPCC 2007). It is even expected that temperature rise over the next century would probably be greater than that observed in the last 10,000 years. As a direct consequence of warming temperature, the hydrologic cycle will undergo significant impact with accompanying changes in the rates of precipitation and evaporation. Predictions include higher incidences of severe weather events, a higher likelihood of flooding, and more droughts. Sea level rise as a result of global warming has an impact on the increasing inundation on coastal area.

In addition to these impacts, the economic, social, and environmental consequences will be enormous in Africa where populations are particularly vulnerable. Awareness of these manifestations and adaptation strategy are key concerns for the continent for the coming years, especially in many domains such as agriculture, water, soils, and vegetation.

Coastal region and in particular low-lying estuary system have retained our attention for this study due to the fact that these low gradient areas would be the most affected areas as they represent the environmentally most sensitive areas.

As defined by Pritchard (1967), according to settings and mixing process (the type of river water and seawater mixing and the degree of salinization), estuaries may be subdivided into two groups. The first group includes normal estuaries, in which freshwater dilutes seawater and water salinity monotonically decreases downstream the river from 10–40 to 0.5–1 ‰ and the water runoff and precipitation exceed evaporation losses. The second group includes reverse, or hypersaline, estuaries, in which the salinity of estuarine water substantially exceeds the salinity of seawater; water evaporation losses exceed freshwater river runoff and precipitation. The first group is widespread in the world, and processes occurring are relatively well understood (Pritchard 1967; Ketchum 1983; McDowell and O'Connor 1983). The processes of mixing of river water and seawater in such estuaries are usually subdivided into three types: (1) complete mixing through the depth and weak density stratification of waters; (2) partial (moderate) mixing and moderate

stratification of waters; and (3) a saltwater wedge and intense stratification of waters (Mikhailov and Isupova 2008). Reverse estuaries are less common in the world and are poorly studied, peculiarities of the processes in reverse estuaries, including the processes occurring in the basin of the Caspian Sea (the Kayak Bay) as well as in west Africa and Australia (Wolanski 1986; Pagès et al. 1987). Under extremely arid condition, particularly during the dry season and drought lasting for many years, considerable deficiency of freshwater may occur in the river mouth reach.

Estuaries systems in semiarid and arid regions are characterized by highly variable seasonal river discharge; they represent in fact the most vulnerable zones with regard to climate variability and climate change. In some regions, with the persistence of drought periods and their consequence of negative water budget (induced by high evaporation effects), seawater may intrude into these systems and salinity will rise monotonically to hypersalinity with distance from the mouth (Ridd and Stieglitz 2002). Such a process has occurred in a coastal river in Senegal, namely the Saloum, actually a tide-influenced inverse estuary. In the Saloum estuary, saltwater contamination constitutes a serious problem. Evaporation and tidal inundation cause salt concentrations in the groundwater to rise above the normal seawater value. Ridd and Sam (1996), Sam and Ridd (1998) found that water inundating the salt flats returns to the estuary with a greater salinity by dissolving salt crystals. The intermittent and reverse flows of the Saloum River due to the runoff deficit caused saltwater advance up to 60 km upstream, contaminating surface waters, groundwater, and large areas of agricultural soils located in these zones. Salinity in the Saloum River showed a gradual upstream increase from 36.7 % at the mouth to more than 90 % at Kaolack (Pagès and Citeau 1990).

In arid and semiarid regions, soils salinity and saltwater intrusion are one of the major threats to agriculture (Ghassemi et al. 1995). Under most global warming scenarios, rate of coastal erosion will accelerate in the twenty-first century (Zhang et al. 2004; IPCC 2007). Remote sensing techniques have potential for mapping and monitoring the degree and extent of salinization. Thus, quantifying and monitoring their spatial distribution are very important for management purposes. Change detection is the process of identifying differences in the state of an object or phenomenon by observing it at different times (Singh 1989). It is one of the major applications of remotely sensed data obtained from Earth-orbiting satellites because of repetitive coverage at short intervals and consistent image quality (Anderson 1977; Ingram et al. 1981; Nelson 1983). Change detection using images has been traditionally performed by comparing the classification of multitemporal data sets or by image processing techniques such as differencing and rationing. Change detection is useful in such diverse applications as land-use change analysis, monitoring of shifting cultivation, assessment of deforestation, changes in vegetation phenology, seasonal changes in pasture production, damage assessment, crop stress detection, disaster monitoring snow-melt measurements, daylight analysis of thermal characteristics, and other environmental changes (Singh 1989).

In change detection processes (Singh 1989; Coppin and Bauer 1994; Lu et al. 2003, 2004a, b; Coppin et al. 2004; Pu et al. 2008; Pan et al. 2011; Datta and Deb 2012; Petropoulos et al. 2012), time series images acquired from different dates are compared to analyze the spectral difference, caused by land-use/land-cover change (LULC) over time while trying to normalize other conditions to similar levels during that period. Therefore, it is necessary to confirm the estuary dynamic for mapping and monitoring the land-cover change with different techniques. Satellite remote sensing has been widely applied and recognized as a powerful and effective tool for detecting land-use and land-cover changes. However, according to the change's indices adopted and the methods of detection applied, results obtained show significant differences that can be evaluated both quantitatively (importance of the changes over time) and qualitatively (types of changes observed). Works of Smits et al. (1999) and Coppin et al. (2004) identified ten types of detection methods that are based on different techniques of image processing, including image subtraction, crossing classifications, principal component analysis (PCA: statistical analysis multivariate), vector calculating change, or neural networks.

In this study, two change detection techniques are evaluated: a classification method and a normalized remote sensing technique—normal difference vegetation index (NDVI) differencing method, focusing on a comparison between the two techniques and also on the determination of the threshold of the NDVI differencing method. Both techniques are common and effective in change detection of LULC (Gong and Howarth 1992; Kontoes et al. 1993; Foody 2004; San Miguel-Ayanz and Biging 1997; Aplin et al. 1999; Stuckens et al. 2000; Franklin et al. 2002; Pal and Mather 2004; Gallego 2004; Lu et al. 2004a, b; Pu et al. 2008; Datta and Deb 2012). Classification and NDVI differencing change detection methods were adopted in this study to analyze land-cover changes associated with salinization.

Study Area

The Saloum estuary system, located approximately between longitudes 14°01′ and 16°56′ W and latitudes 13°31′ and 14°57′ N (Fig. 1), shrank after the last pluvial episode in around 10,000 BC and represents one of the largest African reverse estuaries. It consists of an extensive network of fossil, dried secondary channels (so-called thalwegs) stretching north and eastward. The terrain in the study area is generally flat with altitudes ranging from below sea level in the estuarine zone to about 40 m above mean sea level (a.m.s.l.) inland; the longitudinal slope of the river course is correspondingly low as well as the shallow bathymetry of the river. The climate is Sudano-Sahelian type with a long dry season from November to June and a 4-month rainy season from July to October. The regional annual precipitation, which is the main source of freshwater recharge to the superficial aquifer, increases southward from 600 to 1,000 mm. The average temperature is 28–29 °C, and the average annual evaporation varies from 1,500 to 2,500 mm (source: meteorological data). The geomorphology consists of a gently sloping plain that extends toward the coast, ranging in elevation from 0 m in the estuary system to 40 m a.m.s.l. inland (Barusseau et al. 1985; Diop 1986). Sand dune deposits occur near the coast with an altitude of 1 m in the northern part and between 2 and 8 m a.m.s.l. in the southern part of the region. The hydrologic system of the region is characterized by the river Saloum, its two tributaries (Bandiala and Diomboss), and numerous small streams locally called "bolons." Downstream, it forms a large low-lying estuary bearing tidal wetlands, a mangrove ecosystem, and vast areas of denuded saline soils called "tan" locally.

Methodology

Landsat data were selected to generate time series of land-cover changes in Saloum estuary. The regular revisit times and spatial resolution of the Landsat mission are well suited for regional, national, and global land-use changes. Four images were selected for this study dated October 17, 1984, October 31, 1992, November 01, 1999, and November 26, 2010, respectively. Accordingly, the study period covered about the last three decades.

The methodology applied in this work consists of three major steps: (1) collect and clean training samples; (2) automatic classification for LULC (land-cover–land-cover) mapping; and (3) NDVI differencing analysis. The images were selected with respect to resolution, number of bands, and season. Although the four scenes were already georeferenced to the UTM Zone 28 North and WGS 84 projection, they were geomatching. The outputs of the second and

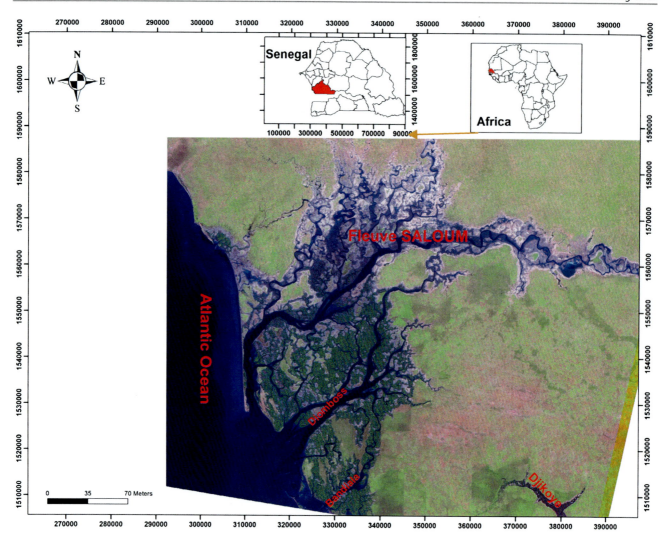

Fig. 1 A location map of the study site

third steps were combined to get the final conclusions. In this way, combination of the thematic information and the NDVI was possible in order to infer the nature of change, between-class or within-class change (Lunetta and Elvidge 1998), or simply an error.

Detecting Outliers and Cleaning Training Sample

A suitable classification system and a sufficient number of training samples are prerequisites for a successful classification (Lu and Weng 2007). Training samples are usually collected from fieldwork or from fine spatial resolution aerial photographs and satellite images, and sampling of sufficient number and their representativeness is critical for image classifications (Landgrebe 2003; Mather 2004). Different collection strategies, such as single pixel, seed, and polygon, may be used, but selecting sufficient training samples becomes difficult to perform when the landscape is complex and heterogeneous as they would influence classification results, especially for classifications with fine spatial resolution image data (Chen and Stow 2002). This problem would be complicated if medium or coarse spatial resolution data are used for classification, because a large volume of mixed pixels may occur (Lu and Weng 2007).

Despite precautions made in the training samples' collection, it is sometimes difficult to identify the most sensitive reference land-cover class for some observations, even resorting to ancillary data (Carrão et al. 2008). Thus, the original training sample, i.e., the sample of pixels that was directly collected by the analyst, contained unusual training units. According to Johnson and Wichern (1998), unusual observations are those that are either too large or too small compared to the others. Thus, in order to identify these anomalies, it is necessary to apply a statistical procedure based on the distance of each training unit to its mean class.

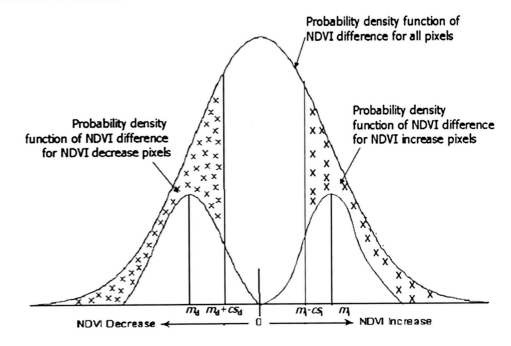

Fig. 2 Illustration of the two assumptions (adapted from Pu et al. 2008)

Let $\Omega = \{w_1, w_2, \ldots, w_k\}$ be the set of class labels, μ_j the mean, and Σ_j the variance–covariance matrix of the jth class of Ω. Assuming that each land-cover class can be modeled by a multivariate normal distribution, it is possible to compute the squared Mahalanobis distance, for a given training unit t assigned to the jth class of Ω, by Eq. (1):

$$d_j^2(t) = (t - \mu_j)^T \Sigma_j^{-1}(t - \mu_j). \quad (1)$$

Under these assumptions, the d_j^2 is modeled by a chi-square random variable with k degrees of freedom, χ_k^2, when the number of observations in each land-cover class is greater than 30 (Johnson and Wicher 1998). Thus, we can develop the following test to identify anomalous training units (Johnson and Wicher 1998). For every class w and for every training unit t of w, if $d_j^2(t)$ is greater than $\chi_k^2(\alpha)$, where α is the significance level of the test, we reject the hypothesis that t is a standard observation in class w; otherwise, t is accepted and kept in the training sample. We have fixed the significance level at 2.5 %.

In practical applications, the class mean and variance–covariance matrix are not a priori known. Thus, we need to estimate them. To that end, we have estimated the class mean and variance–covariance using their standard maximum likelihood estimators, given by the following equations (Johnson and Wicher 1998):

$$\hat{\mu}_j = \frac{1}{n_j} \sum_{t \in w_j} t \quad (2)$$

$$\hat{\Sigma}_j = \frac{1}{n_j - 1} \sum_{t \in w_j} (t - \hat{\mu}_j)(t - \hat{\mu}_j)^T \quad (3)$$

where n_j is the number of training units in the jth class of Ω.

Supervised Image Classification

In recent years, many advanced classification approaches have been widely applied for image classification. Many factors, such as different sources of data, classification system, availability of classification software, and spatial resolution of the remotely sensed data, must be taken into account when selecting a classification method for use. Different classification methods have their own merits, and for the classification, we resort to the linear discriminant classifier (LDC). The LDC is a parametric classifier based on the homoskedasticity assumption, i.e., we assume that each land-cover class is modeled by a multivariate normal distribution and each of these distributions has an equal variance–covariance matrix. The LDC has many advantages over more sophisticated classification algorithms, due to the fact that it does not need as many training units comparing to the maximum likelihood classifier (MLC) or support vector machines (Hastie et al. 2009). It is simple in computational and operational terms and is reasonably robust (Kuncheva 2004), in that the results are good even when the classes do not have normal distributions.

Fig. 3 Classification result showing the land cover in Saloum estuary in October 1984 (**a**), October 1992 (**b**), November 1999 (**c**), and November 2010 (**d**)

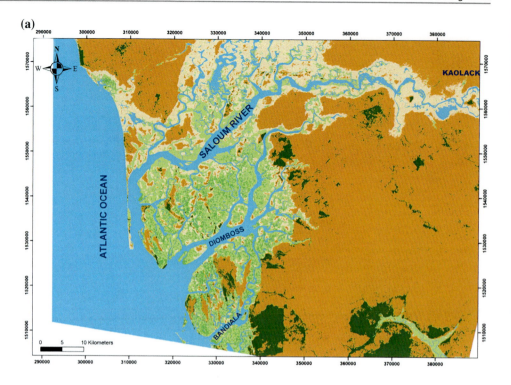

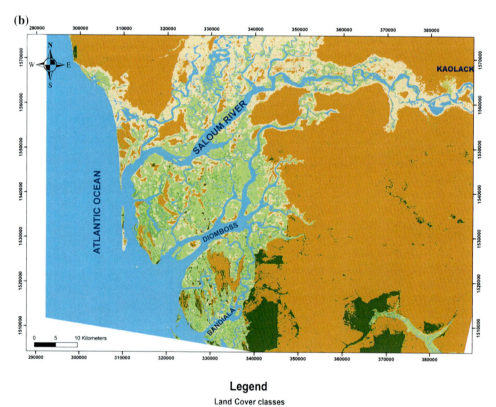

Fig. 3 continued

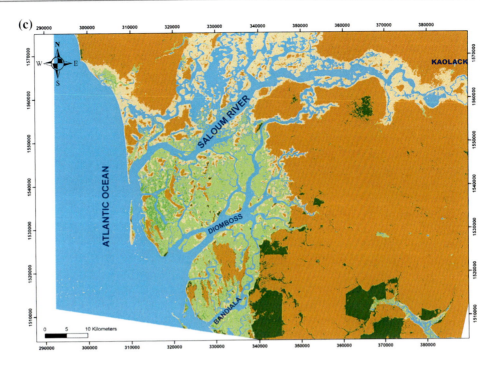

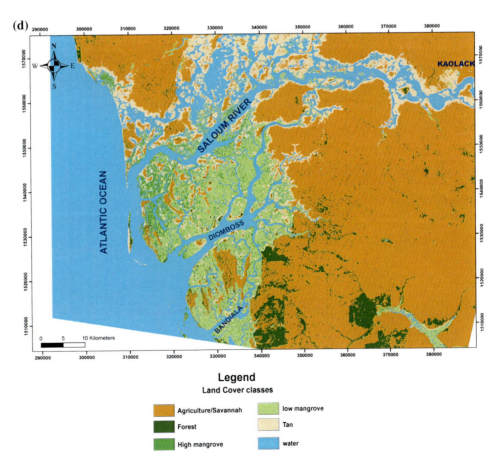

Table 1 Change matrices generated (%) through overlay analysis between the four single-date classification results, with the classification method

		Water	High mangrove	Low mangrove	"tan"	Savanna/rain-fed agriculture	Forest
		1992					
1984	Water	**97**	0	2	1	0	0
	High mangrove	2	**42**	55	0	0	0
	Low mangrove	5	4	**83**	6	1	1
	"tan"	1	0	7	**83**	9	0
	Savanna/rain-fed agriculture	0	0	0	1	**98**	1
	Forest	0	0	1	0	45	**54**
		1999					
1992	Water	**96**	0	2	2	0	0
	High mangrove	5	**21**	72	1	1	1
	Low mangrove	28	3	**61**	7	1	0
	"tan"	11	0	5	**78**	7	0
	Savanna/rain-fed agriculture	0	0	0	2	**96**	2
	Forest	0	0	2	1	26	**70**
		2010					
1999	Water	**92**	1	5	2	0	0
	High mangrove	10	**44**	43	1	1	1
	Low mangrove	12	14	**66**	4	3	1
	"tan"	18	0	7	**63**	12	1
	Savanna/rain-fed agriculture	0	0	0	1	**95**	3
	Forest	0	0	2	0	42	**56**

The bold represent the no change percentage.

From the pixel $x \in \mathbb{R}^k$ and the number of bands k, the label to be assigned to x is given by the class that maximizes the LDC discrimination function (Kuncheva 2004; Hastie et al. 2009). The LDC discrimination function is given, according to Kuncheva (2004), by Eq. (4):

$$h_i(x) = \frac{1}{2} \hat{\mu}_i^T \hat{\Sigma}^{-1} \hat{\mu}_i + \hat{\mu}_i^T \hat{\Sigma}^{-1} x \qquad (4)$$

where $\hat{\Sigma}$ is the common variance–covariance matrix, estimated by the weighted average of the separately estimated class variance–covariance matrix, i.e.,

$$\hat{\Sigma} = \sum_{i=1}^{k} \frac{n_i}{n} \hat{\Sigma}_i \qquad (5)$$

where n_i is the number of training units assigned to the ith class of Ω and n is the total number of training units.

The Landsat data of 1984, 1992, 1999, and 2010 are classified into six spectral classes using the LDC. Savanna and rain-fed agriculture are merged in the same class. This classification algorithm is a supervised parametric classifier, i.e., it requires a training sample in order to classify the pixel from a given image and it assumes that each land-cover class

behaves according to the normal statistic distribution. Thus, in this respect, it resembles the MLC. The difference is that the LDC is based on an additional assumption which is the homoskedasticity hypothesis, in which the classifier assumes that each class has equal variance, and thus, all covariance matrices are equal for every land-cover class of the nomenclature. Although the homoskedasticity hypothesis tends to be unrealistic, the literature has shown that this classifier behaves in a robust way even when there are deviations from the hypothesis of normality and homoskedasticity (Kuncheva 2004). The LDC has several advantages, in that it requires less training samples than the MLC and also it is easy to fine-tune and robust to noisy data (Hastie et al. 2009). In this sense, the LDC is a preferable classification algorithm for land-cover mapping (Carrão et al. 2008), especially when the image analyst does not have a reliable reference database to collect representative training samples. For the classification, a set of training sites and ground truth data were required. A sample set of 50 training sites was established. They characterize the six typical land-cover classes occurring in the study area. The sample plots were digitized on screen, and then, a supervised LDC was applied using a stack of the six (without band 6) original bands of the Landsat image and the remote sensing technique (NDVI) to generate a land-cover map.

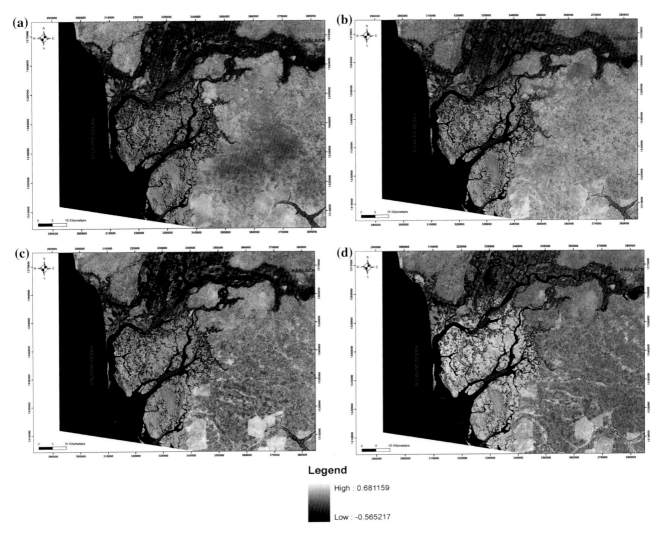

Fig. 4 NDVI images of the four Landsat images: **a** October 1984, **b** October 1992, **c** November 1999, and **d** November 2010. The NDVI images were calculated with the NIR band, the red band

The overall accuracy and a kappa analysis were used to perform a classification accuracy assessment based on error matrix analysis. Using the simple descriptive statistics technique, overall accuracy is calculated by dividing the total correct by the total number of pixels in the error matrix. The kappa analysis is a discrete multivariate technique used in accuracy assessments (Jensen 1996), and it yields a KHAT statistic (an estimate of kappa) which is a measure of agreement or accuracy (Congalton and Green 1993). It is a measure of overall statistical agreement of an error matrix, which takes non-diagonal elements into account, and it is recognized as a powerful method for analyzing a single error matrix and for comparing the differences between various error matrices (Congalton 1991; Smits et al. 1999; Foody 2004).

The next step was to generate a cross-tabulation using GIS technique that combines the information of two types of raster files into a contingency matrix. The procedure consists of counting pairs of categorical values of two given variables in order to produce a categorical frequency distribution.

NDVI Differencing Method

The NDVI differencing method employs NDVI to differentiate images for mapping pixel change in the land-cover types. It is a popular vegetation index differencing used for change detection. For the NDVI differencing method, the NDVI image for each year was first computed according to Tucker (1978) using the NIR and RED bands (Eq. 6):

$$\text{NDVI} = \frac{\text{NIR} - \text{RED}}{\text{NIR} + \text{RED}} \quad (6)$$

NDVI is derived from differences in reflectance of the red (pigment absorption) and near-infrared (scattering from cellular structure), with values ranging from -1 to $+1$. Negative values refer to an absence of vegetation, while

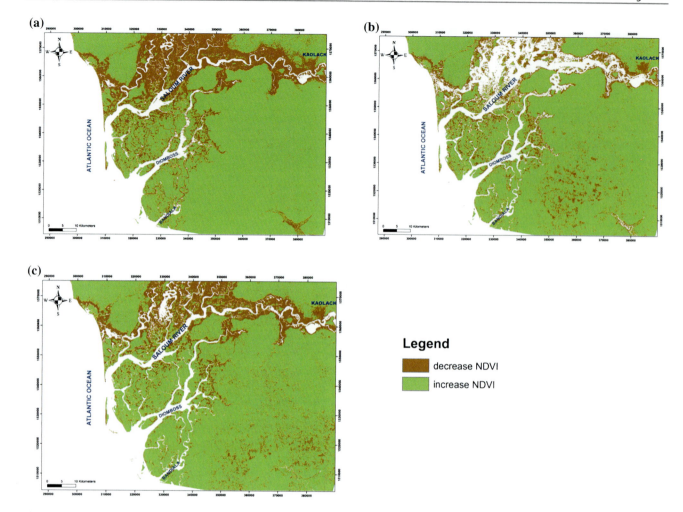

Fig. 5 Maps show areas of NDVI increase and areas of NDVI decrease between 1992 and 1984 (**a**), 1999 and 1992 (**b**), and 2010 and 1999 (**c**)

positive values are related to biomass variables, indicating leaf cover or productivity (Wang et al. 2003; Filella et al. 2004; Pettorelli et al. 2005; Zinnert et al. 2011).

Due to differences in atmospheric and land surface conditions and phenological stages, among other factors, during the acquisition, the satellite images exhibit differences in spectral behavior (Pu et al. 2008). The purpose of change detection is to extract the true LULC changes. Prior to that, it is important to normalize the images in order to identify the changes caused by other factors. It is worth noticing that the radiometric normalization does not completely correct the spectral behavior (Pu et al. 2008), i.e., the procedure will normalize the spectral values to a similar level for the two images acquired in different dates.

In this study, the normalization procedure was based on the work of Pu et al. (2008). The normalization was not done over the NIR and RED bands, but rather over the NDVI. This procedure required less time, due to the reduced number of samples necessary to collect when compared with normalizing the NIR and RED separately. In the normalization procedure, the NDVI values from one date are assumed to be in a linear relation with the NDVI values from the other date. That is, it is possible to correlate using $y = ax + b$, where x is the pixel value of one image, y is its correspondent value, and a and b are coefficients determined by least-square linear regression (LSLR). The x image is usually called the reference image, and the y image is called the subject image (Lunetta and Elvidge 1998). To compute the LSLR parameters, it is necessary to collect a sample of pixel values. In this work, we have applied the pseudo-invariant feature (PIF) method described by Schott et al. (1988), to collect the samples. PIF are pixels that do not represent changes in their spectral response over the period of time in analysis. The PIF method is based on two poles, namely very dark pixels and very bright pixels. Typically, the dark PIFs can be found in deepwater pixels and the bright sets on surfaces with very little or no vegetation, like barren soil and rock (Lunetta and Elvidge 1998). These two poles are the basis to sample pixel values for normalization. Once the coefficients are determined, it becomes possible to apply the linear function to compute the predicted NDVI and then the difference between two

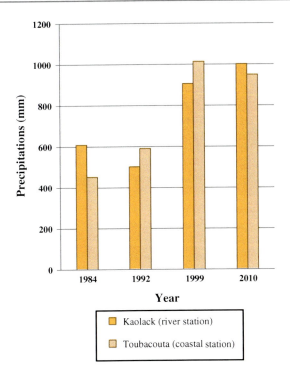

Fig. 6 Average mean annual precipitation (MAP) for 5 years prior to each date showing decreasing trend over time

dates, as suggested by Pu et al. (2008). In the present study, the NDVI differences between 1984 and 1992, 1992 and 1999, and 1999 and 2010 were computed.

After the computation of NDVI differences, the threshold values that will define the change/no-change areas must to be computed. In this step, it is assumed that the difference in NDVI presents a normal distribution centered on zero (Lunetta and Elvidge 1998; Pu et al. 2008). In practice, this is usually not the case in that the NDVI differences have mean value very close but different to zero. Additionally, it is also assumed that NDVI increasing and decreasing parts present a normal distribution (Pu et al. 2008). Under these two assumptions, Pu et al. (2008) determined the threshold values that can be used to identify the change/no-change areas by computing the no-change interval: $[m_d + cs_d, m_i + cs_i]$, where m_d is the mean value and s_d is the standard deviation of the decreasing component; similarly for m_i, s_i (Fig. 2). The constant c can be determined using methods based on the kappa value or accuracy assessment (Pu et al. 2008). However, in this study, the NDVI differences presented a very small standard deviation value. This fact implies that the extreme values in the no-change interval tend to be very near to the global mean value. Thus, the threshold values were set as equal to the NDVI difference mean value. This raises the problem of how to detect no-change areas. To overcome this difficulty, the LULC maps were used, so those pixels with equal thematic label were considered stable over time.

Under the two assumptions, the NDVI difference is centered on zero, and there is a value m_d, the mean value of the NDVI decreasing component, and a value m_i, the mean value of the NDVI increasing component. The no-change area is defined using the interval $[m_d + cs_d, m_i + cs_i]$.

Results and Discussions

Land-Cover Change

Classification

Using the classification method described above, the area of land-cover types in each of four study images was obtained and regional characterization of land-cover and land-cover changes was understood for Saloum estuary over the twenty-five-year period from 1984 to 2010. The results reveal that substantial changes took place during this time. The confusion matrices and kappa values were calculated from test samples for the four single-date classification results (Fig. 3): October 1984, October 1992, November 1999, and November 2010. The accuracy derived from the November 2010 image is evidently lower than that from 1984, 1992, and 1999 images (kappa value 0.79 vs. 0.90 or 0.87 or 0.85, or overall accuracy of 78 vs. 90 or 87 or 84 %). The lower classification accuracy might be due to the poorer quality of 2010 raw image strips in comparison with the other. Table 4 presents three change matrices that reflect the change directions and percentages of land-cover types based on the single-date classification results. The change matrices were calculated by overlaying the four single-date classification maps. The water area has increased in 1992 and 1999, and mangrove (high and low) and "tan" were lost by immersion because the break of the Sangomar spit in 1987. From 1984 to 1992, 55 % of high mangrove shifted to low mangrove and 6 % of low mangrove degraded to denudate soil ("tan"). These changes and conversions increased from 1992 to 1999, and 72 % of high mangrove transformed to low mangrove and 7 % of low mangrove and 3 % of savanna shifted to "tan" (Table 1). In 1999, due to high precipitation (Fig. 4) and sea level rise, the water surface also increased by 15 %.

NDVI Differencing

Over the NDVI images (Fig. 5), areas with gray reflect higher NDVI. From Fig. 4, the area reflecting higher NDVI (areas with gray) on the 2010 image is larger than that on the other three images. For NDVI index normalization between an NDVI image pair, the three linear regression equations are as follows:

Fig. 7 Interannual variation in salinity and rainfall at Kaolack locality from 1927 to 2012

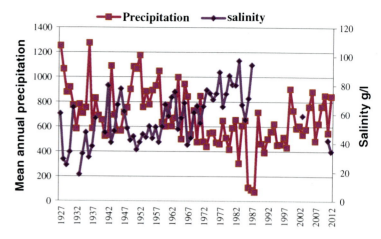

$$y = 1.47x - 0.43$$
$$R^2 = 0.98 \text{ for NDVI image pair 1992 } (y) - 1984 \ (x) \quad (7)$$

$$y = 0.73x + 0.29$$
$$R^2 = 0.98 \text{ for NDVI image pair 1999 } (\cdot) - 1992 \ (x) \quad (8)$$

$$y = 1.7x + 0.02$$
$$R^2 = 0.98 \text{ for NDVI image pair 2010 } (y) - 1999 \quad (9)$$

For the NDVI image pairs 1992 (y) − 1984 (x) and 2010 (y) − 1999 (x), the slope coefficients, respectively, 1.4 and 1.7 reflect that the NDVI for the unchanged pixels increases. For the NDVI image pair 1999 (y) − 1992 (x), the NDVI for the unchanged pixels decreases due to precipitation increases in 1999, which is reflected in the 0.73 slope coefficient. These equations were used to predict the 1992 NDVI, the 1999 NDVI, and the 2010 NDVI images, respectively, from the 1984, 1992, and 1999 NDVI images. The three NDVI difference images were generated by subtracting the predicted NDVI from the actual NDVI.

In the period 1984–1992, 73.1 % of the study area showed an increasing NDVI, in 1992–1999 81 %, and in 1999–2010 79.5 %. Crossing these results with precipitation values for the same periods (Fig. 6), we can see that the increase in NDVI is caused by the increase in rainfall levels. Although the rainfall levels increased in those periods, the location of the areas that show decreasing NDVI is located in zones occupied by tan around the Saloum River due to the high salinity of the water river (Figs. 3 and 6). In fact, the land-cover maps show that the majority of the transitions that show NDVI are transition to decreasing tan.

Naturals Factors of Estuary's Dynamic

Climate change has been particularly evident in west Africa in the last 30 years. Increased drought has led to a significant decrease in freshwater flow as well as an increase in the salinity level in estuary system. This is the case for the inverse estuary of the Sine Saloum where river waters with salinities much higher than seawater salinity occur (Fig. 7). In this region, the climate is characterized by an extended dry season, cool from November to March and warm from April to June, and by a short wet and warm season from July to October. Since the 1920s, the annual rainfall has decreased in this region with variable magnitude and drought period is much pronounced in recent decades (Pagès and Citeau 1990). The combined effects of reduced freshwater inputs, intense evaporation, and a low slope in the lower estuary have resulted in an overall high salinity and an inversion of the salinity gradient upstream the Saloum River course. This characteristic occurs to a lesser extent in the Diomboss branch.

The intermittent and the reverse flows of the Saloum River due to the runoff deficit caused saltwater advance up to 60 km upstream, contaminating surface water, groundwater, and large areas of agricultural soils located in these zones. Salinity in the Saloum River showed an upstream gradual increase from 36.7 ‰ at the mouth to higher than 90 ‰ at Kaolack (Pagès and Citeau 1990).

The chemical and isotopic data of water sampled in the Saloum River estuary (Faye et al. 2003, 2004, 2009; Dieng 2012 unpublished) revealed that high salinity is induced by seawater advance through tide dynamic. During the dry season (December to May) where maximum air temperatures and evaporation occur, seawater advance may reach 90 km inland (Kaolack locality) and salinity as high as 60 g/l (Dieng 2012 unpublished). Consequences of this high salinity are contamination of the shallow groundwater resources, large areas of arable land with formation of saline barren soils at the vicinity of the estuary system and the economically valuable mangrove ecosystem which plays a vital role to the majority of this rural population. The saltwater contamination constitutes a serious problem in this region.

Higher salinity content in the upstream estuary has also consequences on vegetation and soil resources in that the mudflat has been affected by these hydroclimatic variations. The morphopedological modifications induced by a high

Fig. 8 Evolution of the Sangomar sand spit between 1984 and 2010

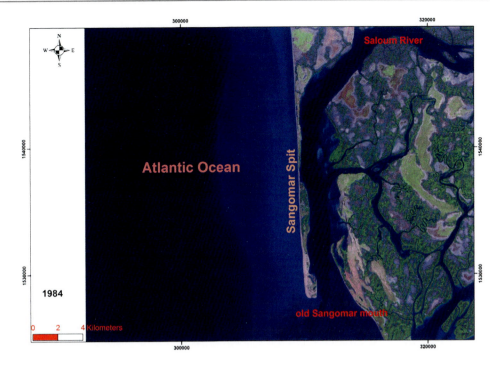

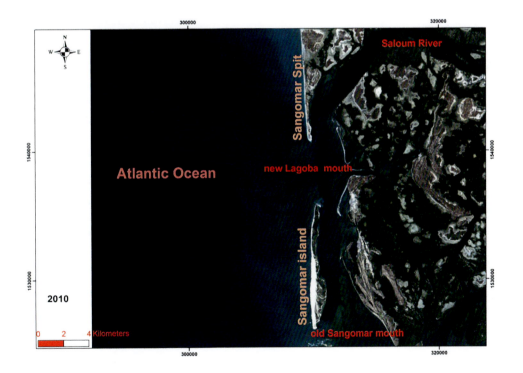

salinization and acidification of soils have led to a gradual degradation, transformation, and disappearance of the mangrove to tan, and the most affected areas are located from the mid-to-upstream parts of the Saloum River (Figs. 3, 4 and 5). In addition to that, the Sangomar spit break which occured in 1987 (Diaw et al. 1991; Dieye et al. 2013) may contribute to the significant changes downstream (mangroves, land loss) since opening of the spit channel (4.96 km currently) favored direct hydraulic connection between the sea and the Saloum River (Fig. 8).

Conclusion

Uses of Landsat image allow us to monitor dynamic changes in land-cover and land-use over a large area. The image processing approach with GIS techniques can provide valuable spatial data for both quantitative and qualitative studies of the land-cover changes. This study showed that significant changes in land cover occur within the whole estuary system. These changes reflect environmental degradation, such as land desertification, salinization, and vegetation degradation, which are caused by salinity increase. Comparisons revealed that conversion of mangrove to "tan" cover was closely linked to precipitation and breaching of the Sangomar sand spit. In addition, these results show that significant changes in land cover occur in the study area, reflecting in this way environmental degradation and land desertification caused by the advance of salty bare lands.

Acknowledgements We thank ESA for providing us with Landsat imagery, TIGER capacity building, and IGP (Portuguese Geographic Institute) for giving us training job to enhance our knowledge in remote sensing and GIS.

References

Anderson JR (1977) Land use and land cover changes. A framework for monitoring. J Res Geol Surv 5:143–153

Aplin P, Atkinson PM, Curran PJ (1999) Per-field classification of land use using the forthcoming very fine spatial resolution satellite sensors: problems and potential solutions. In: Atkinson PM, Tate NJ (eds) Advances in remote sensing and GIS analysis. Wiley, New York, pp 219–239

Barusseau JP, Diop EHS, Saos JL (1985) Evidence of dynamics reversal in tropical estuaries, geomorphological and sedimentological consequences (Saloum and Casamance Rivers, Senegal). Sedimentology 32(4):543–552

Carrão H, Araújo A, Caetano M (2008) Land cover classification in Portugal with intra-annual time series of MERIS Images. In: Proceedings of the 2nd MERIS/AATSR user workshop Frascati, Italy, 22–26 Sept 2008

Chen D, Stow DA (2002) The effect of training strategies on supervised classification at different spatial resolution. Photogram Eng Remote Sens 68:1155–1162

Church JA, White NJ (2006) A 20th century acceleration in global sea-level rise. Geophys Res Lett 33:L01602

Congalton RG (1991) A review of assessing the accuracy of classifications of remotely sensed data. Remote Sens Environ 37:35–46

Congalton RG, Green K (1993) A practical look at the sources of confusion in error matrix generation. Photogram Eng Remote Sens 59:641–644

Coppin PR, Bauer ME (1994) Processing of multitemporal landsat TM imagery to optimize extraction of forest cover change features. IEEE Trans Geosci Remote Sens 32:918–927

Coppin PI, Jonckheere K, Nackaerts BM, Lambin E (2004) Digital change detection methods in ecosystem monitoring; a review. Int J Remote Sens 25:1565–1596

Datta D, Deb S (2012) Analysis of coastal land use/land cover changes in the Indian Sunderbans using remotely sensed data. Geo-spat Inf Sci 15(4):241–250

Diaw AT, Diop N, Thiam MD, Thomas YF (1991) Remote sensing of spit development: a case study of Sangomar sand spit, Senegal. Z Geomorph Berlin-Stuttg 81:115–124

Dieye EHB, Diaw AT, Sané T, Ndour N (2013) Dynamique de la mangrove de l'estuaire du Saloum (Sénégal) entre 1972 et 2010. Eur J Geogr. http://cybergeo.revues.org/25671; doi: 10.4000/cybergeo.25671

Diop ES (1986) Estuaires holocènes tropicaux. Etude géographique physique comparée des rivières du Sud du Saloum (Sénégal) à la Mellcorée (République de Guinée). Thèse Doctorat es Lettres Université L Pasteur Strasbourg. Tome I, p 522

Faye S, Cissé Faye S, Ndoye S, Faye A (2003) Hydrogeochemistry of the Saloum (Senegal) superficial coastal aquifer. Environ Geol 44:127–136

Faye S, Maloszewski P, Stichler W, Trimborn P, Cissé Faye S, Gaye CB (2004) Groundwater salinization in the Saloum (Senegal) delta aquifer: minor elements and isotopic indicators. Sci Total Environ J Springer. doi:10.1016/j.scitotenv.2004.10.00,17p

Faye S, Diaw M, Ndoye S, Malou R, Faye A (2009) Impacts of climate change on groundwater recharge and salinization of groundwater resources in Senegal. In: Groundwater and climate in Africa proceeding of the Kampala conference. IAHS Publ 334, June 2008

Filella I, Penuelas J, Llorens L, Estiarte M (2004) Reflectance assessment of seasonal and annual changes in biomass and CO_2 uptake of a Mediterranean Shrubland submitted to experimental warming and drought. Remote Sens Environ 90(3):308–318

Foody GM (2004) Thematic map comparison: evaluating the statistical significance of differences in classification accuracy. Photogram Eng Remote Sens 70:627–633

Franklin SE, Peddle DR, Dechka JA, Stenhouse GB (2002) Evidential reasoning with Landsat TM, DEM and GIS data for land cover classification in support of grizzly bear habitat mapping. Int J Remote Sens 23:4633–4652

Gallego FJ (2004) Remote sensing and land cover area estimation. Int J Remote Sens 25:3019–3047

Ghassemi F, Jakeman AJ, Nix HA (1995) Salinization of land and water resources. Human causes, extent, management and case studies. Sydney University of New South Wales Press Ltd.

Gong P, Howarth PJ (1992) Frequency-based contextual classification and gray-level vector reduction for land-use identification. Photogram Eng Remote Sens 58:423–437

Hastie T, Tibshirani R, Friedman J (2009) The elements of statistical learning: data mining, and prediction. Springer series in statistics, Springer. ISBN-10 0387848570

Ingram K, Knapp E, Robinson JW (1981) Change detection technique development for improved urbanized area delineation. Technical memorandum CSC/TM-81/6087. Silver Spring, Computer Science Corporation, MD, USA

IPCC (2007) The physical science basis. In: Solomon S et al (eds) Contribution of working group I to the fourth assessment report of the intergovernmental panel on climate change. Cambridge University Press, New York, pp 747–845

Jensen J R (1996) Introductory digital image processing. Prentice Hall, New Jersey, USA

Johnson RA, Wichern DW (1998) Applied multivariate statistical analysis. Prentice Hall, pp 816

Ketchum BH (1983) Estuarine characteristics, estuaries and enclosed seas. Elsevier Science Publication Comp, Amsterdam, pp 1–14

Kontoes C, Wilkinson G, Burrill A, Goffredo S, Megier J (1993) An experimental system for the integration of GIS data in knowledge-based image analysis for remote sensing of agriculture. Int J Geogr Inf Syst 7:247–262

Kuncheva L (2004) Combining pattern classifiers methods algorithms, 1st edn. Wiley-Interscience. ISBN-10: 0471210788

Landgrebe DA (2003) Signal theory methods in multispectral remote sensing. Wiley, Hoboken

Lu D, Weng (2007) A survey of image classification methods and techniques for improving classification performance. Int J Remote Sens 28(5):823–870

Lu D, Mausel P, Brondizio E, Moran E (2003) Change detection techniques. Int J Remote Sens 25:2365–2407

Lu D, Mausel P, Batistella M, Moran E (2004a) Comparison of land-cover classification methods in the Brazilian Amazon Basin. Photogram Eng Remote Sens 70:723–731

Lu D, Valladares G, Li GS, Batistella M (2004b) Mapping soil erosion risk in Rondonia, Brazilian Amazonia: using RUSLE, remote sensing and GIS. Land Degrad Dev 15:499–512. doi:10.1002/ldr.634

Lunetta R, Elvidge C (1998) Remote sensing change detection: environmental monitoring and application. Taylor and Francis, Sleeping Bear Press, Inc.

Mather PM (2004) Computer processing of remotely-sensed images: an introduction, 3rd edn. Wiley, Chichester

McDowell DM, O'Connor BA (1983) Gidravlika prilivnykh ust'ev rek (hydraulics behaviour of estuaries). Energoatomizdat, Moscow

Mikhailov VN, Isupova MV (2008) Hypersalinization of river estuaries in West Africa. Water Resour 35(4):367–385

Nelson RF (1983) Detecting forest canopy change due to insect activity using landsat MSS. Photogram Eng Remote Sens 49:1303–1314

Overpeck JT, Otto-Bliesner B, Miller GH, Mush DR, Alley RB, Kiehl JT (2006) Paleoclimatic evidence for future ice-sheet instability and rapid sea-level rise. Science 311(5758):1747–1750. doi:10.1126/science.1115159

Pagès J, Citeau J (1990) Rainfall and salinity of a Sahelian estuary between 1927 and 1987. Hydrol J 113(1–4):325–341

Pagès J, Debenay JP, Lebrusq JI (1987) L'environnement estuarien de la Casamance. Rev Hydrobiol Trop 20(3–4):191–202

Pal M, Mather PM (2004) Assessment of the effectiveness of support vector machines for hyperspectral data. Future Gener Comput Syst 20:1215–1225

Pan W, Xu H, Chen H, Zhang C, Chen J (2011) Dynamics of land cover and land use change in Quanzhou city of SE China from landsat observations. Electr Eng 40371107:1019–1027

Petropoulos G, Partsinevelos P, Mitraka Z (2012) Change detection of surface mining activity and reclamation based on a machine learning approach of multitemporal landsat TM imagery. Geocarto Int. doi:10.1080/10106049.2012.706648

Pettorelli N, Vik JO, Mysterud A, Gaillard JM, Tucker CJ, Stenseth NC (2005) Using the satellite-derived NDVI to assess ecological responses to environmental change. Trends Ecol Evol 20(9):503–510

Pritchard DW (1967) What is an estuary: physical viewpoint, estuaries. American Association Advanced Science Publication, Washington, DC

Pu R, Gong P, Tian Y, Miao X, Raymond IR, Carruthers GL, Anderson GL (2008) Using classification and NDVI differencing methods for monitoring sparse vegetation coverage: a case study of saltcedar in Nevada, USA. Int J Remote Sens 29(14):3987–4011

Ridd PV, Sam R (1996) Profiling groundwater salt concentration in mangrove swamps and tropical salt flats. Estuar Coast Shelf Sci 43:627–635

Ridd PV, Stieglitz T (2002) Dry season salinity changes in arid estuaries fringed by mangroves and saltflats. Estuar Coast Shelf Sci 54:1039–1049

Sam R, Ridd PV (1998) Spatial variations of groundwater salinity in a mangrove salt flat System Cocoa Creeks Australia. Mangroves Salt Marshes 2:121–132

San Miguel-Ayanz J, Biging GS (1997) Comparison of single-stage and multi-stage classification approaches for cover type mapping with TM and SPOT data. Remote Sens Environ 59:92–104

Singh A (1989) Digital change detection techniques using remotely-sensed data. Int J Remote Sens 10:989–1003

Smits PC, Dellepiane SG, Schowengerdt RA (1999) Quality assessment of image classification algorithms for land-cover mapping: a review and a proposal for a cost-based approach. Int J Remote Sens 20(8):1461–1486

Stuckens J, Coppin PR, Bauer ME (2000) Integrating contextual information with per-pixel classification for improved land cover classification. Remote Sens Environ 71:282–296

Tucker CJ (1978) A comparison of satellite sensor bands for vegetation monitoring. Photogram Eng Remote Sens 44(11):1369–1380

Wang J, Rich PM, Price KP (2003) Temporal responses of NDVI to precipitation and temperature in the central great plains, USA. Int J Remote Sens 24(11):2345–2364

Wolanski E (1986) An evaporation-driven salinity maximum zone in Australian tropical estuaries. Estuar Coast Shelf Sci 22:415–424

Zhang K, Douglas BC, Leatherman SP (2004) Global warming and coastal erosion. Clim Change 64(1–2):41–58

Zinnert JC, Shiflett SA, Vick JK, Young DR (2011) Woody vegetative cover dynamics in response to recent climate change on an Atlantic coast barrier island: a remote sensing approach. Geocarto Int 26(8):595–612

Studies and Transactions on Pollution Assessment of the Lagos Lagoon System, Nigeria

Babajide Alo, Kehinde Olayinka, Aderonke Oyeyiola, Temilola Oluseyi, Rose Alani, and Akeem Abayomi

Abstract

The Lagos Lagoon system is a brackish coastal lagoon—the largest in the West African coast with a large series of estuaries—located between longitude 3°23′ and 3°40′E and between latitude 6°27′ and 6°48′N. It is a shallow expanse of water (0.3–3 m deep), 50 km long and 3–13 km wide and separated from the Atlantic Ocean by a narrow strip of barrier bar complex. This report is on the levels of pollution and nutrients status of the Lagos Lagoon system including physicochemical properties, pesticides organochlorines (OC), polyaromatic hydrocarbons (PAHs), heavy metal species and nutrients observed between 2002 and 2008. Watersheds of the highways on the lagoon had higher concentrations of nutrients (phosphorus and nitrates) relative to other locations on the Lagoon. The western part of the Lagoon was found to have higher concentrations of Cd, Cu, Pb and Zn than the other points. Lagos Lagoon and the adjoining creeks show high anthropogenic input of PAHs and other persistent organic pollutants (POPs). The major hydrocarbon index in most samples was at C_{29}, C_{31} and C_{27}, indicating vascular plants sources. Mean PBT levels in water and in sediment increased with time between 2004 and 2007. PBT distribution in the lagoon followed the pattern, sediment > biota > water, though some exceptions occurred where the biota bioaccumulated more PBTs than are found in both sediment and water. The Lagoon biota bioaccumulated organochlorine pesticides above allowable limits and thus pose a high risk to human health. The levels of some pollutants in the Lagoon have negatively impacted on the environmental quality which has indirectly affected the social and economic activities of the dependants and this requires improved management strategies to ameliorate. Indeed with the high population that the estuary/lagoon system supports, consideration for its designation as an international waterbody and its concomitant attention is now paramount.

Keywords

Lagos Lagoon • Heavy metals • Pollution • Nutrients • Polyaromatic hydrocarbons • Organochlorine pesticides • Polychlorinated biphenyls • Hydrocarbons

Introduction

Urbanization and industrialization have led to the release of a variety of pollutants from both point and diffuse sources, and has placed considerable pressure on the aquatic resource. This is of particular concern in developing countries where expansion of urban area may be relatively

---

B. Alo (✉) · K. Olayinka · A. Oyeyiola · T. Oluseyi · R. Alani · A. Abayomi
Analytical and Environmental Research Group, Department of Chemistry, University of Lagos, Akoka, Lagos, Nigeria
e-mail: profjidealo@yahoo.com

S. Diop et al. (eds.), *The Land/Ocean Interactions in the Coastal Zone of West and Central Africa*, Estuaries of the World, DOI: 10.1007/978-3-319-06388-1_6,
© Springer International Publishing Switzerland 2014

unregulated and where environmental protection may be inadequate (Fonesca et al. 2011; Li et al. 2007; Manning 2011; Nriagu 1992; Oyeyiola et al. 2013a). The ocean's ecosystem and the changes in it induced by pollution and climate change are of worldwide concern. To predict how this ecosystem will respond to further global change and what role the ocean's biota will play requires detailed studies of biological, physical and chemical data of marine-related sample matrices (water, sediment and biota). One of the main concerns at present is the pollution of the marine environment by contaminants whose levels are growing at an alarming rate. Pollutants such as potentially toxic metals, persistent organic pollutants such as pesticides, dioxins, polychlorinated biphenyls and PAHs, and nutrients have been found to be a major threat to the marine ecosystem and thus a major global problem. This is because pollutants released in one part of the world can be transported to regions far from their source of origin through the atmosphere, waters and other pathways. Their effect on human health can be felt directly and/or via the food web (Chaney et al. 1996).

The Lagos Lagoon System

Lagos is the Africa's biggest city and the fastest growing metropolis in the world. It is the most heavily industrialized and urbanized city in Nigeria, with much of the nation's wealth and economic activities located there, and 12 % of the total population (150 million). The City of Lagos is currently undergoing an extensive programme of expansion and development with the aim of becoming a 'megacity' and major international focus for trade and industry (Howden 2010; This Day's Special International Project 2007). The urban area is built on a number of floodplains and encompasses a network of marshes, swamps, streams, creeks, rivers and estuaries which receive large quantities of rain water run-off, domestic, municipal and industrial waste effluents, and each of these receptors discharge finally into the Lagos Lagoon. We describe here the entire hydrological system of the Lagos Lagoon System in Nigeria.

The city of Lagos has about 91 waterways (canals) and their tributaries draining the entire city and discharging into the Lagos Lagoon at different locations. Most of the major markets and industries in Lagos (Iddo, Otto, Yaba, Oshodi, Mushin, Idioro, Balogun, Ebute–Ero, Idumota, Abule Egba, Alaba Rago, Alaba International, Mile2 and Mile 12 markets) generate huge quantities of municipal wastes which are either dumped directly into the lagoon or are incinerated at sites that drain into the lagoon at different locations. Open waste incineration, in some cases on the shores of Lagos lagoon, is a very common sight in Lagos. One of such open incinerators is found at Okobaba, where incessant burning

of sawdust and other municipal waste take places near the Third Mainland Bridge—the longest bridge in West Africa which crosses the Lagos lagoon and links the Lagos Mainland in the north of Lagos Megacity to the Lagos Island in the south of the city.

The Lagos Lagoon is a brackish coastal lagoon—the largest in the West African coast—located between longitude $3°23'$ and $3°40'E$ and between latitude $6°27'$ and $6°48'N$. It is a fairly shallow expanse of water (0.3–3 m deep) which is about 50 km long and 3–13 km wide and separated from the Atlantic Ocean by a narrow strip of barrier bar complex. The lagoon borders the forest belt and empties directly into the Atlantic Ocean at the harbour (Fig. 1). During the rainy season, large volumes of fresh water passes through the harbour into the sea. In view of its complexity (linkages to land, freshwater, sea, salinity fluctuations, dynamics), the lagoon is a fragile ecosystem prone to environmental degradation through pollution from industry, household, resource over-exploitation, etc. The loading of pollutants into the lagoon affects the quality of the water. This is reflected in the colour and appearance of the lagoon water, ranging from oily at some locations to grey, slightly yellow, turbid and dark at others. There is a high spatial and temporal variability of constituents (pollutants) in surface sediments throughout the lagoon. This is reflected by the variation in sediment type and colour at different locations of the lagoon.

Lagos lagoon serves the huge population of Lagos, approximately 15 million. The Lagoon finds its use in artisanal fishing, transport and recreational purposes. It is an important habitat for a wide array of fish and marine organisms and is the major source of sea foods for the people of Lagos (Isebor et al. 2006). Some abuses of the Lagoon include serving as a direct dumpsite for industrial, agricultural and municipal wastes, and even a dumpsite for sewage at some locations. Dredging of sediment from the lagoon for sand-filling and land reclamation although illegal is commonplace to satisfy local demand for building constructions. Pollutants from this lagoon may affect the population directly through contact with contaminated sediment or water and via the food chain.

However, despite rapid industrialization in the region, coupled with the importance of this waterbody, currently, there is a dearth of detailed information on the pollution and pollutant status of the Lagos Lagoon system.

Physicochemical Properties of the Lagos Lagoon

Our extensive studies of the physicochemical status of the water in the Lagos Lagoon System have been revealing on the pollution status of the waterbody. Water samples

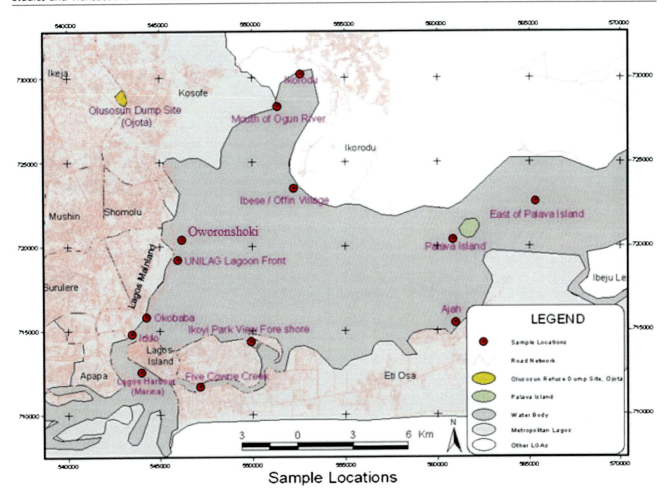

Fig. 1 Map showing the Lagos Lagoon

Table 1 Physico-chemical properties of water samples from the Lagos Lagoon

	Sampling points					WHO limit
Parameter	Q	R	S	COW	PAL	
Dissolved oxygen (mg O_2/l)	4.7–6.3	3.7–5.4	3.4–5.0	6.2–6.5	6.1–6.4	–
BOD (mg O_2/l)	11.8–30.4	16.8–24.4	22.2–34.2	16.2–28.6	14.2–20.6	0
Total hardness (mg $CaCO_3$/l)	191.8–2,057.3	146.8–1,783.8	149.1–916.3	65.8–102.8	21.5–96.2	–
Total Alkalinity (mg $CaCO_3$/l)	61.5–01.7	74.9–9.0	80.3–91.0	42.7–60	32.3–40	–
Acidity (mg $CaCO_3$/l)	[a]–9.3	3.0–11.7	2.5–12.8	5.5–7.0	5.1–6.5	–
Chloride (mg/l)	798.9–14,819	1520.6–11,752	283.5–12,722.7	587.2–5,541	432.8–1,318.2	250
Sulphate (mg/l)	303.6–2,432.9	343.4–22,430	32.0–2,343	43.2–1,220	87.6–93.9	–
Phosphate (mg P/l)	0.2–0.4	0.1–0.3	0.2–0.3	0.1	0.1	1.0
Conductivity (μS cm^{-1})	24 000–410,000	21,000–315,000	3,000–330,000	2,500–170,000	1,300–60,000	–
pH	6.9–8.2	6.9–7.8	6.9–7.5	7.0–7.2	7.2–7.4	6.5–8.5
Total solids (mg/l)	3,100–30,870	3,910–25,630	2,080–23,780	1,720–11,080	980–3,070	500
Total dissolved solids (mg/l)	2,000–28,580	3,070–24,060	1,500–21,850	850–10,840	240–1,220	500

[a] Not detected

Q Iddo, *R* Oko-Baba, *S* Oworonshoki, *COW* five cowrie, *PAL* Palaver Island

Table 2 Concentration of nutrients in the sediments of Lagos Lagoon System

Sampling location	Total PO_4^{3-}–P (Mean ± SD) (mg kg^{-1})	Bioavailable PO_4^{3-}–P (Mean ± SD) mg kg^{-1}	Percent bioavailable (%)	NO^{2-}/NO^3 (Mean ± SD) (mg kg^{-1})	P:N Ratio
Akoka	326.6 ± 15.2	4.65 ± 0.76	1.42	11.24 ± 0.38	0.41
Okobaba	619.3 ± 33.4	7.45 ± 0.26	1.2	0.92 ± 0.06	8.1
Iddo	1,203.9 ± 171.6	20.94 ± 0.62	1.74	0.71 ± 0.08	29.49
Ijora	253.5 ± 100.8	4.52 ± 0.52	1.78	6.56 ± 0.12	0.69
Leventis	65.9 ± 31.8	9.48 ± 0.84	14.38	4.43 ± 0.12	2.14
Cowry Creek	174.5 ± 96.0	8.90 ± 0.78	5.1	9.85 ± 0.32	0.9
Falomo	637.2 ± 4.2	17.03 ± 0.28	2.67	0.56 ± 0.08	30.36
Mean	468 ± 387	10.4 ± 6.2	4.0 ± 4.8	4.90 ± 4.47	10.3

collected from the Lagos Lagoon between January 2004 and March 2005 were analysed, and all physicochemical parameters were determined. Table 1 gives the results for physicochemical properties of the water samples collected from different parts of the Lagos lagoon. Points Q, R and S are Iddo, Oko-Baba and Oworonshoki, respectively, by (Fig. 1) and are close to human activities, while Cow (Five Cowrie Creek) and PAL (Palavar Island) are points on the Lagos Lagoon that are far from industrial or domestic activities.

The dissolved oxygen (DO) at points Q, R and S appeared to be independent of tide or seasons and is suitable for aquatic life to survive. This is because in spite of pollution of the water by sewage and other domestic wastes, the impact of the pollution is not felt because of the relatively large volume of water in the Lagos Lagoon which facilitates dilution of the pollutants. The highest DO values were observed at COW and PAL which are far from human and industrial activities.

The biochemical oxygen demand (BOD) values for points Q and S were the highest. This was probably because of the untreated sewage discharged into the Lagos Lagoon at point Q, and industrial effluents and domestic wastes at point S which may contain high level of organics.

The chloride content of the water is expectedly high. This is due to the presence of chloride salts in the sea water which flows into the Lagos lagoon when the tide is high. The pH range at all the points fall within the WHO limit for unpolluted water (6.5–8.5).

Generally, points Q, R and S were more polluted than COW and PAL in terms of the DO, BOD and pH values of the samples collected. This may be because domestic wastes are discharged into the Lagos lagoon at these points.

Nutrients Status in the Sediments of Lagos Lagoon System

Abayomi et al. (2011) reported on the potential of the Lagos lagoon sediment to act as a sink for anionic nutrient from highway run-off and roadside soils contiguous to the lagoon system. Concentrations of total phosphorus ranged from 73 ± 20 to 622 ± 514 mg kg^{-1} in the wet season and 170 ± 10 to $1,320 \pm 480$ mg kg^{-1} in the dry season. The bioavailable concentration had a range of 2.57 ± 0.64 to 9.4 ± 5.1 mg kg^{-1} and 4.8 ± 0.7 to 22.0 ± 0.9 mg kg^{-1} for the wet and dry season, respectively. The concentrations of bioavailable NO^{2-}/NO^{3-} in the roadside samples ranged from 0.53 ± 0.64 to 12.35 ± 32.88 mg kg^{-1} for the wet season and 0.39 ± 0.08 to 21.35 ± 0.50 mg kg^{-1} for the dry season. In the Lagoon sediment, phosphorus had a mean concentration range of 468 ± 387 mg kg^{-1} total phosphorus and 10.4 ± 6.2 mg kg^{-1} bioavailable phosphorus for the sampled locations (Table 2). Watersheds of the highways on the lagoon had higher concentrations of these nutrients relative to other locations on the Lagoon which confirm roadside as soils as sources of nutrient input into the Lagos Lagoon.

Average Phosphate Concentration in the Lagos Lagoon

Results from the study showed a mean bioavailable phosphate concentration range of 0.046–0.91, 0.046–0.453 and 0.048–0.492 mg L^{-1}, respectively, for the top, middle and bottom strata of the lagoon. Mean total phosphate concentrations were in the range of 0.054–0.513, 0.041–0.961 and

0.058–1.71 mg L^{-1} for the top, middle and bottom strata, respectively (Figs. 2, 3).

In most cases, the levels observed were higher than the 0.04 PO$_4^{3-}$—P limit and above this concentration algal bloom starts to occur (Wallace et al. 2010).

Potentially Toxic Elements in Sediments of the Lagos Lagoon System

In a 2006 study, our group collected surface sediments from the Lagos Lagoon, Nigeria, and three adjoining rivers/estuaries (Odo-Iyaalaro River, Shasha River and Ibeshe River) and these were analysed for their physicochemical properties, pseudo-total concentration, fractionation pattern as well as the ecotoxicological implication of the potentially toxic metals (PTM): Cd, Cr, Cu, Pb and Zn.

According to Oyeyiola et al. (2013a), the Odo-Iyaalaro River was observed to be the most polluted river, with highest concentrations of 42.1, 102, 185, 154 and 1040 mg kg^{-1} of Cd, Cr, Cu, Pb and Zn, respectively. The month of March, which is the peak of the dry season, had particularly higher concentration of metals compared with samples collected from the same point in other months. This was attributed to the accumulation of these PTMs in sediment during the dry season, which are then remobilised during the rainy season. In their study, it was reported that previous workers observed seasonal influence on sediment metal concentrations in tropical systems, but findings are contradictory. Similar to their study, Bahena-Manjarrez et al. (2002) observed the highest metal concentrations in the dry season in the Coatzacoalcos River, Mexico, and this was supported by Alagarsamy (2006) who reported the lowest metal concentrations during the monsoon in the Mandovi estuary, India.

Oyeyiola et al. (2013b) also found potentially toxic elements (PTE) concentrations in sediments from the Shasha River to decrease with decreasing distance from the Lagos Lagoon. This variation was said to correlate with the location of industries and their waste-disposal systems. The average concentrations of PTE in the Shasha River were lower than in the Odo-Iyaalaro River. The Ibeshe River was the least contaminated, apart from a site affected by Cu from the textile industry. The concentration of Cu was also found to decrease downstream with increasing distance into the Lagos Lagoon.

They also observed in their studies that the western part of the Lagoon had higher concentrations of the studied metals than the other parts. This was in agreement with researchers (Otitoloju et al. 2007) who also observed higher concentration of metals in the western part of the lagoon. These may be because there are a lot of bridges around these area as compared with the other parts of the lagoon

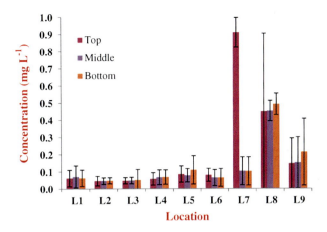

Fig. 2 Average concentration of bioavailable phosphate in the water of the Lagos Lagoon

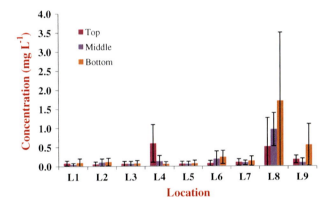

Fig. 3 Average concentration of total phosphate in the water of the Lagos Lagoon

and also there is a lot of human activities. In our study, it was observed that Cr concentration in the Lagos Lagoon was generally high even though the range is small. This is probably an indication that Cr in the Lagos Lagoon is geogenic. There was no difference were observed between concentrations of metals in rainy and dry seasons in the lagoon, nor were there marked seasonal differences in grain size distribution, as had been observed in some of the tributary rivers. This was to be expected because according to Lund-Hansen et al. (1999), re-suspension rate was observed to be much more higher than sedimentation rate in shallow coastal lagoons and the depth of Lagos Lagoon waterbody means that deposition, as well as re-suspension, is likely to occur throughout the year.

Due to lack of legislation in Nigeria governing acceptable levels of PTE in sediments, it was not possible to assess or contextualise the current findings within a local regulatory framework, so the consensus-based probable effect concentrations (PEC) recommended by MacDonald et al. (2000) and Dutch sediment guideline (Grimwood and

Table 3 Descriptive statistics of physicochemical properties and PTM concentration in sediments of dry and rainy seasons

Parameter	Dry			Rainy		
	Minimum	Maximum	Mean	Minimum	Maximum	Mean
Odo-Iyaalaro						
pH	2.8	5.6	4.0	3.6	5.1	4.2
OM (%)	0.8	21	7.8	1.4	15.7	8.5
CEC	2.0	14.6	7.6	4.2	14.2	9.6
Cd (mg kg^{-1})	<0.9	42.1	9.6	<0.9	12.8	4.8
Cr (mg kg^{-1})	<17.5	102	31.0	<17.5	28.9	22.9
Cu (mg kg^{-1})	15.6	185	60.9	15.8	105	55
Pb (mg kg^{-1})	<15.8	154	56.7	21.6	108	63.5
Zn (mg kg^{-1})	53.1	1,040	292	18.7	377	147
Shasha River						
pH	4.3	7.3	5.6	4.2	7.7	6.2
OM (%)	0.1	8.2	2.4	0.2	7.4	2.1
CEC	1.6	9.8	3.9	1.6	10.2	3.8
Cd (mg kg^{-1})	< 0.9	1.2	0.9	<0.9	1.7	1.0
Cr (mg kg^{-1})	<17.5	56.7	27.8	<17.5	140	40.6
Cu (mg kg^{-1})	< 2.8	78.5	31.3	<2.8	106	32.0
Pb (mg kg^{-1})	< 15.8	189	38.6	<15.8	202	45.5
Zn (mg kg^{-1})	1.7	467	101	13	641	104
Ibeshe River						
pH	3.6	7	5.1	3.2	7.3	4.5
OM (%)	0.3	7.2	2.3	0.2	7.6	2.4
CEC	1.6	4.2	2.5	1.8	5.2	3.0
Cd (mg kg^{-1})	<0.9	<0.9	0.9	<0.9	<0.9	0.9
Cr (mg kg^{-1})	<17.5	48.3	27.1	<17.5	54.5	31.3
Cu (mg kg^{-1})	<2.8	332	62.0	<2.8	115	23.3
Pb (mg kg^{-1})	<15.8	26.3	16.2	<15.8	21.8	15.8
Zn (mg kg^{-1})	4.6	158	35.9	6.0	57.5	21.3
Lagos Lagoon						
pH	4.3	6.6	5.4	4.1	7.0	5.6
OM (%)	0.1	12.6	4.2	0.1	10.8	3.7
CEC	1.8	11.2	5.0	2	11.0	4.9
Cd (mg kg^{-1})	<0.9	2.1	1.2	<0.9	<0.9	<0.9
Cr (mg kg^{-1})	34.4	51.7	44.7	23.8	51.7	35.8
Cu (mg kg^{-1})	<2.8	33.7	20.6	<2.8	43.0	18.8
Pb (mg kg^{-1})	<15.8	39	28.4	<15.8	39.2	25.6
Zn (mg kg^{-1})	1.3	190	103	1.3	246	118

Dixon 1997) were used. Of the 103 sediments studied, 11 exceed the PEC for Cd (all in the Odo-Iyaalaro); two exceed the PEC for Cr; four exceed the PEC for Cu (one from Odo-Iyaalaro and three from Ibeshe River); three exceed the PEC for Pb and four exceed the PEC for Zn. Forty of the sediments were above the Dutch sediment guideline for Cu and four (three from Odo-Iyaalaro during the dry season) exceed the guideline for Zn.

In order to determine the mobility and bioavailability of the metals studied, BCR sequential extraction technique was used. In the Odo-Iyaalaro, cadmium and zinc showed a similarity in their fractionation pattern. They were mostly

Table 4 Ratios indicating predominant sources of POPs in Lagos lagoon

	Water (ng/mL)	Sediment (ng/g)	Crayfish ng/g	Shrimp (ng/g)	Blue crab (ng/g)	Tilapia ng/g	Megalops (ng/g)
∑PAHs	0.13	346.94	74.2	99.06	264.61	80.64	88.11
∑Lower PAHs	0.13	210.95	74.2	99.06	264.61	80.64	88.11
∑Higher PAHs	0	136	0	0	0	0	0
∑Low chlorinated PCBs	0	11.78	0.94	0.97	2.23	14.02	16.09
∑High chlorinated PCBs	0	3.25	0.33	0.78	44.15	17.32	29.85
∑PCBs	0	15.04	1.27	1.75	46.38	31.34	45.93
∑Ocs	0.02	2.79	21.03	8.41	69.4	39.94	32.09
PCBs/p,p'DDE	0	8.14		0.65	2.09	1.95	3.96
∑PAHs/∑PCBs		23.07	58.44	56.61	5.71	2.57	1.92

observed to be associated with the acid exchangeable fraction, while chromium and copper were observed to be associated with the reducible and oxidizable fractions, and lead with the reducible and residual fractions. Sediments with higher pseudototal metals contents often contained higher proportions of metals in more labile forms (released earlier in the BCR procedure). This is of environmental concern because it means that, where overall PTM concentrations are highest, the potential for remobilisation and uptake into the food chain is also greatest.

Ecotoxicological Risk Assessment of Sediments

In another study by Oyeyiola and Alo et al., the risk associated with the Lagos Lagoon system sediments was assessed using the risk assessment code (RAC). This was based on the percentage of acid exchangeable fraction of metals (i.e. the most mobile) determined with a sequential extraction procedure as per procedures developed by Davutluoglu et al. (2011), Jain et al. (2007), Perin et al. (1985). Based on these values, Cd was found in the Odo-Iyaalaro to be mostly associated with the very high-risk group. The other metals studied varied from no risk to medium risk, while zinc varied from high risk to very high risk in most of the sites in all the waterbodies.

The Hankanson potential ecological risk index was also used in the assessment (Cai et al. 2011; Hankanson 1980). In the absence of background values of metals in sediment in Nigeria, the background value of metals in sediment presented by the US National Oceanic and Atmospheric Administration (NOAA) (SQuiRTs 2008) was used, and the toxic-response factor presented by Cai et al. (2011) was also used. Cadmium was placed in the very high-risk category,

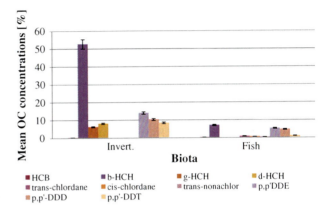

Fig. 4 Mean % OCs in fish and invertebrates

thus posing the highest ecological risk in the Odo-Iyaalaro River (Table 3). The highest RI value was also observed during the dry season (March). Based on the ecological risk assessment, the researchers observed a need for strengthening environmental pollution control in other to prevent ecological risks from metals.

Persistent Organic Pollutants Sources and their Impacts on Lagos Lagoon, Nigeria

Studies in our laboratories have shown that industrial and other anthropogenic sources predominated over agricultural sources of Persistent Organic Pollutants (POPs) in the Lagos Lagoon (Alani et al. 2011). Also pyrolytic sources of PAHs were found to dominate natural or petrogenic sources in the Lagoon. This result agreed with Pazdro (2002), who reported that the occurrence of PAHs in the environment is

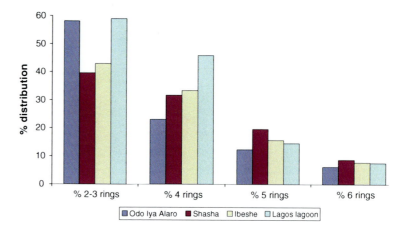

Fig. 5 Percentage distribution of grouped PAHs in all the sampling points

mainly due to combustion and pyrolysis of fossil fuels and to release into the environment of petroleum products. Environmentally, socially and economically, POPs have impacted the Lagos lagoon negatively. Exposure to POPs in the Lagos lagoon has negatively impacted human health by increasing the risks of most modern diseases among the inhabitants of Lagos (Table 4).

Bioaccumulation of Organochlorine Pesticide Residues in Fish and Invertebrates of Lagos Lagoon, Nigeria

Our studies have provided baseline information on the level of OCs in the Lagos Lagoon System including their bioaccumulation and the risk they pose to humans that depend on fish and invertebrates from the Lagos lagoon for food. The most bioaccumulated OCs in fish were beta-HCH (22.72–0.90 ng/g d. w.) and p,p'DDE (16.04–0.44 ng/g d. w.). The most bioaccumulative OCs in the invertebrates were still beta-HCH (24.50–16.10 ng/g d. w.) and p,p'DDE (22.20–1.85 ng/g d. w.). This agreed with the report that p,p'DDE was found to be more stable and persistent (refractory) than either p,p'DDT or p,p'DDD and underwent strong biomagnification with transfer along food chain (Walker 2009). The sum OCs of 55.22 ng/g d. w. in crab eggs, 63.90 ng/g d. w. in agaza (*Caranx hippos*) and 69.40 ng/g d. w. in young blue crabs (*Callinectus amnicola*) revealed these biota as the most contaminated. Beta-HCH and p,p'DDE were identified as the dominant OC in the Lagos lagoon. Consumption of crab eggs, mature crabs, young blue crabs (*Callinectus amnicola*) and Agaza (*Caranx hippos*), and some other seafoods from the Lagos lagoon could pose a high risk of OC health effects on humans as these biota bioaccumulated the contaminants above allowable limits (Fig. 4).

PAH Distribution in Sediments of the Lagos Lagoon System

Three trans-urban waterbodies of the Lagos lagoon system; Odo Iya alaro, Ibeshe and Shasha creeks that receive domestic, municipal and industrial effluents, and eventually empty into the Lagos Lagoon, were studied (Alani et al. 2012a). Sediment samples were collected bimonthly from 21 sampling points for a period of one year, covering both rainy and dry seasons of the year. The distribution of the PAHs in the sediment samples showed large variations in the sites investigated. The concentration of total PAHs ($\sum$PAHs) ranged from 9.76 to 6,448.66 μg/kg and showed a strong influence from anthropogenic inputs. In general, naphthalene, fluorene, phenanthrene, fluoranthene, pyrene, benzo (a) anthracene, chrysene, benzo (b) fluoranthene, and benzo (a) pyrene were the dominant PAHs found in the sediments.

The total PAH concentration for the Odo Iya alaro ranged from 268 to 6,449 μg/kg, Shasha creek ranged from 127 to 3,509 μg/kg, Ibeshe creek ranged from 56 to 2,996 μg/kg, while the Lagos lagoon ranged from 10 to 1,044 μg/kg. The percentage distribution of PAHs was predominantly 2–3 ringed PAHs ranging from 42.35 to 100 %. Samples obtained during the dry seasons of the year were also found to contain higher levels of total PAHs. Significantly higher total PAH concentrations were found at sampling points close to the main metropolis of Lagos compared with points in the lagoon which are far from the city, supporting the conclusion that urbanized and industrialized areas are major sources of PAH contamination in sediments.

The percentage distribution of grouped PAHs shown in Fig. 5 indicated a similar trend in all the sampling points. 2- and 3-ringed PAHs had the highest percentage distribution while the 6-ringed PAHs had the least. This indicated that the sediment samples in the Lagos Lagoon and the adjoining creeks had similar anthropogenic input of PAHs.

Table 5 Results of PAH analysis compared with standard pollution criteria of PAH components for sediment matrix

	Concentration (µg/kg) (Xu et al. 2007)		This study concentration (µg/kg)			
	ERL	ERM	Odo Iya alaro	Shasha	Ibeshe	Lagos lagoon
Naphthalene	160	2,100	446.26	164.45	143.86	138.56
Acenaphthene	16	500	105.09	29.99	45.65	14.76
Fluorene	19	540	41.68	46.47	50.96	36.03
Phenanthrene	240	1,500	138.91	99.53	129.95	148.50
Anthracene	853	1,100	99.51	71.44	170.21	57.98
Fluoranthene	600	5,100	268.60	229.67	164.28	295.16
Pyrene	665	2,600	198.13	153.21	136.72	241.94
Benzo(a)Anthracene	261	1,600	124.62	86.84	137.02	84.39
Chrysene	384	2,800	101.76	61.95	84.17	41.95
Benzo(b)Fluoranthene	NA	NA	176.87	88.42	94.65	70.36
Benzo(k)Fluoranthene	NA	NA	75.90	72.63	59.21	60.37
Benzo(a)Pyrene	430	1,600	73.14	54.61	88.43	44.93
Dibenzo(a,h)Anthracene	NA	NA	43.56	36.45	35.33	29.84
Benzo(g,h,i)Perylene	63.4	260	106.85	77.72	80.93	49.53
Indeno(1,2,3-c,d)Pyrene	NA	NA	100.10	66.42	84.06	61.49
∑PAH	4,000	44,792	2,334.16	1,213.31	1,449.82	880.43

Source identification of PAHs

Sediment samples collected near sewage outlets, cities and the harbour appeared to have extremely high concentrations of total PAHs. These suggest that PAHs accumulated in the sediments of the Lagos lagoon and the adjoining creeks came from different sources such as sewage discharge from nearby human activities, untreated industrial effluents and fuel combustion emissions.

The measured concentrations of PAHs were compared with the effects range low (ERL) and the effects range median (ERM) values which are used for assessment of aquatic sediment using biological thresholds with a ranking of low-to-high impact values (Long et al. 1995). Table 4 lists the thresholds for sediments from the Lagos lagoon and the adjoining creeks. In the sediments of the Odo Iya alaro, the results showed that the average total PAH concentrations at all sites were below the ERL and ERM (Table 5).

It is well established that PAH congener distribution generally varies with the production source as well as the composition and combustion temperature of the organic matter. Molecular indices based on ratios of selected PAH concentrations may be used to differentiate PAHs from pyrogenic and petrogenic origins. Four specific PAH ratios were calculated for the studied samples: phenanthrene/anthracene, fluoranthene/pyrene, benzo (a) anthracene/chrysene, and indeno (1,2,3c,d) pyrene/(Indeno (1,2,3c,d) pyrene + benzo (g,h,i) perylene).

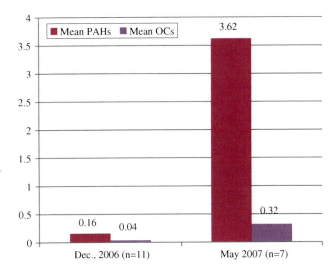

Fig. 6 Mean concentrations of PBTs (ng/mL) in December 2006 and May 2007 (PCB levels in Lagos Lagoon water during this study were below detection limit)

Results from the analysis of sediments of the Odo Iya alaro creek showed that Phe/Anth ratio ranged from 0.47 to 4.26. The ratios were mostly <1 except for a few sites. Flu/Pyr ratio ranged from 0.17 to 2.74. Most of the sampling sites had values less than 1, indicating pyrolytic origin or coal combustion. The BaA/Chr ratios were greater than 1 in

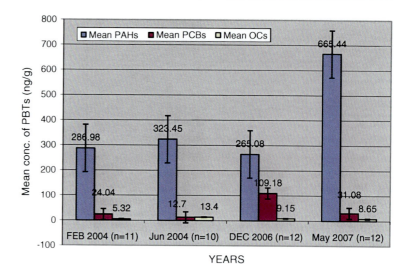

Fig. 7 Mean concentrations of PBTs in sediments from February 2004 to May 2007

most of the samples and could also be indicative of pyrogenic input of the PAHs. The pyrolytic PAH inputs at the Odo Iya alaro creek are also further confirmed by the IP/BgP + IP ratio. The values for this ratio ranged from 0.32 to 0.8 which is typical of fossil fuel combustion, coal, grass and wood combustion, all of which are pyrolytic sources. The results indicated that PAHs in the Odo Iya alaro had presumably undergone similar environmental processes independent of the sampling sites. A similar trend was observed in other sediments of the remaining sampling sites.

N-Alkane Distribution in Sediment Samples from Lagos Lagoon

The concentrations of total aliphatic hydrocarbons (C_9–C_{38}) in sediment samples from the Lagos Lagoon have been found to range from 14.89 to 148.29 µg/g dry weight, and their distribution showed large spatial variations at various sampling points. The distribution of both n-alkanes and PAHs shows great spatial variations in the sediments, which could be attributed to temporal and localized inputs of contamination sources. By examining the distribution indexes, it has been confirmed that the aliphatic hydrocarbons were mainly from anthropogenic to biogenic sources, while the distribution of PAHs came from both pyrolytic and petrogenic sources. The highest hydrocarbon concentrations in sediments were, in general, found in the areas associated with high anthropogenic impact and port activities.

The major hydrocarbon index in most samples was at C_{29}, C_{31} and C_{27}, indicating that vascular plant sources played a very important role (Wang et al. 2006). A lower molecular weight Cmax around C_{21} and C_{23} was also observed in several samples. CPI index ranged from 0.65 to 2.97 which indicated that the presence of n-alkanes in the Lagos lagoon sediments and the adjoining creeks was from both anthropogenic to pyrogenic sources. The sediment samples that had CPI values close to 1 were taken from areas associated with high anthropogenic impact and port activities such as Tin Can Island and Marina.

The temporal distribution of PBTs in the Lagos Lagoon is shown in Figs. 6 and 7.

Pattern of PBT Signatures Across Different Media in Lagos Lagoon, Nigeria

Our studies on the PBT signatures of the Lagos Lagoon System (Alani et al. 2012b) showed highest mean PAHs of 0.080 ng/ml in the lagoon water at the mouth of the Ogun River, a fluvial source, reflecting the effect of runoffs on the PAH load in the lagoon. In coastal areas, direct deposition of atmospheric PAHs may be relatively minor compared with fluvial inputs, but in open ocean areas it can dominate (Kowalewska and Konat 1997). The levels of PAHs in the lagoon water were also determined by Anyakora et al. (2004). The highest mean OCs of 0.069 ng/ml was obtained at Okobaba, a slum residence by the shore of the lagoon, where the use of pesticides for the control of insect vectors is relatively high. PCBs were not detected in water from the lagoon. In the sediment, highest mean PAHs of 68.251 ng/g d. w. was obtained at Okababa, highest mean OCs of 11.859 ng/g d. w. was obtained at Aja, while highest mean PCBs of 1.331 ng/g d. w. was obtained at the mouth of the Ogun River, confirming the effect of runoff load through fluvial sources. In the invertebrates, the highest mean PAHs of 18.659 ng/g d. w. was found in crayfish; the highest mean PCBs of 17.070 ng/g d. w. was found in crayfish; and

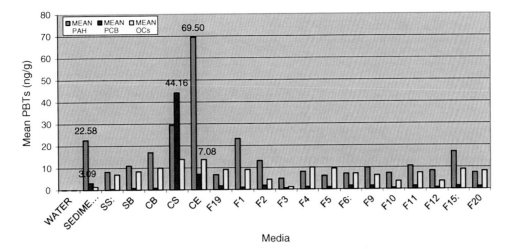

Fig. 8 Mean PBTs in different media of Lagos lagoon in December 2006 (Alani 2011), *Note SS* Crayfish, *SB Macrobranchium Vollenloevensi* (Shrimps), *CB Callinectus amnicola* (adult crabs, had eggs), *CS Callinectus amnicola* (young crabs), *CE* Crab eggs, *F1 Caranx hippos*(Agaza), *F2 Mugil cephalus* (mullet), *F3 Sphyraena barracuda* (barracuda), *F4 Saratherodon melanotheron* (Tilapia), *F5 Tilapia guineensis* (Tilapia), *F6 Edmalosa fimbriata* (Bonga), *F9 Tarpon Atlanticus* (megalops), *F10 Scomberomorus Tritor* (mackerel) (Ayo), *F11 Lutjanus agennes* (African red snapper), *F12 Pomadasys Jubelini* (Grunter), *F15 Chrysicthys Nigrodigitatus* (Catfish) (Obokun), *F20 Lutjanus Dentatus* (African *brown* snapper)

the highest mean OCs of 13.880 ng/g d. w. was found in young crabs. These patterns are most likely linked to their feeding habits, habitat, lipid contents metabolism ability and physiological characteristics among other factors. In the fish fillet tissues, the highest mean PAHs of 28.996 ng/g d. w. was found in agaza; the highest mean PCBs of 1.925 ng/g d. w. was found in mullets; and the highest mean OCs of 9.986 ng/g d. w. was found in tilapia (Fig. 8).

Conclusion

This paper has given an insight into the levels of pollution of the Lagos Lagoon with respect to PTE, nutrients and various other organic pollutants. The levels of some of the pollutants presently in the Lagoon has directly and indirectly affected the environmental, social and economical benefits derivable from the Lagoon. Indeed with the services expectation of the waterbody and the high population that the estuary/lagoon system supports, consideration of its designation as an international waterbody and its concomitant global attention is now paramount. It is important that strategies to ensure industries around the lagoon treat their effluents before discharging into the lagoon are needed. Also there is the need for the education of all stakeholders on the effects of indiscriminate dumping of municipal waste and burning of biomass around the Lagoon. Regulatory bodies must be encouraged to enforce existing regulations on municipal waste disposal and effluent discharge into this lagoon system.

References

Abayomi A, Nimmo M, Williams C, Olayinka KO, Osuntogun B, Alo B, Worsfold P (2011) The contribution of roadside soil to phosphorus loading in the eutropic Lagos Lagoon, Nigeria. J Environ Monit 13:1884–1889

Abayomi AA, Olayinka KO, Osuntogun B, Alo BI (2006) Urban highway runoff in Nigeria II: effects of sheet flow on Roadside Soils in Lagos Metropolis. J Sci Res Dev 10:119–128

Alagarsamy R (2006) Distribution and seasonal variation of trace metals in surface sediments of the Mandovi Estuary, west coast of India. Estuar Coast Shelf Sci 67:333–339

Alani RA, Drouillard KG, Olayinka KO, Alo BI (2012a) Bioaccumulation of polycyclic aromatic hydrocarbons (PAHs) in fish and invertebrates of Lagos Lagoon, Nigeria. J Emerg Trends Eng Appl Sci (JETEAS) 3(2):287–296. ISSN: 2141-7016)

Alani RA, Drouillard KG, Olayinka KO, Alo BI (2012b) Modeling and risk assessment of persistent, bioaccumlative and toxic (PBT) organic micropollutants in Lagos lagoon, Nigeria. In: Book of proceeding of the chemical society of Nigeria 35th annual international conference, Owerri, 17–21 Sept 2012

Alo BI, Abayomi AA, Osuntogun B, Olayinka K (2007) Urban highway runoff I: heavy metals in sheet flow from the main expressway in Lagos, Metropolis. J Appl Sci 7(19):2800–2805

Alo B, Orgu B, Abayomi A (2010) Low sub-surface harmattan season hypoxia events in the Lagos Lagoon, Nigeria. Eur J Sci Res 40(2):279–286

Anyakora C, Ogbeche KA, Uyimadu J, Olayinka K, Alani RA, Alo BI (2004) Determination of polynuclear aromatic hydrocarbons in water samples of the Lagos Lagoon. Nig J Pharm 35:35–39

Bahena-Manjarrez JL, Rosales-Hoz L, Carranza-Edwards A (2002) Spatial and temporal variation of heavy metals in a tropical estuary. Environ Geol 42:575–582

Cai J, Cao Y, Tan H, Wang Y, Luo J (2011) Fractionation and ecological risk of metals in urban river sediments in Zhongshan City, pearl River Delta. J Environ Monit 13:2450–2456

Chaney RL, Ryan JA, O'Connor GA (1996) Organic contaminants in municipal biosolids: risk assessment, quantitative pathways analysis and current research priorities. Sci Total Environ 185:187–216

Davutluoglu OI, Seckin G, Ersu CB, Yilmaz T, Sari B (2011) Metal content and distribution in surface sediments of the Seyan River, Turkey. J Environ Manage 92:2250–2259

Ekpo BO, Oyo-Ita OE, Wehner H (2005) Even-n-alkane/alkene predominances in surface sediments from the Calabar River, SE Niger Delta, Nigeria. Naturwissenschaften 92:341–346

Fonseca EM, Baptista Neto JA, Fernandez MA, McAlister J, Smith B (2011) Geochemical behaviour of heavy metals in different environments in Rodrigo de Freitas Lagoon Brazil-RJ/Brazil. Anais da Academia Braisileria de ciencias 83:457–469

Grimwood MJ, Dixon E (1997) Assessment of risks posed by List II metals to Sensitive Marine Areas (SMAs) and adequacy of existing environmental quality standards (EQSs) for SMA protection. Report to English Nature

Hankanson L (1980) An ecological risk index for aquatic pollution control: a sedimentological approach. Water Res 14:975–1001

Heemken OP, Theobald N, Wenclawiak BW (1997) Comparison of ASE and SFE with Soxhlet, sonication and methanolic saponification extractions for the determination of organic micropollutants in marine particulate matter. Anal Chem 69:2171–2180

Howden D (2010) Lagos: inside the ultimate mega-city. The independent Africa. http://www.independent.co.uk/news/world/africa/lagos-inside-the-ultimate-megacity%201945246.html

Isebor CE, Awosika A, Smith SV (2006) Preliminary water, salt, and nutrient budgets for Lagos Lagoon, Nigeria. http://nest.su.se/MNODE/Africa/Lagos/lagosbud.htm

Jain CK, Malik DS, Yadav R (2007) Metal fractionation study on bed sediments of Lake Nainital, Uttaranchal, India. Environ Monit Assess 130:129–139

Kowalewska G, Konat J (1997) Distribution of polynuclear aromatic hydrocarbon PAHs in sediments of the southern Baltic Sea. Oceanologia 39(1):83–104

Li Q, Wu Z, Chu B, Zhang N, Cai S, Fang J (2007) Heavy metals in coastal wetland sediments of the Pearl River Estuary, China. Environ Pollut 149:158–164

Long ER, MacDonald DD, Smith SL, Calder FD (1995) Incidence of adverse biological effects within ranges of chemical concentrations in marine and estuarine sediments. Environ Manage 19:81–97

Lund-Hansen LC, Petersson M, Nurjaya W (1999) Vertical sediment fluxes and wave-induced sediment resuspension in a shallow-water coastal lagoon. Estuaries 22:39–46

MacDonald DD, Ingersoll CG, Berger TA (2000) Development and evaluation of consensus-based sediment quality guidelines for fresh water ecosystems. Arch Environ Contam Toxicol 39:20–31

Mackay D, Fraser A (2000) Bioaccumulation of persistent organic chemicals: mechanisms and models. Environ Pollut 100:375–391

Manning WJ (2011) Urban environment: defining its nature and problems and developing strategies to overcome obstacles to sustainability and quality of life. Environ Pollut 159:1963–1964

Nriagu JO (1992) Toxic metal pollution in Africa. Sci Total Environ 121:1–37

Olajire AA, Altenburger R, Küster E, Brack W (2005) Chemical and ecotoxicological assessment of polycyclic aromatic hydrocarbon—contaminated sediments of the Niger Delta, Southern Nigeria. Sci Total Environ 340:123–136

Oluseyi TO, Olayinka KO, Alo BI, Smith RM (2012) Distribution and source identification of Polycyclic Aromatic Hydrocarbons (PAHs) in surface sediments of the Lagos Lagoon System. Nig J Chem Soc Nigeria 37(1):92–98

Oluseyi TO, Olayinka KO, Alo BI, Smith RM (2011) Comparison of extraction and clean-up techniques for the determination of polycyclic aromatic hydrocarbons in contaminated soil samples. Afr J Environ Sci Technol 5(7):482–493

Otitoloju AA, Don-Pedro KN, Oyewo EO (2007) Assessment of potential ecological disruption based on PTM toxicity, accumulation and distribution in media of Lagos Lagoon. Afr J Ecol 45:454–463A

Oyeyiola AO, Olayinka KO, Alo BI (2004) Heavy metals in water and sediments of two Rivers in Lagos State, Nigeria. In: Proceedings of the chemical society of Nigeria annual 2004 conference, pp 24–27

Oyeyiola AO, Olayinka KO, Alo BI (2005) Speciation of heavy metals in sediment of a polluted tropical stream, using the BCR sequential extraction technique. J Nigeria Environ Soc 4:23–35

Oyeyiola AO, Olayinka K, Alo BI (2006) Correlation studies of heavy metals concentration with sediment properties of some rivers surrounding the Lagos lagoon. Niger J Health Biomed Sci 5:118–122

Oyeyiola AO, Olayinka K, Alo BI (2007) Speciation of heavy metals in sediments of a polluted tropical stream using the BCR sequential extraction technique. J Niger Environ Soc 4:139

Oyeyiola AO, Davidson CM, Olayinka KO, Oluseyi TO, Alo BI (2013a) Multivariate analysis of potentially toxic metals in sediments of a tropical coastal lagoon. Environ Monit Assess 185(3):2167–2177

Oyeyiola AO, Davidson CM, Olayinka KO, Alo BI (2013b) Fractionation and ecotoxicological implication of potentially toxic metals in sediments of three urban rivers and the Lagos Lagoon, Nigeria, West Africa. Manuscript in Preparation

Pazdro K (2002) Persistent organic pollutants in sediments from the Gulf of Gdańsk. Institute of Oceanology, Polish Academy of Sciences, Sopot

Perin G, Craboledda M, Lucchese M, Cirillo R, Dotta L, Zanette ML, Orio AA (1985) Heavy metal speciation in the sediments of northern Adriatic Sea—a new approach for environmental toxicity determination. In: Lekkas TD (ed) Heavy metals in the environment, vol. 2. pp 454–456

Silva BO, Adetunde OT, Oluseyi TO, Olayinka KO, Alo BI (2009) Comparison of some extraction methods and clean up procedures for the 16 priority EPA PAHs. J Sci Res Dev 13:129–143

SQuiRTs (2008) NOAA screening quick reference tables, NOAA. http://response.restoration.noaa.gov/book_shelf/122_NEW-SQuiRTs.pdf

This day's Special International Summit, Ney York, USA (2007) Lagos city project. New city on the Atlantic. http://www.africanloft.com/lagos-mega-city-project-new-city-on-the-atlantic/

Walker CH (2009) Organic pollutants. An ecotoxicological perspective. CRC Press, New York

Wallace DWR, Law CS, Boyd PW, Collos Y, Croot P, Denman K, Lam PG, Riebesell U, Takeda S, Williamson P (2010) Ocean fertilization. A scientific summary for policy makers. IOC/UNESCO, Paris (IOC/BRO/2012/2)

Wang X-C, Sun S, Ma H-Q, Liu Y (2006) Sources and distribution of aliphatic and polyaromatic hydrocarbons in sediments of Jiaozhou Bay, Qingdao, China. Marine Pollut Bull 52:129–138

Xu J, Yu Y, Wang P, Guo W, Dai S, Sun H (2007) Polycyclic aromatic hydrocarbons in the surface sediments from Yellow River, China. Chemosphere 67(1408):1414

Estuarine and Ocean Circulation Dynamics in the Niger Delta, Nigeria: Implications for Oil Spill and Pollution Management

Larry Awosika and Regina Folorunsho

Abstract

The Niger Delta is a fan-shaped sedimentary environment located between the Benin River in the west and the Imo River in the east (longitudes 5°4′00″E and 7° 40′00″E), extending a few kilometres south of the village of Aboh at the point where the Niger bifurcates into several rivers with estuaries opening into the Gulf of Guinea. Within the estuaries, circulation patterns are dictated mainly by the tidal regimes and fluvial flows superimposed on the ebb and flood currents. The ebb tidal currents are usually stronger than the flood tidal currents, with water fluxes during the ebb tide almost twice those of the flood tides. Offshore circulation is predominantly along-shelf and is oscillatory in nature with a fortnightly component. The across-shelf circulation, which is more subdued, is both tide and wind dependent. Understanding the circulation patterns both in the estuaries and in the open ocean has implications for oil spill management and the resulting health of the estuaries, which serve as breeding grounds for a wide variety of fish and floral species.

Keywords

Bathymetry • Continental shelf • Gulf of Guinea • Longshore currents • Littoral drift • Ocean currents • Niger Delta • Nigeria • Oil spill • Oscillating currents

Introduction

The Niger Delta is a fan-shaped sedimentary environment located between the Benin River in the west and the Imo River in the east (longitudes 5°4′00″E and 7° 40′00″E), extending a few kilometres south of the village of Aboh at the point where the Niger bifurcates into the Nun and Forcados Rivers (Fig. 1). Barrier islands with beach ridges rim the seaward margin of the Delta for about 480 km.

Allen (1965) recognized 20 major barrier islands in the Niger Delta separated by deep tidal channels (estuaries). The widths of the barrier islands range from a few 100 m to 12 km with lengths varying from about 5–37 km (Allen 1965). The Niger Delta is composed of seven major environments (Table 1). The flood plain of freshwater swamp ecological zone extends from the Forcados-Nun bifurcation of the Niger to the levee-indented margin of the mangrove swamp zone. River channel deposits, point bars, levees, back swamp and cut-off channel deposits develop in this zone (Whiteman 1982).

The barrier islands separate the mangrove swamps from the open sea. In the central part of the delta, the mangrove swamps are 8–16 km across which interlace with lower freshwater swamps. On the flanks of the delta, the mangrove swamps are 32–40 km wide. The mangrove swamp ecological zone is characterized by mudflats, tidal saline and

L. Awosika (✉) · R. Folorunsho
Nigerian Institute for Oceanography and Marine Research,
Victoria Island, Lagos, Nigeria
e-mail: lfolaawo@gmail.com

R. Folorunsho
e-mail: rfolorunsho@yahoo.com

S. Diop et al. (eds.), *The Land/Ocean Interactions in the Coastal Zone of West and Central Africa*, Estuaries of the World, DOI: 10.1007/978-3-319-06388-1_7,
© Springer International Publishing Switzerland 2014

Fig. 1 Map showing the Niger Delta, estuaries, and offshore bathymetry

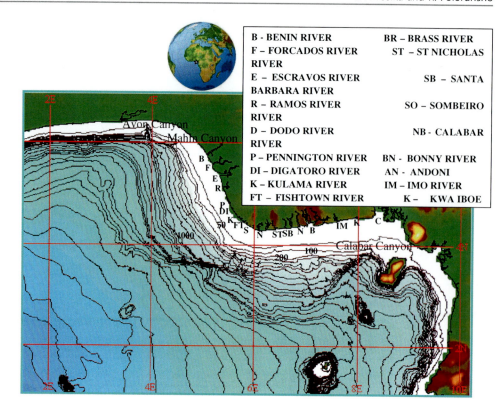

Table 1 Estimated areas occupied by selected environment, in present day Niger Delta (After Whiteman 1982)

Environment	km²
Onshore and offshore total	160,930
Onshore 'fossil' delta complex	128,744
Upper and lower floodplains	5,219
Mangrove swamps	3,058
Barrier Islands	1,117
River mouth bars	993
Delta front platform	2,795
Pro delta slope	4,846
Open shelf (to shelf break)	6,214
Continental slope (Shelf break to 1,80 m)	9,941
Total (Onshore to Offshore to 1,280 m)	323,857

brackish waters, numerous creeks and creeklets, marshes, and mangrove. The barrier bar complex consists of the modern active beach and sand ridges developed in relation to older strand lines. Intertidal coastal plain extends behind the barrier bars and comprises tidal flats and swamps with dense mangrove vegetation which alternate with tidal distributary channels and lagoons. The plants are mainly halophytic red mangrove (*Rhizophora racemosa*) characterized by breathing roots, growing above the surface of the organically rich but oxygen-depleted muds. Barrier beach islands, commonly broken through by tide, separate the swamps from the open sea. The edge of the tidal swamps moves seaward as the back of the barrier bar is eroded. The sub-aerial delta is nearly eight times larger than the sub-aqueous delta, probably as a result of the relatively high near-shore wave action and the presence of strong littoral currents.

Continental Shelf

The continental shelf off the Niger delta can be divided into an inner shelf comprising of bathymetric configurations parallel to the coastline with depths ranging from 0 to 45 m (Fig. 1). The middle shelf depths range from 45 to 80 m are indented with gullies and terraces (Allen 1965). The outer shelf, with depths above 80 m, is also terraced and has gradual slopes. There are other smaller gullies, especially off the nose of the Niger delta within the middle to outer shelf and the slope. The Avon and Mahin canyons are located west of the Niger Delta while the Calabar canyon (Fig. 1) is east of the Delta (Allen 1965; Burke 1972; Awosika and Ibe 1994).

Oceanographic Conditions

Oceanographic parameters that control circulation patterns in estuaries and ocean consist of waves, tides and longshore currents, with wind and salinity modifying these oceanographic processes.

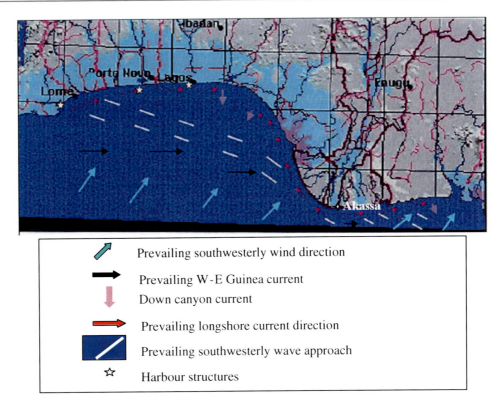

Fig. 2 Characteristics of oceanographic parameters offshore Nigeria (After Awosika et al. 2000)

Waves

Waves affecting the Niger Delta are wind-generated and approach the coastline from a south-westerly direction. The wind systems consist of the onshore south westerlies and offshore north easterlies. Swells generally approach the coastline from the south/south-westerly direction, with oblique angles of approach of not less than 3° and not more than 15°.

Tides

Tides within the Niger Delta are semi-diurnal with two inequalities. Swells produced by the south-westerly winds approach the Niger Delta coast in a south-westerly direction (Fig. 2). Elevated water, close to the shore during the two daily high tides, generally assists the waves to progress further inshore. Stronger tidal currents, exceeding 5 m/sec at the inlet/ocean intersection, have been found to occur during ebb tides. Flood and ebb streams usually run for about 3 h after High and Low water, respectively. The active distributaries are tidally dominated and often display typical bell-shaped river mouths.

Longshore Currents

The dominant longshore drift cells responsible for nearshore circulation and transport in the nearshore zone (Fig. 2) consist of the north-westerly longshore drift of the western Niger Delta and the west-to-east longshore drift (Awosika et al. 1994, 1996). Between Akassa point and the Benin River, the Niger delta runs NW–SE. The coastal configuration along this part of the Niger results in the south-westerly waves breaking obliquely to the coast with the breaking waves opening to the north-west. This generates a longshore drift in a north-westerly direction which is fed by successive distributaries of the Niger delta between Akassa Point and Benin River (Fig. 2). East of Akassa is the east-to-west littoral longshore currents. This longshore drift cell continues eastwards to the Calabar estuary (Fig. 2).

Ocean Currents

Ocean currents off the Niger Delta consist of the Guinea current, which is an extension of the North Equatorial Counter current said to attain speed of 0.3 m/sec with some reversals (Fig. 3). The Guinea current runs above an

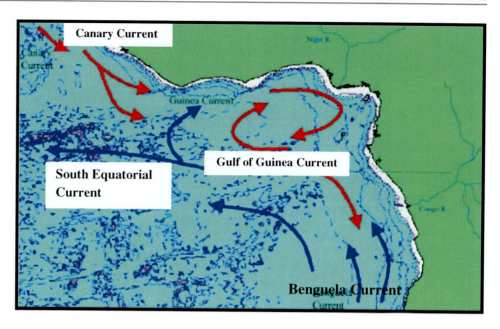

Fig. 3 Gulf of Guinea showing prevailing ocean currents

undercurrent which is thought to be a westward-flowing extension of the northern branch of the equatorial undercurrent which splits into two branches after impinging on the African continent at Sao Tome Island. Due to the fact that the equatorial undercurrent carries cool and highly saline water, the thermocline beneath the Guinea current is particularly intense. The other important surface current in the Gulf of Guinea is the South Equatorial Current (SEC).

Estuarine Circulation

Circulation within the Niger Delta estuaries is driven mainly by the tides from the sea, modified by the morphologic and bathymetric configuration of the estuaries and fluvial flow. Many of the estuaries consist of sand bars, which in most cases are ephemeral and hence modify circulation patterns. Typical current and water fluxes in some Niger delta estuaries (Forcados in the west of the Niger Delta and Brass in the east of the Niger delta) are shown in Figs. 4, 5, 6 7, 8, 9, 10, 11; Table 2.

The river estuaries in the western part of the Niger Delta enter the ocean at a westward oblique angle. The Forcados river estuary, which is largest in the east, exhibit current speeds of between 0.35 m/s to a high of 1.65 m/s with water discharge in the ebb condition almost double that of the flood (Table 2). River estuaries in the eastern Niger Delta enter the sea almost due south, with Brass being typical. The Brass river estuary, though smaller than the Forcados, has faster currents and total discharge during ebb condition than during flood conditions.

The rainy season is characterized by more persistent and strong winds as well as a larger tidal range compared with the dry season when met-ocean forces are subdued. Fresh water input into the estuaries during ebb tides results in dilution of the water, increase in the volume of water discharge and increased current speed (Table 2). While seasonal variations affect the tidal range at the river mouth during dry and wet seasons, the currents and estuary water fluxes vary with the seasons. Significant tidal incursions exist in the Niger delta estuaries during rainy and dry seasons. All these produce salt wedge limits within the estuarine zones which are not fixed but change as a result of variations in tidal spring-neap cycles, atmospheric forcing, and river discharge. During rainy season, the salt wedge moves seawards while during the dry season and high tide, the salt wedge moves inland. Variations in the salt wedge could develop a velocity difference between the two layers of fresh and salt water with the shear forces generating internal waves at the interface resulting in mixing the seawater upward with the freshwater.

The tidal variations superimposed by the freshwater discharge create plume areas in the ocean off the mouth of the estuary. These plumes, which are heavily laden with sediments (Fig. 12), are more prevalent especially during low tides when fresh water input extends farther out into the sea. These plumes are in the form of a fan that protrudes from the estuary into the coastal ocean.

Ocean Circulation

Ocean circulation off the Niger Delta estuaries is usually governed by tides, wind stress, bathymetry and salinity gradient. According to Awosika and Folorunsho (2005), among others, two major circulation patterns have now been documented offshore of the Niger Delta. The across-shelf patterns are driven and in phase with the tides, while the along-shelf patterns alternate east to west almost fortnightly.

Estuarine and Ocean Circulation Dynamics in the Niger Delta, Nigeria

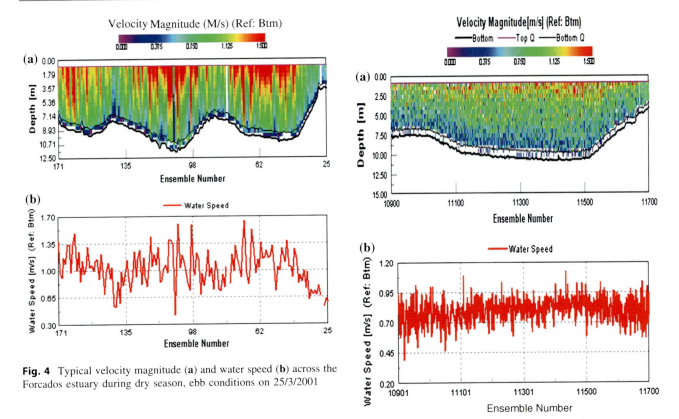

Fig. 4 Typical velocity magnitude (**a**) and water speed (**b**) across the Forcados estuary during dry season, ebb conditions on 25/3/2001

Fig. 6 Typical velocity magnitude (**a**) and water speed (**b**) across the Forcados estuary during wet season, ebb conditions on 15/8/2001

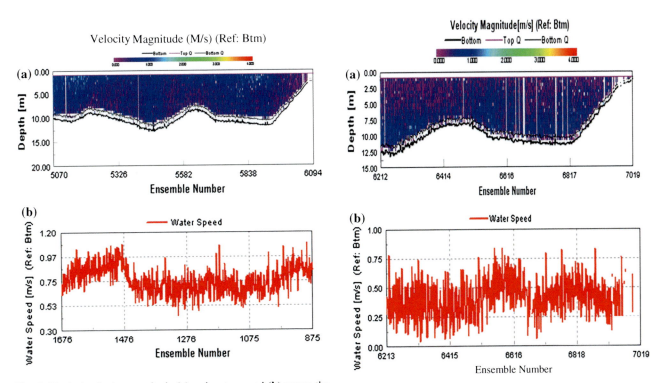

Fig. 5 Typical velocity magnitude (**a**) and water speed (**b**) across the Forcados Estuary during dry season flood tide conditions on 22/12/2000

Fig. 7 Typical velocity magnitude (**a**) and water speed (**b**) across the Forcados estuary during wet season, flood conditions on 15/8/2001

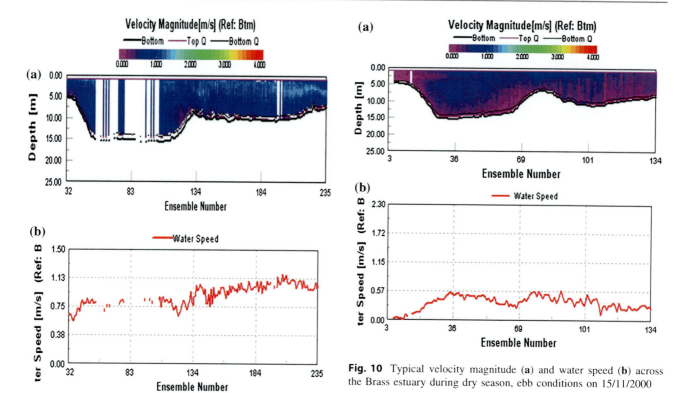

Fig. 8 Typical velocity magnitude (**a**) and water speed (**b**) across the Brass estuary during wet season, ebb conditions on 25/8/2000

Fig. 10 Typical velocity magnitude (**a**) and water speed (**b**) across the Brass estuary during dry season, ebb conditions on 15/11/2000

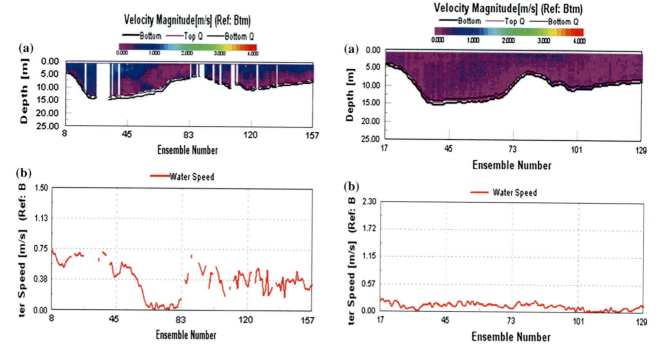

Fig. 9 Typical velocity magnitude (**a**) and water speed (**b**) across the Brass estuary during wet season, flood conditions on 25/8/2000

Fig. 11 Typical velocity magnitude (**a**) and water speed (**b**) across the Brass estuary during dry season, flood conditions on 15/11/2000

Table 2 Summary circulation parameters in Forcados and Brass River Estuaries in the Niger Delta

Estuary	Tidal state	Width of estuary along profile (m)	Average current speed (m/s)	Total discharge (m^3/s)
Forcados	Ebb—dry season	3,484.17	1.0	29,571.10
Forcados	Flood—dry season	3,557.57	0.75	22,391.73
Forcados	Ebb—wet season	3,456.46	0.9	24,225.69
Forcados	Flood—wet season	3,591.34	0.6	11,507.43
Brass	Ebb—dry season	2,450.25	0.38	2,314.43
Brass	Flood—dry season	2,329.46	0.32	9,797.25
Brass	Ebb—wet season	2,441.90	0.95	20,637.63
Brass	Flood—wet season	2,411.99	0.32	9,669.77

Fig. 12 Satellite imagery of part of the Niger Delta showing plumes off Nun, Brass and Sangana estuaries

Across-Shelf Circulation

The across-shelf circulation is principally driven by tides and is in phase (north–south) with the tidal cycle. Acoustic Doppler current profiler (ADCP) station number S2, installed at a location offshore the Benin Estuary between 19/04/2000 and 04/02/2001 in water depth of 15.8 m longitude 004°56′40″E and latitude 05°42′10″N, shows the characteristics of tides and the across-shelf circulation (Fig. 13) (Evan Hamilton Inc 2001).

Records from this ADCP (Fig. 13) show several days of long-period oscillations. Tidal periods are averaged at 0.52 days with oscillation periods of between 13 and 14 days. The semi-diurnal tidal periods (0.5 days) contain only relatively small amounts of current energy, mostly confined within the across-shelf (north–south) component (Fig. 13). Most of the energy in the tidal current resides primarily in the fortnightly and long-period along-shelf (north-western/south-eastern) component of the current.

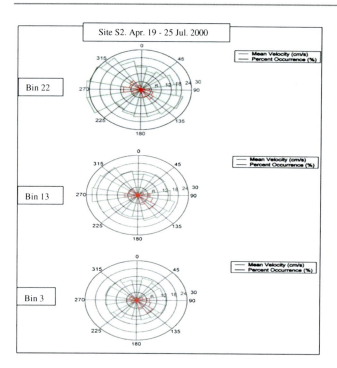

Fig. 13 Current polar plots for station S2 (15.8 m water depth) between 19 April 2000 and 25 July 2000. (Evan Hamilton Inc 2001). Bin 3 is 2.3 m from sea bottom, bin 13 is 7.3 m from seabed, and bin 22 is 11.8 m from sea bed

Mean velocities of 12 cm/s (0.12 m/s) are dominantly east to west (Fig. 13). Percentage occurrence of these directions is often over 20 %. The north–south across-shelf oscillating currents are much more subdued and less frequent. Near-surface currents (Bin 22), as expected, are much stronger and more consistent with regard to the general current directions. Frequency analysis of the data, however, shows energy peaks at several periods including those of daily tides, fortnightly, and approximately 20 days.

The tides, which create currents reaching a maximum of approximately 20 cm/s, provide the most consistent cross-shelf currents and may be responsible for the majority of cross-shelf mixing and stirring of coastal waters. However, these do not appear to be of as much significance as along-shelf net transport.

Along-Shelf Circulation

The along-shelf-period oscillations dominate the along-shelf transport which is mainly east to west. Records of ADCP data (Fig. 14), from station S5 offshore the Kwa Ibo estuary in the eastern Niger Delta, shows near-surface oscillating currents (Bin 26). Reversals along this east–west direction are very easily observable in the feather plots (Fig. 14).

In April 2000, during the onset of the rainy season, the dominant direction is west to east. In July during the rainy season, the west-to-east direction is highly reinforced; however, south-easterly and south-westerly directional components are observed. The dry season months of October to December exhibit a reversal with a dominant westerly component (Figs. 13, 14).

The oscillating along-shelf currents have also been captured by the trajectories of two Davies drifters launched in the Niger Delta in 2000. Drifter number 25,629 was deployed offshore the Sangana River (western Niger Delta) on July 24, 2000, at 11.00 h (GMT) and showed the oscillating pattern in a west–east direction (Fig. 15). The trajectory (Fig. 16) of another Davies drifter (25,620) launched offshore the Nun River (Eastern Niger Delta) showed again an oscillatory pattern with some gyres circulation.

The oscillatory circulation patterns described above seem to be different from those earlier described by several authors like NEDECO (1961), Longhurst (1962), Richardson and Reverdin (1987), Awosika et al. (1994), who described the dominant ocean current in this part of the Gulf of Guinea to be predominantly a west-to-east flowing pattern.

The causes of these reversals have been the subject of debate. Richardson and Reverdin (1987), Longhurst (1962) and Ingham (1970) concluded that the seasonal instability of the North Equatorial Countercurrent (NECC) and the Canary Current affect the seasonal variability of the Guinea Current and hence the oscillatory current patterns. Longhurst (1962), Ingham (1970), and Boisvert (1967) attributed the reversal in current direction during the minima to variation changes in the flow of the NECC, the Canary Current, and the Benguela Current and to the weakening of the easterly winds. Another explanation, proposed by Ingham (1970), was that the apparent reversals are actually caused by cyclonic eddies between the current and the coast.

While no agreement has been reached on the direct causes of the observed oscillating along-shelf surface currents offshore the Niger Delta, the implications of the oscillating currents for management of oil spill and nutrient circulation are apparent.

Implications of Circulation Current Patterns on Oil Spill Management in the Niger Delta

Oil movement in the coastal and marine area is influenced by a combination of dynamic forces, including wind speed and velocity, air and water temperature, tides, and currents. To effectively respond to an oil spill, scientists must gather critical chemical, oceanographic, climatic, and on-scene environmental data. Although various types of remote sensing techniques are available for detecting and mapping oil distribution, it is important to understand the characteristics of surface currents in the area.

Fig. 14 Feather plots of currents from ADCP S5 showing oscillations along the west–east direction (Evan Hamilton Inc 2001) (*Top*—rainy season and *bottom*—dry season)

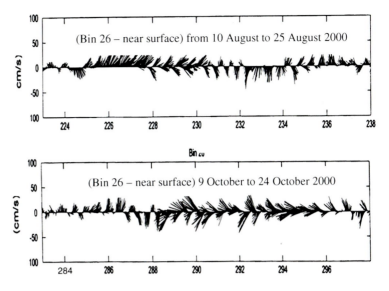

Fig. 15 Trajectory of drifter No 25620 (*purple colour*) offshore the Benin River western Niger Delta (Evans Hamilton Inc 2001)

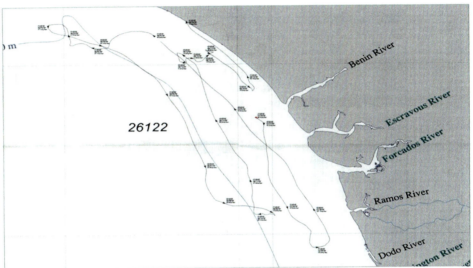

Fig. 16 Trajectory of drifter No 25620 (*purple colour*) offshore the Strand coast and eastern Niger Delta (Evans Hamilton Inc 2001)

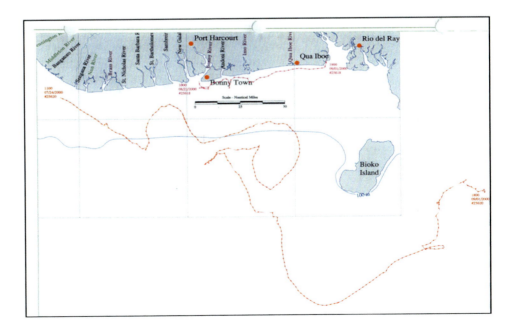

Ocean circulation can transport oil for thousands of miles in months to years. Knowing the trajectory of the spill gives decision-makers critical guidance in deciding how best to protect resources and direct clean-up. However, it is often very difficult to predict accurately the movement and behaviour of an oil spill when data and circulations patterns are sparse or unavailable.

Many oil spills have occurred several times along the Nigerian coast, either through accident or through sabotage. In many cases, forecasting the movement of an oil spill in the Niger Delta is often hampered by insufficient data, particularly in the first few hours of the release. Detailed spill data (location, volume lost, and product type) are often sketchy and environmental data (wind and current observations and forecasts) are often sparse or unavailable. This in many cases has led to loss of time, and inefficient oil spill response efforts. Understanding the nature and patterns of currents and resulting circulation system both in the estuaries and in the open ocean can ensure prompt and effective oil spill management in the Niger Delta.

Conclusion

The Niger Delta and its estuaries have a very complex deltaic environment. It is criss-crossed by several estuaries, creek, and rivers. Circulation patterns in the delta and offshore are hence very complicated. The complicated estuarine circulation and the oscillating surface currents offshore, such as those that have been observed in the Niger Delta, could present great challenges to oil spill response and contingency plans. Hence, understanding the estuarine dynamics and the offshore oscillating currents could greatly enhance oil spill response activities in the Niger Delta. Future studies to determine the residence time and exposure time of water in each estuary particularly from human induced stresses, can enhance the understanding of other parameters such as heavy metals, dissolved nutrients, blooms that may affect the health of estuaries.

References

Allen JRL (1965) Late quaternary Niger Delta and adjacent areas: sedimentary environments and lithofacies. AAPG Bull 49(V.I):289–332

Awosika LF, Folorunsho R (2005) Fortnightly oscillating surface current patterns observed from Davies drifter offshore the Western Niger Delta Nigeria. In: Proceedings of operational oceanography and meteorology for the 21st century. Joint commission for oceanography and meteorology scientific conference, Halifax, NS, Canada

Awosika LF, Ibe AC (1994) Geomorphic features of the Gulf of Guinea shelf and littoral drift dynamics. In: Proceedings of international symposium on the results of the first IOCEA cruise in the Gulf of Guinea, 17–20 May 1994. Pub Centre for Environment and Development in Africa, pp 21–27

Awosika LF, Collins M, Clay AT (1994) Clay mineralogy variations in the Gulf of Guinea and implications for sediment transport dynamics. In: Proceedings of international symposium on the results of the first IOCEA cruise in the Gulf of Guinea, 17–20 May 1994. Pub. Centre for Environment and Development in Africa, pp 99–103

Awosika LF, Dublin-Green CO, Oyewo EO (1996) Nearshore littoral drift cells along the Nigerian Coast: Implications for management of Petroleum industry wastes and oil spills. In: Proceedings of 8th Biennial international seminar on the petroleum industry and the Nigerian Environment, Port Harcourt, Nigeria, 17–21 Nov 1996

Awosika LF, Folorunsho R, Dublin-Green CO, Imevbore VO (2000) Review of the coastal erosion at Awoye and Molume areas of Ondo State. A consultancy report for Chevron Nigeria Limited, 75p

Boisvert WE (1967) Major currents in the North and South Atlantic Ocean between 64°N and 60°S, Technical report Hydrogr. Off. Washington, TR-193, 92p

Burke KCB (1972) Longshore drift, submarine canyons, and submarine fans Am. Assoc Petrol Geol Bull 56:1975–1983

Evan Hamilton Inc (2001) Nigeria OPTS shallow water current measurement survey; data reports 3 and 4

Ingham MC (1970) Coastal upwelling in the northwestern Gulf of Guinea. Bull Mar Sci 20:1–34

Longhurst AR (1962) A review of the oceanography of the Gulf of Guinea. Bull Inst Fr d'Afrique noire 24:633–663

NEDECO (1961) The waters of the Niger Delta. NEDECO (Netherlands Engineering Consultants), The Hague

Richardson PL, Reverdin G (1987) Seasonal cycle of velocity in the Atlantic North Equatorial Countercurrent as measured by surface drifters, current meters, and ship drifts. J Geophys Res 92:3691–3708

Whiteman AJ (1982). Nigeria: its petroleum geology, resources and potentials. Graham and Trotham, London, vol 2, pp 306–361

Morphological Characteristics of the Bonny and Cross River (Calabar) Estuaries in Nigeria: Implications for Navigation and Environmental Hazards

Regina Folorunsho and Larry Awosika

Abstract

The Bonny and Cross River estuaries are two major estuaries along the eastern coast of Nigeria. These estuaries are characterized by deep and shallow channels with semi-diurnal tides that generate tidal currents in phase with the tidal directions. The mouths of these estuaries are used extensively by a wide range of vessels and boats with varying sizes ranging from small dinghies and powerboats to merchant ships and commercial fishing boats. The morphology of these estuaries is being shaped by high tidal oscillations superimposed on waves and high volumes of sediment brought in by the rivers and creeks that flow into their drainage basins. The estuaries serve as navigational ways to the Bonny, Onne, Port Harcourt, Okrika and Calabar ports which are of high socio-economic importance to Nigeria. Navigational and environmental problems facing these estuaries include strong currents, occurrence of sandbars or shoals around the mouth, and erosion.

Keywords

Estuary • River discharge • Tides • Niger Delta • Ports

Introduction

The Bonny and Cross River estuaries are two major estuaries along the eastern coast of Nigeria (Fig. 1). In particular, the Bonny River estuary is within the Niger Delta while the Cross River estuary is within the Strand coast of Nigeria (Fig. 2). These estuaries are strongly influenced by currents, tides and storms surges. The mouths of these estuaries are used extensively by a wide range of vessels and boats with varying sizes ranging from small dinghies and powerboats to merchant ships and commercial fishing boats to the five major ports in the eastern Nigeria. These ports (Bonny, Onne, Port Harcourt, Okrika and Calabar) are essential for economic development in Nigeria.

R. Folorunsho (✉) · L. Awosika
Nigerian Institute for Oceanography and Marine Research,
Victoria Island, Lagos, Nigeria
e-mail: rfolorunsho@yahoo.com

Climate

The Bonny and Cross River estuaries, like other estuaries in the Niger Delta, are affected by the Intertropical Convergence Zone (ITCZ), generally referred to as the Intertropical Discontinuity (ITD). This frontal zone is related to the northward and southward movement of the sun. Associated with the movement of the ITCZ are the warm humid maritime Tropical (mT) air mass with its south-westerly winds and the hot and dry continental (cT) air mass with its north-easterly winds. These air masses determine the dominant seasons (rainy and dry) during any given time and are responsible for the different factors that govern the climate.

Wind speeds during the dry season are generally subdued with a mean monthly speed of between 2 and 4 m/s. Wind directions along the coast are predominantly south-westerly. However, the north-easterly dust-laden harmattan winds affect the coast especially in December and January. During

S. Diop et al. (eds.), *The Land/Ocean Interactions in the Coastal Zone of West and Central Africa*, Estuaries of the World, DOI: 10.1007/978-3-319-06388-1_8,
© Springer International Publishing Switzerland 2014

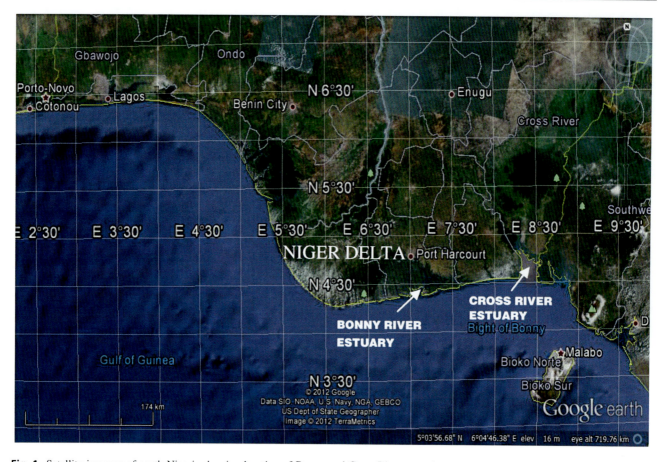

Fig. 1 Satellite imagery of south Nigeria showing location of Bonny and Cross River estuaries

the dry season between November and February, when the cT air mass is dominant, very low mean monthly rainfalls of between 5 and 32 mm are experienced. Temperatures are high to very high throughout the year (averaging 30 °C) because of the abundant and consistent insolation experienced throughout the year. However, because of the maritime location of the Bonny and Cross River estuaries, temperatures are tempered down between November and February. Minimum temperatures of between 24 and 25 °C are observed in July and August. The distribution of relative humidity in the Bonny and Cross River estuaries is determined to a large extent by the two major seasons (wet and dry). During the dry season, a marked difference is noticed with relative humidity less than 60 % in the mornings and less than 40 % in the afternoons.

The rainy season is experienced between March and October when the maritime air mass (mT) predominates. Very high mean monthly rainfalls of between 220 and 460 mm are experienced. High temperatures of about 30 °C are also experienced during this season; however, between July and August relatively low temperatures of between 24 and 26 °C are noted. Maximum wind speeds of between 10 and 14 m/s characterize this season, while minimum wind speeds of about 5 m/s are also experienced. Relative humidity during the rainy season could peak to more than 80 % in the mornings and more than 60 % in the afternoons. Thus, in the Bonny and Cross River areas, the values of relative humidity are very high throughout the year.

Bonny Estuary

Setting and Morphology

The Bonny estuary (Fig. 3) is located on the immediate eastern flank of the Niger Delta between longitudes 7°00′ and 7°15′E and latitudes 4°25′ and 4°50′ (Figs. 1 and 3). The strategic location of the estuary serves as an entrance point to the Port Harcourt, Onne and Okirika ports in Rivers State. Immediately east of the estuary is the Bonny barrier island. The mouth of the estuary is jointly shared by the Cawthrone channel and the New Calabar River. The width of the mouth of the estuary is over 13.8 km and drains a total area of 621,351 km^2. It has an estimated area of 206 km^2 and extends 7 km offshore to a depth of about 7.5 m.

The topography of the area is flat and consists mainly of swamps and flood plains. The dominant island around the

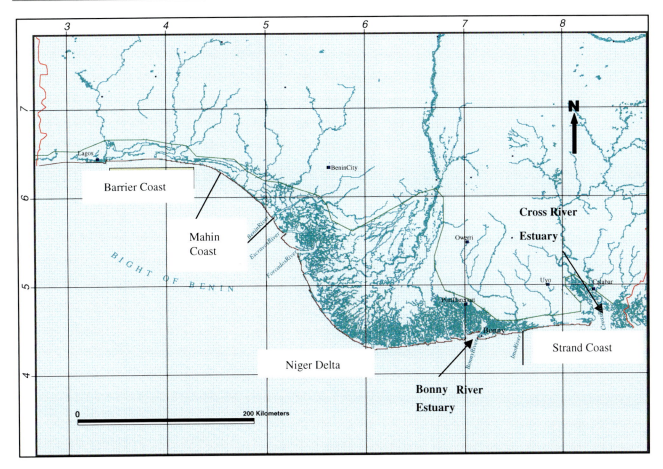

Fig. 2 Location of Bonny and Cross River estuaries within the four Nigerian coastal morphological zones (after Awosika et al. 2000)

estuary is the Bonny Island, one of the twenty barrier bar islands (Allen 1964) fronting the ocean in the Niger Delta. The island is bordered by the Andoni River estuary in the east and the Bonny River estuary in the west. The landscape, consisting of coastal plain sediments, is mostly flat and low-lying in comparison with many other regions in Nigeria. The coastal plain deposits are a mosaic of marine, deltaic, estuarine, lagoonal and fluvio-lacustrine materials. Bonny Island is relatively flat and consists of a series of shallow sand ridges that indicate past regular seaward accretion. Soils are generally sandy or sandy loam and the natural soils are uncontaminated.

Tidal Regime

Characteristically, the area is a typical tidal water zone with little freshwater input but with extensive mangrove swamps, inter-tidal mud flats, and influenced by a semi-diurnal tidal regime. The tidal range in the Bonny River estuary is about 0.8 m at neap tides and 2.20 m during spring tides (NEDECO 1961). Extreme tidal ranges of between 3.5 and 12.5 m are occasionally experienced at Bonny town, a distance of about 15 km from the mouth of the estuary (NEDECO 1961). The Bonny River system has the largest tidal volume of all the river systems in the delta. There is generally a net flux of tidal water up the river, which disperses into various creeks and channels.

Vegetation

The vegetation along the estuary is made up of saline mangroves, shrubs, creepers and climbers. There are extensive brackish wetlands within which the mangrove vegetation thrives. The mangroves occur mainly as thickets and low woodlands. Both the thickets and the woodlands contain varieties of mangrove species such as *Rhizophora racemosa*, *Rhizophora mangle* and *Rhizophora harrisonii* as well as *Avicennia africana*. However, the most prevalent species is *R. racemosa*, which in an undisturbed state can grow to a height of 30 m and above Isebor and Awosika (1993). The root structures play an important role in the stabilization of the sediments. Mangrove ecosystems in the Bonny provide valuable physical habitat for various coastal species such as shorebirds, wintering/migrating birds as well as crabs, shrimps and juvenile stages of fish and shellfish. There is also the presence of the Nympa palm (an introduced

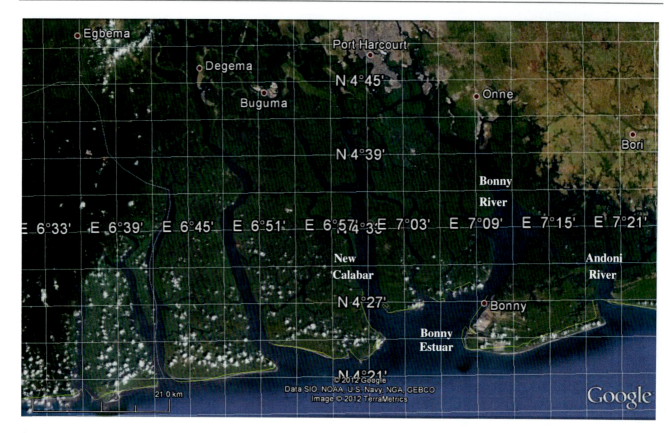

Fig. 3 Map showing Bonny estuary and adjoining river systems

exotic plant species) that is displacing the mangroves along the estuaries and creeks. Other plant species that occur in the area are the grass, *Paspalum* sp., the fern, *Acrostichum* sp., and shrubs, *Ipomoea* sp. The estuary maintains exceptionally high levels of biological productivity and plays important ecological roles of exporting nutrients and providing habitat for aquatic invertebrates and vertebrates. It is also a breeding/nursery ground for aquatic resources.

Socio-Economic Activities and Infrastructures

The Bonny estuary is one of the most environmentally stressed areas in the Niger Delta due to shipping activities associated with four port complexes (Port Harcourt, Onne, Okrika and Bonny) and other boat landings such as Buguma which services oil and gas production. Several oil fields are located in and around the Bonny estuary, including: Orubiri field on Primrose Creek, Onne fields on Ogu Creek; Bomu, Bodo and Bonny fields on the eastern part of Bonny estuary. Also located within the river system are the Nigerian National Petroleum Corporation (NNPC), Petroleum Refinery plant near Okrika, the NAFCON fertilizer plant on Ogu Creek and the Bonny Crude-oil Tank farm, the Eleme Petro chemical plant and other oil- and gas-related installations. The liquefied natural gas (LNG) plant is also located at the mouth of the Bonny River.

Cross River (Calabar) Estuary

Setting and Morphology

Located on Nigeria's south-eastern frontier, the Cross River landscape descends precipitously from the Oban Obudu rugged foothills (1,000–2,000 m) of the Cameroun Mountains on the east, into the Cross River plains (30 m) to the west. The Cross River estuary lies in the eastern extremity of Nigeria and is the largest along the coast (Fig. 4). It has a maximum width of about 25 km at the mouth and narrows northwards to the port of Calabar. Off the mouth of the estuary are several shoals of sandbars like the Tom shot banks. Geologically, the Cross River estuary is a delta of its own, totally independent of the Niger Delta. The Cross River, with a catchment area of 53,590 km^2, delivers more sediment load to the coast than the present Niger Benue

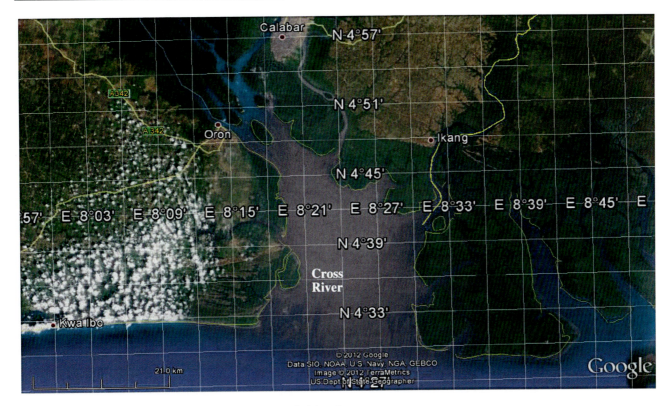

Fig. 4 Satellite imagery showing Cross River estuary and adjoining river systems

drainage system. Structurally, the Cross River plain is underlain by the Ogoja syncline and the Abakaliki uplift wherein lie thick Cretaceous sandstones, marine shales and limestones of the Asu River and the Cross River group.

The Cross River estuary is fed by several smaller rivers which take their sources from the Oban hills in the north. The river flows through swampy rainforest with numerous creeks and forms an inland delta near its confluence with the Calabar River about 20 km wide and 50 km long between the cities of Oron on the west bank and Calabar on the east bank. The delta empties into a broad estuary which it shares with a few smaller rivers. At its mouth in the Atlantic Ocean, the estuary is 24 km wide. The eastern side of the estuary is in the neighbouring country of Cameroun. The Cross River delta has been built out onto the cT shelf with deep submarine canyons (e.g. the Calabar Canyon). The canyons have been incised, with deep sea fans building out onto the ocean floor and providing excellent petroleum traps.

Tidal Regime

The estuary has a tidal range of between 2.1 and 3.0 m. Tides are semi-diurnal with two inequalities. This estuary had the largest tidal range in the Gulf of Guinea reaching a maximum of 3.5 m in Calabar port.

Vegetation

The Cross River estuary is characterized by a broad belt of mangrove swamps fronted by Nympa palms. Mangrove and swamp ecosystems dominate the vegetation type. The mangrove ecosystem is dominated by *Rhizophora* species, while the Nympa palms and swamp vegetation are also abundant. Along the Calabar estuary, Nympa palms are rapidly replacing the mangrove vegetation (Isebor and Awosika 1993).

Socio-Economic Activities and Infrastructure

Socio-economic activities in this area involve fishing, tourism and subsistence agriculture. Commercial activities include ship building, subsistence and large-scale agriculture and lumbering. Considerable activities in support of the oil and gas industry are also carried out within the area. The population of the lower Cross River traditionally uses water transport and Calabar has long had a major seaport, located on the Calabar River about 10 km from its confluence with the Cross River and about 55 km from the sea. The estuary serves as a major navigational way to Calabar Port. The physical modification of the estuaries result from sedimentation processes involving tidal currents, storm surge, exploration and exploitation of oil and gas.

Table 1 Summary of some ADCP records in Bonny and Cross River estuaries (*Source* Field survey)

River estuary	Date of survey	Max surface current vel. (m/s)	Max depth along transect (m)	Total discharge fluxes (m^3/s)
Bonny–peterside	18/12/11	1.81	17.2	16,840.23
Bonny–peterside	19/11/10	2.589	18.64	11,523.63
Calabar estuary	23/12/11	0.216	16.4	12,936.11
Calabar estuary	22/11/10	1.87	15.6	12,481.56

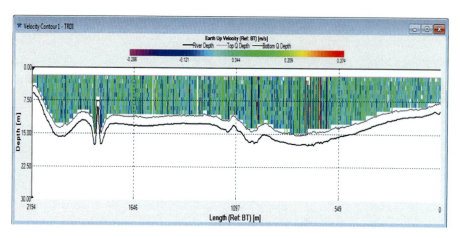
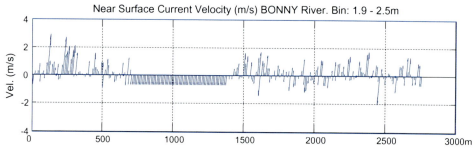
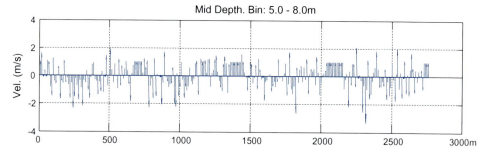
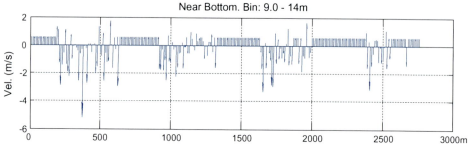

Fig. 5 ADCP current data showing bathymetric configuration and velocity along a profile line in the Bonny estuary (*Source* Field survey)

Fig. 6 ADCP current data showing bathymetric configuration and velocity along a profile line in the Calabar estuary (*Source* Field survey)

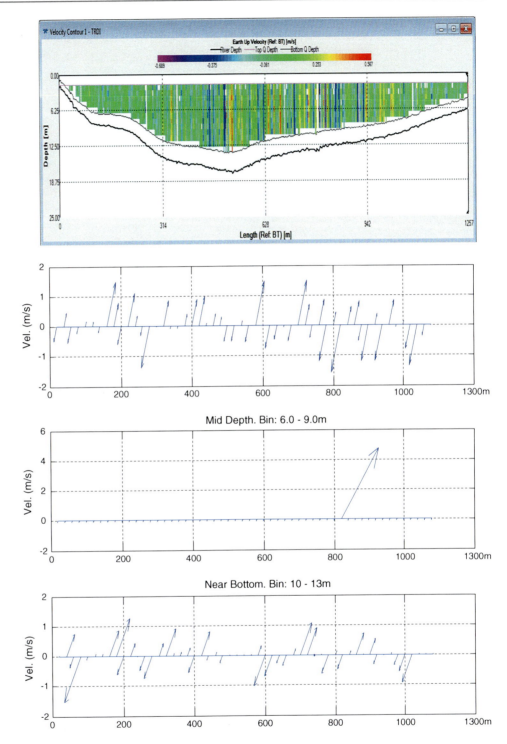

Navigational and Environmental Hazards

Both the Bonny and the Cross River estuaries serve as navigational and socio-economic routes to inland ports (Port Harcourt, Onne, Calabar, Bonny and Okrika) and hence are important for the socio-economic activities in Nigeria. However, these estuaries are plagued with navigational and environmental problems such as fast water currents, sandbars, shoals, bank and shoreline erosion, as discussed below.

Water Currents

Near the mouth of the estuary, the interplay of tidal currents and waves presents navigational hazards that have caused several small boats to capsize and even caused big vessels

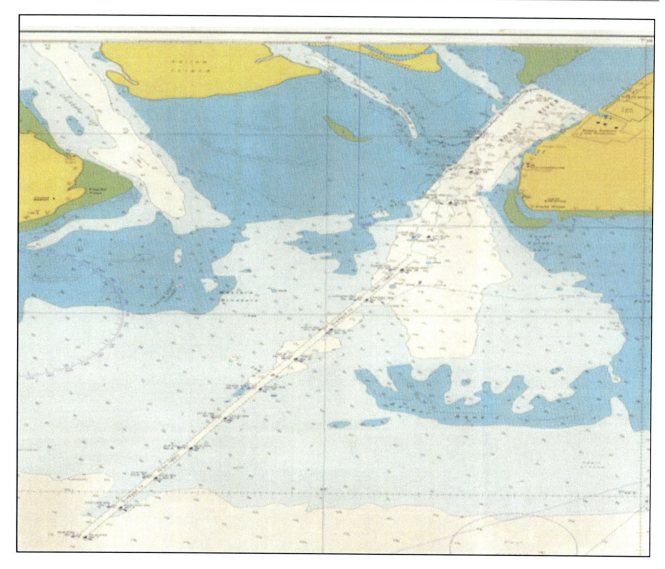

Fig. 7 Navigational chart of Bonny estuary showing several sandbars, sandbanks and other underwater obstructions (from British Admiralty Chart 3286 1991)

to drift and run aground. In the Bonny Estuary, surface current velocity, which ranges between 1.8 and 2.5 m/s had been recorded while the near bottom ranged between 0.9 and over 1.8 m/s (Table 1 and Fig. 5). In the Calabar estuary, currents can exceed 1.87 m/s (Table 1 and Fig. 6). These very fast currents superimposed on wave breakers pose navigational hazards to boats and ships.

Sandbars and Shoals

Both the Bonny and the Cross River estuaries are fed by numerous rivers, creeks and rivulets that carry huge amount of sediments to the estuary and the nearshore zone. During low tide and slack tides, plumes of sediments can be seen in the estuary mouths and the immediate offshore area. The large volumes of sediment reaching the coast and the nearshore zone are worked and shaped by a combination of waves and tides resulting in banks, shoals and ridges. Many of these sandbars and shoals are under water during high tides and hence present navigational problems for vessels. In the Bonny estuary, the main navigational channel from the ocean consists of a south-western north-eastern dredged channel with depths ranging from 10 to 12.5 m. This is the only navigable channel leading the ports of Port Harcourt, Onne, Okrika and Bonny. Surrounding this main channel are sandbars like the Baleur bank towards the east where depths are less than 3 m, and breaker spit towards the west where depths are less than 2.5 m (Fig. 7). Inside the estuary are other isolated banks, sandbars and shoals (Fig. 7). Also

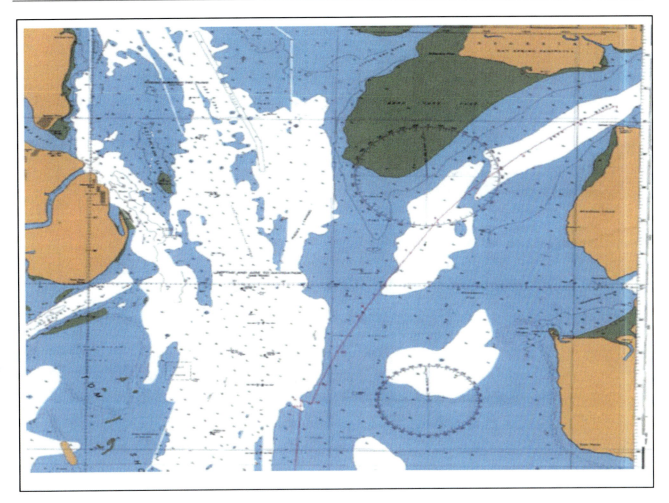

Fig. 8 Navigational chart of Calabar estuary showing several sandbars, sandbanks and other underwater obstructions (from British Admiralty Nautical Chart 3433 2013)

in the Calabar estuary (Fig. 8), the main navigable channel is almost midstream just east of the Tom Shot bank. Depths within the navigable channel are between 6 and 7 m. Even within this channel, there are several sandbars (Fig. 8). Apart from the underwater ridges and shoals, several ship wrecks (some underwater), pose navigational hazards in the Bonny and Cross River estuaries.

Shoreline and Bank Erosion

Shoreline and bank erosion is very prevalent along the banks of the estuaries. Erosion is due to high currents and the construction of several jetties which are perpendicular to the shoreline. These features, coupled with the high currents, cause massive shoreline erosion. Shoreline and bank erosion therefore pose navigational and environmental hazards to ships and vessels as well as berthing facilities. In many cases, jetties have been washed away as result of erosion, leaving the landing jetties isolated from the estuary.

Conclusion

The Bonny and Cross River estuaries are of huge economic importance to the economy of Nigeria. The morphology and accompanying oceanographic and fluvial interactions pose hazards for navigation and hence have adverse impacts on the socio-economic activities in Nigeria. Regular dredging of the navigable channels involves high cost and negatively impacts the health of the estuary. Continuous monitoring and understanding of the oceanographic and fluvial processes in these estuaries should assist in a much more pragmatic management of these estuaries, which serve as important navigational channels for many ports along the Niger Delta.

References

Allen JRL (1964) The Nigerian continental margin: bottom sediments, submarine morphology and geological evolution. Marine Geol 1:289–332

Awosika LF, Folorunsho R, Dublin Green CO, Imevbore V O (2000) Review of the coastal erosion at Awoye and Molume areas of Ondo State. A consultancy report for Chevron Nigeria Limited, p 75

British Admiralty Nautical Chart 3486 Chart Title (1991) Approaches to Bonny River Publication Date: 18/01/1991 Latest Edition date: 10/03/2011

British Admiralty Nautical Chart 3433 Chart Title (2013) Approaches to Calabar Sheet 1 Publication Date: 25/09/1970 Latest Edition date: 14/02/2013

Isebor CE, Awosika LF (1993) The Nigerian mangrove resources; status and management. In: C. D. F. a. M. V. E.S. Diop (ed) Proceedings of a workshop on conservation and sustainable utilization of mangrove forests in Latin America and Africa regions, vol Project PD114/90 (F). ITTO/ISME, Dakar, pp 14–16

NEDECO (1961) The waters of the Niger delta. The Hague, p 317

Status of Large Marine Flagship Faunal Diversity Within Cameroon Estuaries of Central African Coast

Isidore Ayissi, Gordon N. Ajonina, and Hyacinthe Angoni

Abstract

An assessment of the status of large marine flagship faunal species along Cameroon estuaries within the Central African coast was carried out through several surveys, interviews, literature reviews and experience to compile species checklists, causes of their presence (migration, reproduction, feeding, etc.), the conservation status and different threats to species. Results showed that four species of sea turtles were identified and common along Cameroon estuaries: *Dermochelys coriacea, Lepidochelys olivacea, Chelonia mydas* and *Eretmochelys imbricata* for nesting and feeding activities. Eight cetaceans (*Sousa teuszii, Delphinus capensis, Delphinus sp,. Tursiops truncatus, Stenella attenuata* or *S. frontalis, S. coeruleoalba, Megaptera novaeangliae, Physeter macrocephalus*) and one sirenian species (*Trichechus senegalensis*) were found to be common, seasonal or rare. We recorded up to 61 waterbird species represented by 17 families from monthly counts within 20 km of the Sanaga River estuary and associated rivers and lakes in the Douala-Edea Wildlife Reserve between March 1999 and December 2012. The families of Ardeidae, Scolopacidae, Charadriidae and Alcedinidae were top with 12, 10, 8 and 7 species, respectively. Twenty-two (36.1 %) of the 61 species appeared to be resident, while 21(34.4 %) and 16(29.5 %) were seasonal and occasional visitors, respectively. Of particular significance is the high abundance of African Skimmers, Grey Pranticoles, Open-billed Storks and Common green shanks with monthly numbers of up to 811, 583, 336 and 189, respectively. In spite of the existing laws and conservation policies on these threatened species in Cameroon, most are

I. Ayissi (✉)
Cameroon Marine Biology Association (CMBA),
PO. Box 52, Ayos, Cameroon
e-mail: isidoreayissi@gmail.com

I. Ayissi
Specialized Research Center for Marine Ecosystems
(CERECOMA), Institute of Agricultural Research for
Development, P.O. Box 219, Kribi, Cameroon

G. N. Ajonina
CWCS, Coastal Forests and Mangrove Conservation Programme,
PO. Box 52, Mouanko, Cameroon

H. Angoni
Faculty of Science, Department of Plant Biology,
University of Yaounde I, PO. Box 812, Yaounde, Cameroon

I. Ayissi · G. N. Ajonina
Institute of Fisheries and Aquatic Sciences,
University of Douala (Yabassi), Douala, Cameroon

S. Diop et al. (eds.), *The Land/Ocean Interactions in the Coastal Zone of West and Central Africa*, Estuaries of the World, DOI: 10.1007/978-3-319-06388-1_9,
© Springer International Publishing Switzerland 2014

facing many threats. By-catches in gillnets and other fishing gears and the potential for increasing direct takes may be the most severe threats and causes of significant mortality rates. Other threats of varying magnitude of concern include the following: habitat encroachment through coastal development (e.g. port and road construction), over-fishing, chemical and acoustic pollution, ship collisions and ghost nets. The almost complete lack of scientific data on the biology, distribution, stock structure and abundance of sea turtles and cetaceans in Cameroon waters makes it difficult to properly assess the impact of these threats, let alone addressed them. An acceleration of research is urged with the involvement of national Universities and Research Institutes. More faunal surveys are needed to unveil the potentials of the area and the need for the establishment of important relationships between species abundance, site temporal conditions (sandbank dynamics) and socio-economic activities with a view to identifying sustainable wetlands ecosystem utilization options.

Keywords

Cameroon estuaries • Cetaceans • Flagships • Marine mammals • Sea turtles • Waterbirds

Introduction

In international literature, an estuary is defined as a semi-enclosed coastal body of water which has a free connection with the open sea and within which sea water is measurably diluted with freshwater derived from land (Cameron and Pritchard 1963; Pritchard 1967). Estuaries is a partially enclosed permanent water body, either continuously or periodically open to the sea on decadal time scales, extending as far as the upper limit of tidal action or salinity penetration. During floods, an estuary can become a river mouth with no seawater entering the formerly estuarine area or when there is little or no fluvial input an estuary can be isolated from the sea by a sandbar and become a lagoon which may become fresh or hyper-saline.

Estuaries constitute one of most heavily utilized and productive zones on our planet. Their integration processes weave a web of complexity far out of proportion to their occupation of less than 1 % of the planet's surface area Welsh (1984). Estuary management is a complex task, for it deals with the use and care of the interface between the land, rivers and the sea (UNEP/MAP/PAP 1999).

Rapid industrialization and burgeoning population have caused a related increase in the demand for freshwater and a resultant alteration in the flow regime of many of Cameroon's rivers. Estuaries are also at the receiving end of poor catchment practices, such as pollution, erosion, excessive water abstraction and impoundments. Poorly regulated activities have led to the destruction of Cameroon estuaries habitats by physical development such as land reclamation, pollution, deforestation, agriculture, urbanization of Douala and correlated towns, all of these affect the rich and important bio-diversity in this area.

The Cameroon section of the Central Africa coastline hosts an important marine biodiversity, especially sea turtles, marine mammals and water birds. Yet, very little is known about the importance and status of these species. Certain information has been documented on the status of these species along the Cameroon estuary from limited surveys. Atangana (1996) presents the biogeography of these ecosystems in general aspects; Fretey (1998a, b, c, 2001) present the different species of sea turtles occur in this zone and the need of their conservation; Angoni (2005) and Angoni et al. (2010) present the ecology of sea turtles related to their habitats; Ayissi (2000), and Ayissi et al. (2006a, b, 2007, 2013) present also these species but insisting on the need of their conservation by developing alternative sources of income through ecotourism; Ajonina et al. (2002, 2003, 2007), Ajonina and Ayissi (2012) present avifaunal diversity in this area with the aim of the wetlands management. This also includes those undertaken recently under some conservation-based institutions working within the zone including Specialized Research Center of Marine Ecosystems (CERECOMA), Cameroon Marine Biology Association (CMBA), Cameroon Wildlife Conservation Society (CWCS), WWF and other partners who have also carried out some rapid assessments of the status of large marine flagship faunal diversity along the Cameroon coastline.

However, these studies have often been scattered without any clear consolidation to build up any baseline information on these flagship species, especially the evaluation of different impacts necessary for any conservation measures needed for long-term management of these species and their habitats. The present study sets to achieve this aim and drive recommendations in future for sustainable management of Cameroon's estuary.

Methods

Cameroon Coastal Profile and Study Sites

The coastal zone of Cameroon stretches over 402 km (Sayer et al. 1992), from the Nigerian border in the north (Akwayafe River, latitude 4°40′N) to the Equatorial Guinean border in the South (Campo River, latitude 2°20′N), falling between longitude 8°15′E and 9°30′E.

The vegetation of this region belongs to the large set of massive dense humid forest of Cameroon in low and medium altitudes in the coastal forest group, consisting of dense vegetation moist evergreen lowland to *Sagoglottis gabonensis* and *Lophira alata*, biafran subtype; this primitive forest is similar to South American affinities with the humid Amazon rainforest (Letouzey 1968).

Cameroon Continental Shelf

The continental shelf of Cameroon occupies an area of about 10,600 km^2 and gradually descends through 30, 50 and 100 m depths (Bye et al. 1974; Zogning 1986; Morin and Kuete 1989). The northern part has a width of about 25 nautical miles on average, while the southern portion is narrow (15 nautical miles on average). Its relief shows two distinct zones separated by a parallel which passes through the mouth of the Lokoundje River. In the north, the slope is gentle, with a drop in altitude of 130 m. This zone is rocky, with intermittent occurrence of sandbanks. Meanwhile, two major faults have been identified: a reef north of the mouth of the Sanaga River and a series of outliners in the neighbourhood of Macias Nguema Island (Bioko-Equatorial Guinea). This area is favourable for trawling (industrial fishing) (Crosnier 1964). South of this parallel, the relief of the continental shelf is more disjointed; there are many reefs and sandbanks. The interruption of the slope occurs quite early (e.g. at 50 m depth between Campo and Kribi). This area is not suitable for trawling, but is favourable for small-scale fishing. Many corals can be found at 150 m depth.

Coastal Landscape and Hydrology

According to Kramkimel and Bousquet (1987), four characteristics areas can be distinguished within the Cameroon coastal landscape:

- From Campo to the River Nyong mouth

  The coast is high and shows an alternation of rocky outcrops and sandy mud. The main rivers are Ntem, Lobe, Kienke, Lokoundje and Nyong. Their discharges are low, and they transport little alluvium towards the sea. Mangroves are slightly represented; when present, they are in the form of patches on a rocky substrate; this is comparable to the situation described by Villiers (1973) along the Gabonese coast. On the continent, the vegetation is made up of low-altitude Atlantic forest, preceded on the seaward side by patches of a few species of grass which grow on the beaches.

- From River Nyong to Limbe

  The coast is low and is characterized by the presence of estuary and riverine mangroves, separated from the Atlantic forest by a marshy complex of brackish waters. The rivers here are Dibamba, Wouri, Mungo and Sanaga. These waterways have high discharges and transport huge quantities of sediments towards the sea. The Mungo enters the sea through a delta, while other rivers enter together through the estuary (Dibamba, Wouri and Sanaga). The creation of the Douala-Edea Wildlife Reserve has been justified by the great fauna diversity in the area.

- From Limbe to Idenau

  The coast is volcanic and is overhung by Mount Cameroon which has a peak of 4,095 m at the level of Fako. The vegetation is made up of low-altitude mountain forest rich in endemic species. It is characterized by lava flows and the industrial plantations of the Cameroon Development Corporation (CDC). These plantations currently cover more than 90,000 ha. The Mabeta-Moliwe reserve is found here.

- From Idenau to the Nigerian border

  The coast is once more low and marshy; this part of the coast is watered by the mouths of rivers Akwayafe, Ndian, Lokete and Meme which together enter the sea through the Rio-Del-Rey estuary. The vegetation consists of mangroves and swampy species. In the hinterland, the Atlantic forest includes the Korup rainforest and national park.

Geological Characteristics

Sedimentary Basin

The Cameroon coast includes three sedimentary basins of different dimensions. These are the Campo–Kribi basin, the Douala basin and the Rio-del-Ray basin. The Campo–Kribi basin covers an area of 45 km^2 (1–3 km wide and 25 km long). It is situated north of the River Ntem, and its fossils give it great paleogeographic importance. The increase varies between 30 and 100 m. The slope variations measured at sea are a reflection of the situation obtained on land. This can be explained by the existence of many recent faults parallel to the coast and rising several metres above the base. These faults are associated with the formation of the Congo basin, the Lobe waterfall and the Ntem and Bongola rapids. The Douala Rio-del-Rey basin stretches from latitudes 20° to 50° North. It is made up of two sub-basins: the Douala basin in the east (7,000 km^2) and the Rio-Del-Rey basin in the west (2,500 km^2). From south to north, one passes successively through symmetrical geomorphologic settings on both sides of Mount Cameroon: the

Sanaga delta; the "Bouches du Cameroun"; the volcanic horst proper the Rio-del-Rey; and the Niger delta.

The Douala–Rio-del-Rey basin takes the shape of an isosceles triangle with its peak at Yabassi and its side measuring 150 km. The height of the triangle corresponds to the maximum width of the basin (50–60 km). The relief has preserved traces of destructive tectonic activities which calved out the base into steps. The 200 m isobath Douala is at the same distance from the coast (40 km), it is off Kribi–Campo. On the other hand, within the Rio-del-Rey basin, this isobath lies up to 80 km from the beach. The continental shelf in this area is twice as broad as it is in the South-East of Mount Cameroon.

Sediment Dynamics

Sediments deposition leads to the creation of sandy offshore bars whose origin is either marine (effect of the Benguela and Gulf of Guinea currents) or volcanic (Mount Cameroon). The progression of offshore bars and sandy spits parallel to the coast (Souelaba Point), and of various points between Idenau and Bamusso, is caused by: the predominance of the Benguela current over that of the Gulf of Guinea which flows from the west; the low amplitude of tides (2 m on the average); the low charge of coarse detritus material in rivers which flows through a woody hinterland; the build-up of these coastal structures tends to regularize the coastal profile.

Erosion is significant along the volcanic coast of Cameroon. A displacement of the coastline towards the continent has been observed in the South West province. The estuaries and mangroves are characterised by high turbidity which extends right up to 30 km into the sea from Bakassi. This phenomenon is also noticed in the estuaries of "Bouches du Cameroon". The entire eastern part of Rio-Del-Rey basin is blocked by accumulation of mud and fine sand advancing southwards the River Meme. The evolution of the coast will also depend on the quantity and rate of deposition of alluvial material. Between River Akwayafe and Limbe, the offshore currents can reverse direction. This phenomenon can either lead to enlargement of the beaches or otherwise cause erosion as in the case of Bamusso. The portion of the coast between Kribi and Campo consists of crystalline rocks which appear sometimes as isolated out crops in the sea. This rocky portion is characterized by the absence of significant deposits of sand and mud.

Climate and Oceanographic Conditions

The coastal climate in Cameroon, just as in the rest of the Gulf of Guinea, is influenced by the meteorological equator, which is the meeting point between the anticyclone of Azores (North Atlantic) and that of Saint Helen (South Atlantic). This climate results from the combined effect of convergence of the tropical low-pressure zone and the inter-tropical front within the continent. Along the coast, rainfall intensity increases from south to north. Recorded values show average annual rainfall of 3,000 mm in Kribi, 4,000 mm in Douala and more than 11,000 mm in Debundscha. There are two distinct seasons: a long rainy season of more than 8 months and a dry season which generally stretches from November to February. Air temperature is high throughout the year (above 25 °C). The coastal climate is also characterized by monsoon winds of the Guinean type, predominantly south-westerly. These winds cause humidity values to be most always at saturation point. Winds speeds attain exceptional values of 18 m/s (April 1993). In general, average wind speeds recorded over a period of 10 years (1983–1993) vary between 0.5 and 2 m/s.

Cameroon coastal surface waters are warm throughout the year, unlike the coastal waters of other West African countries (Côte d'Ivoire, Ghana, Togo, Benin, etc.) which are characterized by seasonal upwelling. Water temperatures remain always above 24 °C. This warm water layer has a thickness of 20–30 m (Crosnier 1964) depending on the location and the season. It overlies a less warm water layer whose temperature varies between 18 and 20 °C. There is a thermocline between the two water layers which plays an important role in the dynamics of living organisms. Cameroon's coastal waters are generally characterized by low salinity due to high rainfall and a dense river network which supplies freshwater. Lafond (1965) recorded peak salinity values of 20 ‰ at 15 km from Douala port in the dry season and less than 12 ‰ in the rainy season. Tides on the Cameroon coast are of the semi-diurnal type. In general, the amplitude varies between 0.3 and 3 m depending on the location. Their effects are felt in the estuarine complexes. The propagation of the waves and ebb tides are enormous, but poorly known. Olivry (1986), Morin and Kuete (1989) estimate them at $10^6 \, m^3$ for the River Dibamba and 50×10^6 for the River Wouri, in nature are tidal currents which are sometimes violent: 1–1.5 m/s for the flux and up to 2.6 m/s for the reflux. The river flow disturbs this already unstable system by submerging the estuarine complexes. According to observations made by Chaubert and Garraud (1977), sea swells are from the south to south-west sector and of distant origin. Their peculiarity results from the double obstacle constituted by Bioko Island and the widening of the continental shelf at the level of Rio-Del-Rey.

Data Collection Methods

Sea Turtles

From several monitoring of sea turtles species population in the area, we made identification and description of nesting sites and species according to standard KUDU-Program protocols and the different nesting beaches patrolled by teams of eco-watchers according to tidal conditions to observe and identify females laying eggs on the beaches. Biometric data were also collected on sea turtles (curved carapace length and curved carapace width) during patrols and those caught incidentally in nets and observed during the return of fishermen from sea recorded on cards pre-established according to the KUDU-Program standard protocols. We also noted all ringed live turtles with MONEL ECO tags used in the Gulf of Guinea.

The data were also collected in the morning on the tracks and prints of turtles on the beaches; different turtle's nests with some eggs transplanted in hatcheries. Evaluation of human impacts was carried out by surveys through fishermen capture questionnaires and search for carapaces in homes and museums.

Marine Mammals

Beach surveys coverage on foot totalled 784 min and 30.52 km. At least one observer walked along the high waterline so as to maximize distant view, while simultaneously allowing close inspection for cetacean skeletal material among flotsam. Every few minutes, the sea was scanned with 8 × 40 mm binoculars. Prior knowledge of the small-scale geography of coastal stretches is essential for effective beach surveying. Cameroon's beaches are widely interspersed with rocky formations as well as with small and larger freshwater outflows which are often difficult, or time-consuming, to cross or circumvent. At high tide, dense vegetation at the high waterline can obstruct passage.

Five small-boat outings were implemented using both indigenous wooden canoes and a small open fibreglass boat. Duration of visual survey effort was 1,008 min, with 259.1 km distance covered. Traditional canoes are ubiquitous in Cameroon and the most economical way to get onto the water. The main drawbacks are poor stability and velocity, low height above sea level (especially the smaller canoes) allowing adequate view only under optimal sea conditions. A fibre glass boat, equipped with a 40 hp outboard motor, was found to be the most functional and safest platform for inshore work.

Some 18 ports and smaller fish landing sites were visited and checked for evidence of cetacean catches and landings. When direct evidence (carcasses, bones) was lacking, fishermen were interviewed about the presence of cetaceans and by-catches.

Waterbirds

Waterbird census were done monthly (first Thursday and Friday of each month starting at 6.30 a.m) within 20 km of R. Sanaga length starting some 6 km from its mouth, 2 distributaries and a lake (Lake Tissongo) by CWCS Project staff using a 15 hp motorised canoe, binoculars, telescopes, measuring tapes, GPS and identification manuals (Serle et al.

1977; Maclean 1988; Sinclair et al. 1993; Girard 1998). Prior to the beginning of the census in March 1999, the river was divided into five sections following the settlement villages with all sandbanks and vegetated islands mapped and allocated identification numbers including an estimate of their areas. These sites were then monitored (100 % counts) monthly following standard bird census techniques (Bibby et al. 1992; Dodman et al. 1997; Girard 1998). Monthly data collected include the following: species data (waterbird species, other bird species and animals); meteorological data (temperature and rainfall) from CWCS weather station near the river; sandbank status data, i.e. disappearance (area monitoring); and human activity data (wetland use).

Data Analysis

Data were compiled, analysed and presented using simple descriptive statistics, especially frequencies of species and abundance and threat status.

Results and Discussions

Species Status and Indices of Abundance

Sea Turtles

From our surveys, four species of sea turtles were identified and common along Cameroon coastline: *Dermochelys coriacea*, *Lepidochelys olivacea*, *Chelonia mydas* and *Eretmochelys imbricata*. If the first two species are there for nesting activities, the last two are present in this area for feeding activities although *Chelonia mydas* rarely nest (Ayissi et al. 2006c).

Marine Mammals

(1) *Cameroon dolphin*

The Atlantic humpback dolphin *Sousa teuszii* was observed in Cameroon despite the 119 years since the species' discovery. The definition of the "Cameroon Estuary stock" (Van Waerebeek 2003), derived from the species' type location, implied such a premise. This specie was confirmed on 17 May 2011: at 11:05 a.m. we sighted and photographed a small group of about 10 (min. 8–max. 12) Atlantic humpback dolphins near Bouandjo, at N02°28.708′, E09°48.661′. Some individuals showed a strongly developed dorsal hump, while others, thought to be juveniles, had only a faint indication of a hump.

(2) *Humpback whales*

Humpback whales *Megaptera novaeangliae* are seasonally present for calving and breeding in waters of several coastal nations in the Gulf of Guinea, ranging west from (at least) Côte d'Ivoire east to western Nigeria (Van Waerebeek et al. 2001, 2009; Van Waerebeek 2003). Further south, the species occurs also off Gabon and the Republic of Congo (Harmer 1928; Rosenbaum et al. 2004). Until the present survey, no substantiated records of humpback whales existed for Cameroon. Two calves captured incidentally by artisanal fishers were landed and butchered for food. Several other unidentified whales may also have been humpback whales. The presence of calves suggests Cameroon waters may also be part of the calving ground in the northern Gulf of Guinea. Freshly stranded or by-caught whales are flensed and consumed mostly at the community level. Reports from fishers, who (confusingly) call humpback whales "cachalots", suggest that their seasonality coincides with those known from other coastal nations in the northern Gulf of Guinea (Van Waerebeek et al. 2001, 2009). While this will require testing, presumably the same or a closely related Southern Hemisphere humpback whale population is involved.

(3) *Sperm whales*

An entangled sperm whale *Physeter macrocephalus* was stranded at Bakingili (N04°04′17″, E09°02′27″), Southwest Region, in May 2005. It was flensed in situ and served as food for many people from Limbe and Idenau. A second, very large animal was stranded near Kribi in 2009; the weathered skull of a third sperm whale, which was stranded in 1990, was examined at Mpollongue.

(4) *Other marine mammalian species*

Over a distance of 30.52 km and 784 min duration, beach combing effort was implemented on foot. Flotsam at the high waterline was searched but no cetacean skeletal specimens were found. However, after interviews with fishermen, at times, we were shown miscellaneous cetacean bones (primarily vertebrae and ribs) which were documented photographically. Cameroonians often utilize whale bones, especially vertebrae, as ornaments at home. Enquiries resulted in cetacean bones of some 10 specimens. Single whale vertebrae, out of context, are difficult to identify to species because many morphological features overlap among species.

A fisherman who collected teeth from a small whale, referred to as "cachalot", provided the teeth for study. The teeth's pronounced curvedness and their relatively small size (height 49.70–57.80 mm; max breadth 15.05–23.95 mm; max thickness 13.30–16.30 mm) were initially assumed to be from a juvenile *P. macrocephalus*. The pulp cavities were filled to about half tooth length. However, the shape, small size and lack of osteodentine (common in sperm whale, see Boschma 1938) would also concur with a killer whale *Orcinus orca*. A detailed morphological comparison with reference specimens is awaited. Earlier fieldwork yielded photographic evidence for two further species

of Delphinidae, both dead due to fisheries interactions: a long-snouted common dolphin *Delphinus capensis* and a common bottlenose dolphin *Tursiops truncatus*. Interviews repeatedly suggested the occurrence of at least one species of spotted dolphin (*Stenella attenuata* or *Stenella frontalis*). On two occasions (one across the border in Equatorial Guinea), a chunk of a freshly butchered dolphin had called the attention of interlocutors due to its spotted skin. Both spotted dolphin species have been documented from by-catches in Ghana (Van Waerebeek et al. 2007; Debrah et al. 2010) and are likely to occur also in Cameroon. A striped dolphin was observed stranded on a beach, the body has been severed by the time it was photographed but colouration pattern positive identifies it as a striped (*Stenella coeruleoalba*), a new species record for both Cameroon and the Gulf of Guinea (Perrin and Van Waerebeek 2007 in Ayissi et al. 2011a, b). No striped dolphins have been found during extensive dolphin by-catch monitoring in Ghana (Ofori Danson et al. 2003; Van Waerebeek et al. 2009; Debrah et al. 2010). Weir (2009) and Weir et al. (2011) did not sight *S. coeruleoalba* in the Gulf of Guinea, whereas it was fairly frequent offshore Angola.

A larger dolphin locally known as "iowa" may be identifiable with *T. truncatus* or *O. orca*. It is said to exhibit an assertive, fearless behaviour towards people, vessels and fishing gear. Francophone fishermen in Cameroon typically refer to humpback whales as "cachalots", an obvious source of confusion as the true "cachalot" (= sperm whale) also appears to be a frequent visitor of Cameroonian waters. Several fishermen independently mentioned a "Dauphin blanc" (white dolphin), possibly identifiable with a common dolphin (*Delphinus* sp.). "Dauphin blanc" seen to be distinguished from dolphins with darker flank patterns, presumably bottlenose and humpback dolphins.

Waterbirds

Sixty-one (61) waterbird species represented in 17 families have so far been recorded. The families Ardeidae, Scolopacidae, Charadriidae and Alcedinidae have the highest with 12, 10, 8 and 7 species, respectively. The migratory status is also presented

Of the 61 species, 22 (36.1 %) appeared to be resident while 21 (34.4 %) and 16 (29.5 %) were seasonal and occasional visitors, respectively. Of particular significance is the high abundance of African Skimmers, Grey Pranticoles, Open-billed Storks and Common Green Shank with monthly numbers of up to 811, 583, 336 and 189, respectively, close to Glazebrook et al. (1998) counts of 833, 318, 414 and 77, respectively, for River Sanaga during their coastal waterbirds survey of Cameroon Coast in February 1998. According to them, River Sanaga holds nationally significant numbers of Grey Pranticole and African

Skimmer. The seasonal visit of Open-billed Stork coincides with the bivalve extraction activities in the dry season where the birds share in the harvest.

Current and Potential Threats

Sea Turtles

In spite of existing laws and conservation policy on these threatened species, sea turtles are facing several threats such as human predation for local consumption, meat and eggs, selling of carapaces to tourists and gathering of fat for medicinal uses.

The evaluation of impacts of by-catch on sea turtles revealed around 1,241 individuals per year for 13 leatherback and others were green, hawksbill and olive species (Ayissi 2008). Turtle meat is common in the feeding habits of coastal people in Cameroon, but the majority of their catch is not intentional. However, in certain cases, those reptiles are caught intentionally as in Sandje where results include 400 individuals per year by traditional fishermen.

- *Cetacean by-catches*

Nigerian and Ghanaian fishermen occupy a dominant niche among many fisher communities in Cameroon, and customs transfer such as fishing and processing techniques and diet habits, including the consumption of cetacean products, should be expected. Although interviewees frequently denied the occurrence of cetacean by-catches at first, apparently because they feared it was illegal, when the issue was revisited after reinforcing trust with the interviewer, most fishers finally admitted that cetacean by-catches occur with some regularity. Fresh carcasses obtained from such catches and from strandings are utilized in the villages, primarily as food item. Such use of "marine bush meat" is in line with findings for several coastal nations in western Africa, e.g. Ghana, Togo, Nigeria and Guinea (e.g. Clapham and Van Waerebeek 2007; Bamy et al. 2010; Uwagbae and Van Waerebeek 2010; Debrah et al. 2010; Segniagbeto et al. in preparation; Jeff et al. 2010). While there is a lack of material evidence, this can be explained. In other regions where cetacean carcasses are utilized by fishermen (e.g. in Peru), significant quantities of cetacean remains are retrieved from beaches, especially around fishing ports and landing sites. In Ghana, bones of cetaceans are cleaved with machetes and sold attached to the meat. The smoking process of such chunks burns and destroys the bony structures, and little or no recognizable skeletal parts remain after consumption (Debrah et al. 2010), explaining the scarcity of skeletal specimens.

Dolphin meat is consumed freshly cooked or smoked. Stranded or by-caught whales are also flensed and eaten. One case of a sperm whale stranded in Kribi was widely remembered by independent sources who indicated that several people suffered acute gastro-intestinal problems after ingestion, and some were even hospitalized. As elsewhere, teeth of sperm whales are eagerly collected as ivory.

The potential utilization of cetacean carcasses as bait in long-line fisheries, mainly for shark, as reported from Ghana (Ofori-Danson et al. 2003; Debrah et al. 2010), was rarely mentioned by interviewees in Cameroon, and perhaps this practice is indeed uncommon. However, no overhasty conclusions can be drawn as such (illegal) uses, if they occur, are typically shrouded in silence and very hard to ascertain.

- *Direct takes*

One of us (Ayissi) surveyed Japoma and Mbongo (Littoral Region) from 1 to 4 June 2011. Reports from locals indicated that a group of about 12 dolphins were spotted in the Dibamba River with rising tide, near Japoma (N4.0365°, E 9.8196°) and Mbongo (N4.4620°, E8.9840°) in May 2010. Dolphin sightings were suggested to be unusual in the Dibamba River. A few days later one dolphin was found stranded among mangrove roots and was killed by Nigerian fishermen. When additional dolphins became stranded, they suffered the same fate. The village chief mentioned (pers.-comm. to Ayissi, I. 2 June 2011) that two dolphins were butchered in his presence and the meat was distributed among the villagers for personal consumption. The species of dolphin has not yet been identified but *T. truncatus* is considered possible. Some skeletal material that was collected awaits examination.

We recognize the danger in the possible repetition of a global trend documented in a number of developing nations in South America, Africa and Southeast Asia (e.g. Clapham and Van Waerebeek 2007). The consumption of cetacean products initiated with the opportunistic but regular utilization of by-catches can give rise to a larger market demand and ultimately may turn commercial, leading to directed takes of mainly delphinids, especially in situations where important fish stocks are depleted following over-exploitation. The relatively low prices cited by two fishers as typically paid per dolphin suggest the current local market for dolphins is still immature. However, as seen in Ghana, this market can be developed in few years.

- *Over-fishing*

Both humans and marine mammals act as top marine predators and inevitably compete for fish resources. The coasts of Cameroon are characterized by intense fishing effort (Folack and Njifondjou 1995; Ayissi 2008). Besides nationals, thousands of fishermen from Nigeria, being long-term residents, were found to operate from Cameroon, as well as smaller numbers from Togo, Benin and Ghana. A wide variety of fishing arts are practiced by the small-scale fishers, including drift and set gillnets long-lines, purse-seine nets and beach seines. Both multifilament and monofilament nets are widely used, depending on target species and size. In the course of the past few years Ayissi, I. (personal observations) noted an increase in the presence of Asian trawlers (from China, Korea, Japan) off Cameroon's coast, vessels with the reputation of often unsatisfactory adherence to fisheries regulations. Between 1999 and 2009 Chinese pair-trawlers "chalut-boeuf" were deployed on Cameroon's continental shelf. Pair-trawling is well known for its devastating effects on benthic fauna and flora (Liggins and Kennelly 1996).

Little or no recent data are published on catch statistics and the status of fish stocks in Cameroon, but circumstantial evidence suggest that these follow the general trend of fisheries in the Eastern Central Atlantic (FAO area 34), i.e. increasingly overexploited stocks (FAO 2011).

- *Chemical pollution*

Only the lower 20 km of the Sanaga River are navigable, up to Edea, home to the second largest hydropower plant in the country (265 MW). The ALUCAM aluminium smelter in Edea is dependent on the Sanaga for process water and is the single biggest energy consumer in Cameroon (Van der Waarde 2007). The lower reaches of the Sanaga, including its estuary, are sparsely populated ($<20/km^2$) with the local population engaged in benthic bivalve harvesting from the river and fishing. The coastline is mostly inhabited by foreign fishermen from Nigeria, Benin, Ghana and Togo fishing along the coast in larger fishing boats. The aluminium smelting industry produces 500,000 tonnes/year of material in suspension in the Sanaga River (Atangana 1996). The impact on the river's ecology and on its estuary near Mouanko (N 03.58867°, E009.6489°) is unclear. "Red mud", the waste product from the extraction of aluminium from bauxite, is highly contaminating for the environment since it consists of a highly alkaline fine particulate containing heavy metals and other pollutants (White et al. 1997; Pascucci et al. 2009). The question arises about heavy metal toxicity among the fisher communities who subsist on bivalves and other locally extracted sea food. Similarly, the health of top-level marine predators such as small cetaceans, which are known to accumulate contaminants, may be at risk. The coastal beaches are also important breeding grounds for various species of sea turtles (Ayissi 2000; Ayissi et al. 2006a, b).

Cameroon is considered to have abundant offshore natural gas resources. The country's petroleum reserves are located offshore in the Rio-del-Rey basin, offshore and onshore in the Douala and Kribi–Campo basins, and onshore in the Logone-Birni basin in the northern part of the country. Cameroon's only refinery, which is located in the port city of Limbe, had a capacity to produce 45,000 barrels per day (Newman 2006). Tankers, tugboats and other supporting vessels contribute to the heavy vessel traffic around Limbe. Evidence of the hydrocarbon exploration and production industry are ubiquitous. Near Bolondo, on the southern shores of the Cameroon Delta, we found considerable quantities of a tar-like substance (a heavy hydrocarbon fraction) that contaminates the sandy beaches, apparently related to Cameroon's single most important shipping lane which leads to the port of Douala. Locals indicated that the fisher's community of Bolondo had shrunk over the past decade as fishers moved out, blaming declining fish catches. An earlier gravel road that connected Bolondo with Mouanko, no longer maintained, has been reclaimed by the forest. Hence, access to Bolondo is by sea or motorbikes which drive along the beach at low tide. North of the Ntem estuary (Campo), hydrocarbon pollution was seen dispersed through the upper sand layers at several sites along the shore. However, locals claimed pollution had improved from the 1990–2000 period when major timber exploitation along the river transported logs by tugboats down the Ntem River and out to cargo ships anchored in deeper water. The river and estuary was then highly degraded by hydrocarbons but, allegedly since timber has been transported by road, water quality had improved. No documentation was found on this subject.

- *Discarded nets*

On open shores and around ports, we encountered important quantities of various types of abandoned, lost or discarded nets, both monofilament and multifilament. Long after, fishing gear is lost or abandoned at sea by fishers, it continues to ensnare fishes (so-called ghost fishing) and thus harms the marine environment. The drifting gear also causes entanglements of sea turtles and marine mammals (Mac Fadyen et al. 2009). A nationwide awareness campaign might help reduce abandoning of damaged nets and urge fishers to dispose of them on land and/or incinerate them. Alternative uses, disposal methods or recycling should be explored. Nets also pose a hazard to propellers of vessels, especially smaller ones. When visiting Youme I village and spotting discarded nets, our team offered a constructive recommendation in that sense, surprisingly well received, to 12 fishermen including their chief. Several readily acknowledged the problem while the chief announced that they would address it. This spontaneous reaction suggests that with a carefully planned and implemented nationwide educational effort, perhaps with some incentives, this serious environmental problem may not be as intractable as it seems.

- *Shipping and port construction*

Heavy shipping traffic to and from the Gulf of Guinea enters the Cameroon delta via deep-water shipping lanes that lead to the major industrial port of Douala. Concerns are that this waterway may be linked to hydrocarbon pollution and, inevitably, underwater acoustic pollution. Vessel collisions are also expected to constitute a significant risk to the coastal-dwelling population of humpback whales, a threat that has been documented near several West African ports, e.g. in Senegal, Guinea, Côte d'Ivoire, Ghana and

Togo (Félix and Van Waerebeek 2007; Van Waerebeek et al. 2007; Bamy et al. 2010). Near Lolabe, in the South Region, coastal forest clearance, reportedly for new port construction and access roads, was blatantly evident. Of obvious immediate concern is the Cameroon dolphin, our only sighting of which was registered in the South Region. Sediment and detritus run-off may significantly alter and degrade the coastal habitat with a negative impact on littoral biodiversity. Impacts of a fully operational new port evidently could be major. If neritic fish populations decline, and with increased disturbance, near shore-living cetaceans, in particular Cameroon dolphins and (inshore-ecotype) common bottlenose dolphins, are going to be affected. It is worth remembering that no sightings of Atlantic humpback dolphin have ever been reported near a major port in its entire range (Van Waerebeek et al. 2003).

Waterbirds

Apart from the mentioned threats to marine faunal diversity directly or indirectly faced by waterbirds, they are subjected to additional risk of coastal wetlands loss from upland and catchment deforestation, water diversion, encroachment into protected areas and non-protected areas (due to population expansion) (Ajonina and Ayissi 2012).

Addressing the Threats and Future Directions

The almost complete lack of scientific data on the biology, distribution, stock structure and abundance of sea turtles and cetaceans in Cameroon waters makes it difficult to properly assess the impact of these threats, let alone addressed them. An acceleration of research is urged with the involvement of national Universities and Research Centres. More faunal surveys are needed to unveil the potentials of the area and the need for the establishment of important relationships between species abundance, site temporal conditions (sandbank dynamics) and socio-economic activities with the view to identifying sustainable wetland ecosystem utilization options.

In order to achieve management of Cameroon's estuaries with important flagship species encountered, strategies could be set up to deal with the key threats that prevent the achievement of the vision of conservation. It is urgent to develop programmes according national and international laws and policies by action plan with the following goals:

- Identify key threats;
- Action to address key threats;
- Tools and methods required for these programmes;
- Also it is urgent to involve all stakeholders in this area.

Conclusion

Cameroon estuaries hold important marine fauna (birds, mammals and sea turtles); most of them are listed under important conventions and laws. Although this legislation, they are facing many threats from human activities along the area. In the future, many actions involving all stakeholders need to be taken.

This area does not operate in isolation but is connected ecologically with human needs. As a result, certain decisions need to be made at a higher level to ensure overall sustainability (taking into account social equity, economical growth and ecological integrity). It should be better to set up conservation programmes according main strategic objectives and goals need to be achieved for the vision with key actions.

Acknowledgements Thanks are due to all who were involved in these surveys particularly Koen Van Waerebeek from Conservation and Research of West African Aquatic Mammals, Ecological Laboratory (COREWAM), University of Ghana, Legon, Ghana; COREWAM-Senegal, Musée de la Mer/IFAN, Ile de Gorée, Dakar, Senegal; Centro Peruano de Estudios Cetológicos (CEPEC), Lima 20, Peru; Gabriel Segniagbeto from Département de Zoologie et de Biologie Animale, Faculté des Sciences, Université de Lomé, Togo; and Jacques Fretey from IUCN and Chélonée-France. We also express gratitude to all technical fields from Kudù à Tubè, Cameroon Marine Biology Association (CMBA) and CWCS (Cameroon Wildlife Conservation Society (CWCS).

References

Ajonina G, Ayissi I (2012) Linking river hydrodynamics and sedimentology to waterbird numbers from longterm monthly monitoring of waterbirds within the lower Sanaga basin. In: 13th Pan-African ornithological congress, Arusha, Tanzania, p 7

Ajonina GN, Ayissi I, Usongo L (2002) Provisional checklist and migratory status of waterbirds in the Douala-Edea Reserve, Cameroon. Nature et Faune: Biodiversity files, FAO, Rome

Ajonina G, Ganzevles W, Trolliet B (2003) Rapport national du Cameroun, pp 110–117. In: Dodman T, Diaguna CH African waterbird census/les dénombrements d'oiseaux d'eau en Afrique 1999, 2000 et 2001.Wetlands International Global Series N° 16 Wageningen, 368 p

Ajonina G, Chi N, Skeen R, Van de Waarde JJ (2007) (ed) Waterbird census of coastal Cameroon and Sanaga River, January–March 2007, WIWO report 83, Beek-Ubbergen, 114 p

Angoni H (2005) Biologie et Ecologie des Tortues Marines en Rapport avec les Ecosystèmes côtiers. Conservation et Aménagement. Thèse de Doctorat 3è Cycle. Université de Yaoundé I, 125 p

Angoni H, Amougou A, Bilong BCF, Fretey J (2010) La tortue marine au Cameroun, genre Lepidochelys: nidification, biométrie de Lepidochelys olivacea (Eschscholtz, 1829) (Reptilia, Cheloniidae) dans la réserve de faune de Campo (Sud Cameroun). Int J Biol Chem Sci 4(3):649–656

Atangana ER (1996) Biogéographie des écosystèmes côtiers et marins. Rapport Plan National de Gestion de l'Environnement, 34 p

Ayissi I (2000) Projet d'un Ecotourisme <<Tortues Marines>> dans la Région d'Ebodjé. (Unité Technique Opérationnelle de Campo-Ma'an). Mémoire DESS, Université de Yaoundé I. 57 p

Ayissi I (2008) Rapid gillnet bycatch survey of Cameroon, University of Yaoundé I (Unpublished report)

Ayissi I, Ajonina GN, Usongo L, Fretey J (2006a) Etude Préliminaire des tortues marines dans la Réserve de Faune de Douala-Edéa (Cameroun) en vue de l'établissement d'une stratégie de conservation. In: Proceedings. 2 Congrès International sur la Conservation des Chéloniens, Saly-Sénégal, pp 195–198

Ayissi I, Angoni H, Amougou A, Fretey J (2006b) Ecotourism and Natural Resources (Case study of sea turtles in Campo-Ma'an National Park in Cameroon). In : Proceedings of 26th international congress on sea turtles in Creta, Greece, pp 211–212

Ayissi I, Ajonina GN, Usongo L, Fretey J (2006c) Etude préliminaire sur les tortues marines dans la Réserve de Faune de Douala-Edéa (Cameroun) en vue de l'établissement d'une stratégie de conservation. Chelonii 4:195–198

Ayissi I, Angoni H, Amougou A, Fretey J (2007) Impacts of artisanal fishing on natural resources case study on sea turtles along Cameroon coastline (West-Africa). In: sea turtles in Myrtle Beach (South-Carolina, USA)

Ayissi I, Folack J, Van Waerebeek K, Segniagbeto G (2011a) The rediscovery of the Cameroon humphack dolphin (Sousa teuszii) after 119 years of silence. Journées d'excellence scientifique et l'innovation. Yaoundé 30 Nov. 3 Déc. 2011. Yaoundé-Cameroon

Ayissi I, Van Waerebeek K, Segniagbeto G (2011b) Report on the exploratory survey of cetaceans and their status in Cameroon. Document UNEP/CMS/ScC17/Inf.10. Presented to 17th Meeting CMS Scientific Council, Bergen, 17–18 Nov 2011. http://www.cms.int/bodies/ScC/17th_scientific_council/Inf_10_Rpt_Cameroon_cetaceans_exploratory_survey_Eonly.pdf

Ayissi I, Aksissou M, Tiwari M, Fretey J (2013) Caractérisation des habitats benthiques etde ponte des tortues marines autour du parc national de Campo-Ma'an (Cameroun). Int J Biol Chem Sci 1525-IJBCS

Bamy IL, Van Waerebeek K, Bah SS, Dia M, Kaba B, Keita N, Konate S (2010). Species occurrence of cetaceans in Guinea, including humpback whales with Southern Hemisphere seasonality. Marine Divers Rec 3(e48):1–10. doi:10.1017/S1755267210000436

Bibby CJ, Burgess ND, Hill DA (1992) Bird census techniques. Academic Press, London 257 pp

Boschma H (1938) On the teeth and some other particulars of the sperm whale (Physeter macrocephalus L.) Temminckia 3:151–278

Boye M, Baltzer F, Caratini C (1974) Mangrove of the Wouri estuary. In: International symposium of biology and management of mangrove, Honolulu, pp 435–455

Cameron WM, Pritchard DW (1963) Estuaries. In: Hill MN (ed) The sea 2. Willey, New-York, pp 306–324

Chaubert G, Garraud P (1977) Oceanography data collection, storage and dissemination in Cameroon. Sixth Regional workshop on Global Oceanography data archaeology and rescue project, Accra, Ghana, 22-25 April 1977. p 7

Clapham P, Van Waerebeek K (2007) Bushmeat, the sum of the parts. Mol Ecol 16:2607–2609

Crosnier A (1964) Fonds de pêche le long des côtes de la République Fédérale du Cameroun. Cah.ORSTOM, N° special, 133 p

Debrah JS, Ofori-Danson PK, Van Waerebeek K (2010) An update on the catch composition and other aspects of cetacean exploitation in Ghana. IWC Scientific Committee document SC/62/SM10, Agadir, Morocco, June 2010

Dodman T, de Vaan C, Hubert E, Nivet C (1997) African waterfowl census 1997. Les Dénombrements Internationaux d' Oiseaux d'eau en Afrique. Wetlands International, Wageningen, The Netherlands, 260 p

Félix F, Van Waerebeek K (2007) Whale mortality from ship strikes in Ecuador and West Africa. Latin Am J Aquat Mammals 4(1):55–60

Folack J, Njifondjou O (1995) Characteristics of marine artisanal fisheries in Cameroon (Caractéristiques de la pêche artisanale maritime au Cameroun). IDAF Newsletter/ La letter de DIPA 28:18–21

Fretey J (1998a) Quelques Notions sur l'identification et la Biologie des tortues marines. Les différentes espèces. Atelier Guadeloupéen sur les tortues marines, 93 p

Fretey J (1998b) Statues des tortues marines en Afrique Centrale-Afrique de l'ouest. (5) Le Cameroun. Rapport UICN, 152 p

Fretey J (1998c) Tortues Marines de la façade atlantique d'Afrique. Rapport de mission UICN, 253 p

Fretey J (2001) Biogéographie et Conservation des tortues marines de la cote Atlantique d'Afrique. CMS. Technical series publication N° 6. Bonn, 428 p

Girard O (1998) Echassiers, Canards et limicoles de l'Ouest Africain. Office National de la Chasse, Paris, France, 136 p

Glazebrook J, Mbella F, O'kah ME, Njie F, West R (1998) Coastal waterbirds in Cameroon. Phase 1-Feb 1998, A Wetlands International Project, 40 p

Harmer SF (1928) The history of whaling. In: Proceedings of the Linnean Society of London, vol 140: 1927–1928

Jeff EM, Tara MC, Lewison RL, Read AJ, Bjorkland R, Mc Donald SL, Crowder LB, Aruna E, Ayissi I, Espeut P, Joynson-Hicks C, Pilcher N, Poonian C, Solarin B, Kiszka J (2010) An interview-based approach to assess marine mammal and sea turtle captures in artisanal fisheries. Biol Conserv 143:795

Kramkimel JM, Bousquet B (1987) Mangrove d'Afrique et de Madagascar: les mangroves du Cameroun. CEE, SECA, Luxembourg, pp 127–137

Letouzey R (1968) Etude phytogéographique du Cameroun. Encyclopédie biologique, 69. Eds. P. Lechevalier, Paris, 511 p

Liggins GW, Kennelly SJ (1996) By-catch from prawn trawling in the Clarence River estuary, New South Wales, Australia. Fish Res 25:347–367

Mac fadyen G, Huntington T, Cappell R (2009) Abandoned, lost or otherwise discarded fishing gear FAO: Fisheries and Aquaculture, Technical paper 523, Rome

Maclean GL (1988) Roberts' birds of Southern Africa. New Holland Publishers, London 848 p

Morin S, Kuete M (1989) Le Littoral Camerounais: Problèmes morphologiques. Trav. Labo. Goegr. Phys. appliquée. Inst. Geogr. Univ. Bordeaux III, N° 11:pp 5–53

Newman HR (2006) The mineral industries of Cameroon and Cape Verde. U.S. Geological Survey Minerals Yearbook, 2006, 3 p

Ofori-Danson PK, Van Waerebeek K, Debrah S (2003) A survey for the conservation of dolphins in Ghanaian coastal waters. J Ghana Sci Assoc 5(2):45–54

Olivry JC (1986) Fleuves et rivières du Cameroun. Collection monographies hydrologiques. ORSTOM Mém. 9:733

Pascucci S, Belviso C, Cavalli RM, Laneve G, Misurovic A, Perrino C, Pignatti S (2009) Red mud soil contamination near an urban settlement analyzed by airborne hyperspectral remote sensing. In: Geoscience and remote sensing symposium 2009 IEEE International, IGARSS, pp 893–896

Pritchard DW (1967) What is an estuary: physical viewpoint. In: Lauff GH (ed) Estuaries. American Association for the Advancement of Science, Washington, DC, pp 3–5

Rosenbaum HC, Pomilla C, Leslie M, Best PB, Collins T, Engel MH, Ersts PJ, Findlay KP, Kotze PJH, Meyer M, Minton G, Barendse J, Van Waerebeek K, Razafindrakoto Y (2004) Mitochondrial DNA diversity and population structure of humpback whales from their wintering areas in the Indian and South Atlantic Oceans (Wintering Regions A, B, C and X). Paper SC/56/SH3 presented to IWC Scientific Committee, July 2004, Sorrento, Italy

Sayer JA, Harcourt CS, Collins NM (eds) (1992) The conservation atlas of tropical forest, Africa. Macmillan Publishing Ltd., London

Segniagbeto GH, Van Waerebeek K, Bowessidjaou EJ, Okoumassou K, Ahoedo K An annotated checklist of the cetaceans of Togo, with a first specimen record of Antarctic minke whale *Balaenoptera bonaerensis* Burmeister, 1867 in the Gulf of Guinea. (In preparation)

Serle W, Morel GJ, Hartwig W (1977) Birds of West Africa. Harper Collins, London 351 p

Sinclair I, Hockery P, Tarboton W (1993) Birds of Southern Africa. New Holland, London, p 426

UNEP/MAP/PAP (1999) Conceptual framework and planning guidelines for integrated coastal Area and River Basin Management.Split, priority Action Programme. United Nations Environmental Programme, Mediterranean Action Plan, Priority Actions Programme: United Nations Environment Programme

Uwagbae M, Van Waerebeek K (2010) Initial evidence of dolphin takes in the Niger Delta region and a review of Nigerian cetaceans. IWC Scientific Committee document SC/62/SM1, Agadir, Morocco, June 2010. p 8

Van der Waarde J (2007) Integrated river basin management of the Sanaga River, Cameroon. Benefits and challenges of decentralised water management. Unpublished, UNESCO Institute of Hydraulic Engineering, Delft, The Netherlands. [www.internationalrivers.org/files/IRBM%20Sanaga.pdf]

Van Waerebeek K (2003) A newly discovered stock of humpback whales in the northern Gulf of Guinea. CMS Bulletin 18:6–7

Van Waerebeek K, Tchibozo S, Montcho J, Nobime G, Sohou Z, Sohouhoue P, Dossou C (2001) The bight of Benin, a North Atlantic breeding ground of a Southern Hemisphere humpback whale population, likely related to Gabon and Angola substocks. Paper SC/53/IA21 presented to the Scientific Committee of the International Whaling Commission, London, July 2001, 8 p

Villiers JF (1973) Plantes camerounaises imparfaites connues. *Octoknema olinklagei.* Eng (Octonemacées) Ann De la Fac Sci Cameroun, 18:35–42

Van Waerebeek K, Baker AN, Félix F, Gedamke J, Iñiguez M, Sanino GP, Secchi E, Sutaria D, Van Helden A, Wang Y (2007) Vessel collisions with small cetaceans worldwide and with large whales in the Southern Hemisphere, an initial assessment. Latin Am J Aquat Mammals 6(1):43–69

Van Waerebeek K, Ofori-Danson PK, Debrah J (2009) The cetaceans of Ghana: a validated faunal checklist. West Afr J Appl Ecol 15:61–90

Weir CR (2009) Distribution, behaviour and photo-identification of Atlantic humpback dolphins *Sousa teuszii* off Flamingos, Angola. Afr J Mar Sci 31(3):319–331

Weir C, Van Waerebeek K, Jefferson TA, Collins T (2011) West Africa's Atlantic humpback dolphin: endemic, enigmatic and soon endangered? Afr Zool 46(1):1–17

White C, Sayer J, Gadd G (1997) Microbial solubilization and immobilization of toxic metals: key biogeochemical processes for treatment of contamination. FEMS Microbiol Rev 20:503–516

www.fao.org/newsroom/common/ecg/1000505/en/stocks.pdf [Accessed 14 July 2011]

Zogning A (1986) Les formations superficielles latéritiques dans la région de Douala; Morphologie générale et sensibilité aux activités humaines. Séminaire Régional sur les latérites, Douala, pp 289–304

Morphology Analysis of Niger Delta Shoreline and Estuaries for Ecotourism Potential in Nigeria

O. Adeaga

Abstract

Estuaries provide numerous goods and services needed for human development and socioeconomic sustenance, housing about 60 % of the world's population with highly fragile natural endowment of luxuriant diverse ecological types. A beneficial link through long-term biodiversity conservation with local, social, and economic development in estuaries therefore remains a well-planned ecotourism development. Such development is necessary since disturbance of the dynamic 'steady state' of estuaries may result in total modification of their morphology or losing entire ecosystems. This study examines the shoreline morphology of the Niger Delta and major estuaries for better understanding of the natural forcing within the estuary systems and their potential for ecotourism development planning and management beyond the twenty-first century.

Keywords

Morphology • Shoreline • Estuaries • Ecotourism • Niger delta

Introduction

Natural luxuriant flora and fauna, and the natural resources of estuaries, commonly found on the numerous sedimentary coasts around the world, provide numerous goods and services needed for human development and socioeconomic sustenance, housing about 60 % of the world's population. Estuaries are semi-enclosed coastal bodies of water with free connection to the open sea within which sea water is diluted with freshwater derived from land drainage (Pritchard 1967). These water bodies receive sediments from both fluvial and marine sources and contain facies influenced by tides, waves, and fluvial processes (Dalrymple et al. 1992), stretching from the sea inlet to the upper limit of tidal rise influence within river valleys (Fairbridge 1980).

Estuaries link marine waters (subtidal and intertidal), freshwaters, and terrestrial ecosystems and usually contain wetlands formed at the margins of the land and sea features. Thus, an estuary functions as a transient open system in a dynamic 'steady state' through exchanges of energy, water, and sediment with the surrounding systems (catchment and open sea). External environmental inputs and the system constraints therefore change as the estuary develops.

Estuary development involves interplay between its different components as it attains a dynamic steady state, with mixing stratification over space and time scales within the estuary system. The system dynamic pattern therefore depends on the relationship between accretion, water movement, and sediment transport, influenced by long-term average sediment supply transport (from inland or coastal origin), its characteristics (direction and magnitude), and abrupt changes in the estuarine morphology (Milliman 1991; Pethick 1994).

O. Adeaga (✉)
Department of Geography, University of Lagos, Lagos, Nigeria
e-mail: oadeaga@yahoo.com

S. Diop et al. (eds.), *The Land/Ocean Interactions in the Coastal Zone of West and Central Africa*, Estuaries of the World, DOI: 10.1007/978-3-319-06388-1_10,
© Springer International Publishing Switzerland 2014

Hence, estuaries are often characterized by fine sediment deposition (sand and mud) and include extensive intertidal areas (saltmarshes, mudflats, and sand flats), and semidiurnal or diurnal tidal regimes (Plate 1). Pritchard (1967) proposed four classes of estuaries based on their geomorphological status, while Hume and Herdendorf (1988) developed a 16-type classification scheme based on the geomorphology and oceanographic characteristics of the estuary, tides, and the catchment hydrology. The Defra (2002) estuaries classification was based on sediment origins, and Hayes (1975) classified estuaries of tidal ranges of less than 2 m, between 2 and 4 m, and above 4 m into microtidal, mesotidal and macrotidal estuaries, respectively. Dyer (1997) pointed out that when the ratio between river flow and tidal flow is equal to or greater than one, the estuary is highly stratified, at about 0.25 is partially mixed, and well mixed at values less than 0.1, based on the salinity structure involving the mixing process between fresh and salt water (Seminara et al. 2001). Dilution processes within estuaries take place at different mixing levels due to many forcing mechanism like tides, tidal currents, and wave-induced motion. At the mixing point, a salinity gradient is formed and the occurrence of internal waves on the interface generates a density flow gradient which tends to increase flood currents in the deepest parts and decrease flow near the surface and the shallow cross-sectional area of the river channel.

Distortions to the dynamic 'steady state' of an estuary may therefore severely alter the geomorphic and geological characteristics and resilient behavior of the system, with the possibility of total modification of its morphology or losing the entire ecosystem (Pethick 1994; French 2006). Hence, this study examines the shoreline morphology of the Niger Delta shoreline and major estuaries for better understanding of the natural forcing within the estuary systems and their potential for ecotourism development planning and management beyond the twenty-first century.

Ecotourism entails the ability of human beings to utilize the natural resources for the promotion of tourism without being destructive to the ecosystem. It therefore entails benefiting from the enormous provision of the numerable natural resource environment without destroying the livelihood of indigenous communities. It is therefore beneficial not only as a means of conserving the estuary-rich biodiversity and extensive ecosystem environment but also to the economy of the host community (Moelry 1990; Mowforth and Munt 2009). It is a promising means of advancing social, economic, and environmental objectives of sustainable development of the numerable natural resource.

> Box 1: Overview of Estuary and Ecotourism
> - Estuaries are semi-enclosed coastal bodies of water with free connection to the open sea.
> - Estuaries are commonly found on the numerous sedimentary coasts, providing numerous goods and services.
> - Estuaries classification can be based on geomorphological and oceanographic characteristics, sediment deposition, and tidal regimes and dilution processes.
> - Ecotourism entails the ability of human beings to utilize the natural resources for the promotion of tourism without being destructive to the ecosystem.
> - There is need to examine the Niger Delta shoreline morphology and their potential for ecotourism development.

Niger Delta Shoreline Morphology System

The Nigeria coastline region of length approximately 859 km consists of a narrow low-lying topography coastal strip of land and stretches inland for a distance of about 15 km in Lagos to about 150 km in the Niger Delta and about 25 km east of the Niger Delta. This coastline lies within Latitude 4°10′N and 6°20′N and Longitude 2°45′ and 8°32′ in the eastern section (Fig. 1). Tides along the entire Nigerian coast are semidiurnal and from a southwesterly direction with two inequalities. The tidal range increases progressively eastwards from 1 m at Lagos to about 3 m at the Calabar estuary. Though the tidal range is relatively small, the spring tidal range reaches on average 1.8 m across the coastline (Abam 1999). This intense tidal activity can be destructive and significantly modify the coastline morphology characteristics, as evident along the Mahin coast during a spring tidal range of 1.5 m (Nwilo and Onuoha 1993).

Sediments near the shore of the coastline area are composed of coarse to fine sand except off the Mahin mud coast, which is void of sand but with sediment grades varying from fine sand to silt; Nigerian coast are semidiurnal and from a southwesterly direction with two inequalities to mud at the outer shelf, with finer-grained sand beaches occurring along the flanks of the Niger Delta (Plate 2). The coastal region is divided into four main geomorphic zones (from east to west) namely the stranded coast/estuary complex, the Niger Delta, the transgressive Mahin mud coast, and the barrier beach lagoon coast complex (Awosika et al. 2000).

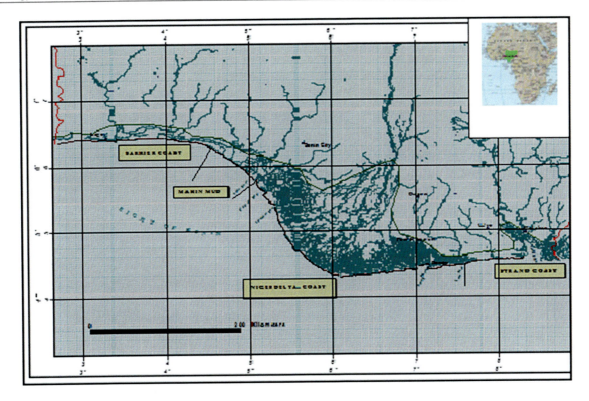

Fig. 1 Nigeria coastal area geomorphologic zones (Awosika et al. 2000)

Box 2: Nigerian Coastal Region
- Nigeria coastline region covers a total length of approximately 859 km.
- Nigerian coastal region is divided into four main geomorphic zones (from east to west) namely the stranded coast/estuary complex, the Niger Delta, the transgressive Mahin mud coast, and the barrier beach lagoon coast complex.
- Nigerian coast is semidiurnal and from a southwesterly direction with two inequalities.

Based on large- and small-scale geomorphic differences and numerous physical characteristics (i.e. beach sediment grain size and coastal processes), the coastline was classified into six geomorphic units (Fig. 2). The Niger Delta consists of the western delta flank, arcuate delta, and eastern flank categories.

The Niger Delta is a vast sedimentary basin with a complex river network and a fragile ecology in which fresh and saline water ecosystems maintain a dynamic equilibrium. The delta covers about 50 % of the total length of the Nigerian coastline, stretching from the Benin River estuary for about 450 km eastward, and terminates at the mouth of the Imo River estuary. The region has a delta flank coast (western and eastern flank) on either side accounting for 115 km of Nigerian coastline and characterized by semidiurnal tidal regimes with tidal amplitude of about 1.2 m and higher ebb flow velocities than flood flow. For example, ebb flow velocity in excess of about 5,000 m^3/sec above flood flow was recorded at the Bonny River (NE-DECO 1961). The geomorphic unit has 21 major river mouths/tidal inlets that intersect the coast, breaking it up into a series of barrier islands of which 16 are within the arcuate delta region. This natural delta receives its sediments, which mainly comprise medium to coarse unconsolidated sands, silt, clay, shale, and peat, from the suspended and traction load of the Niger and Benue Rivers and their tributaries.

The Niger Delta is characterized by a sandy shoreline backed by extensive mangrove swamp and Barrier Island separated by tidal channels. This region has the longest barrier island of about 35 km (Ramos–Dodo Island), of which the widest is Focados–Ramos Island which is 10 km wide. In the eastern flank, the barrier islands are narrower and appear as remnants of beach ridges, due to wave action and tidal erosion in and around the associated creek network. The barrier islands on the delta are generally 15–20 km in length and 3–5 km wide and are better developed on the western than on the eastern flank. Based on the shoreline morphology deposition (Fig. 3), the delta region is vulnerable to sandy coast erosion and horizontal recession due to the white beach sand and sandy ridge substrates with grades of fine- to very fine-grained,

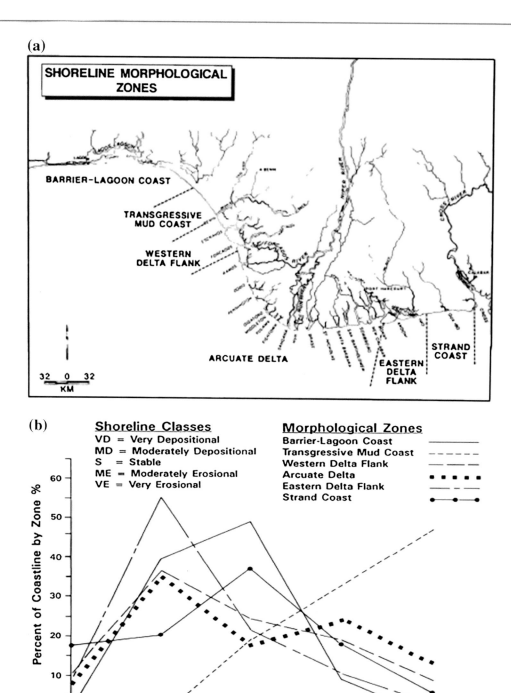

Fig. 2 Nigeria shoreline morphological zones (a) and coastal processes (b) (Sexton and Murday 1994)

moderately well-sorted sand. While the western flanks are less susceptible to erosion due to the dark and dark brownish organic and peaty clay of high plasticity (Abam 1999), the eastern flank of brown sandy clay soil is more vulnerable. The western delta flank extends from the Benin River to east of the Forcados River estuary, while the eastern delta flank extends from east of the Bonny/New Calabar Estuary entrance to immediately east of the Imo River.

The salinity intrusion within the Niger Delta depends strongly on the spatial extent of the diurnal tidal range within the region, which diminishes inland (Abam 1999), with more inundation in the western and arcuate delta part of the delta with settlements such as Port Harcourt, Bonny,

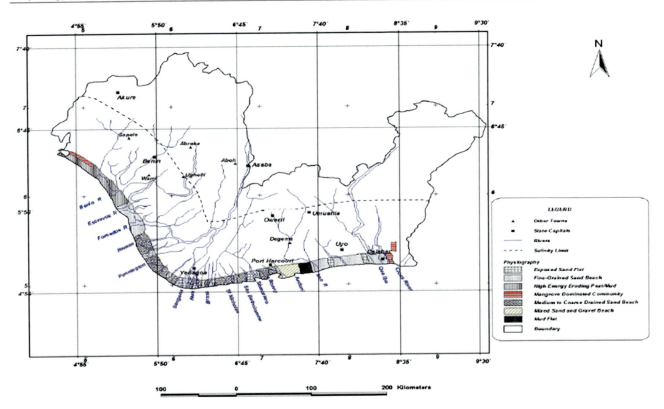

Fig. 3 Niger Delta shoreline morphology (Ogba and Utang 2010)

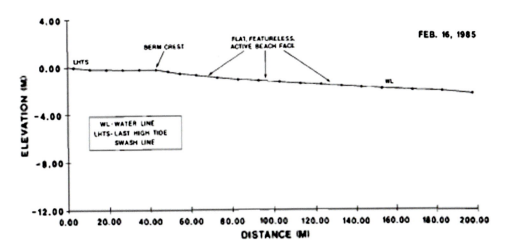

Fig. 4 Typical delta flanks beach profile plot (Sexton and Murday 1994)

Brass, Forcados, Nembe, Opobo, Koko, Degema, Burutu, and southern part of Warri, which are severely under the influence of tides and salinity, but not the entire coastal area (100 km from the coast).

It should be noted that the western side of the delta has significant active freshwater and sediment delivery from the upstream Niger–Benue river system, while the eastern side is influenced primarily by marine processes (waves and tides) (Sexton and Murday 1994). Erosion processes are prominent in the eastern flank of the delta as a result of reworking by marine processes, while the western flank is more constructional or depositional and fluvial dominated due to the active river systems (Sexton and Murday 1994). The post-Kainji dam therefore subjected the region to strong erosional stress due to reduction in water and sediment delivery into the Niger Delta region (Abam 2001).

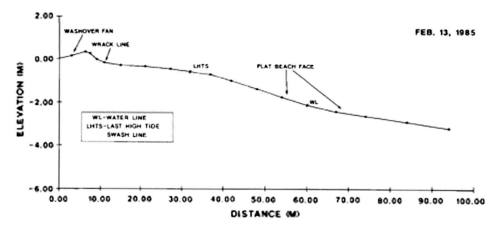

Fig. 5 Typical arcuate delta flanks beach profile plot (Sexton and Murday 1994)

Niger Delta Beach Profile

The Niger Delta flanks (Fig. 4) consist of very fine-grained, mica-rich sand with prominent anti-dunes on active beach faces. These beaches usually have moderate (less than 1 m) wave heights, and the very wide intertidal beach faces heights often greater than 175 m with gentle slope profiles ranging from 1:50 to 1:90. Beaches on the western delta flank are slightly finer grained and flatter, while the tidal amplitudes for the eastern delta and western delta flanks are 2.5 and 1.5 m, respectively.

The Niger arcuate delta (Fig. 5) covers 284 km of Nigeria's coastline (not including the delta flanks) and is composed of fine- to medium-grained, well-sorted sand with barrier island beaches having moderate intertidal beach faces (50 m) with steeper slopes in relation to the delta flanks of 1:15–1:20.

> Box 3: Nigerian Coastal Region
> - The Niger Delta consists of the western delta flank, arcuate delta, and eastern flank categories.
> - The western side of the delta has significant active freshwater and sediment delivery, while the eastern side is influenced primarily by marine processes (waves and tides).

Niger Delta Estuaries

The Niger Delta estuarine waters cover an area of about 680 km² (Fig. 6). The Bonny/New Calabar river systems form about 39 % of the total area (Sott 1966). The Niger Delta area is the richest part of Nigeria in terms of natural resources with large deposits of petroleum products (oil and gas) (Braide et al. 2004). The Niger Delta is a tidal river estuary of fluvial erosion origin and a fully developed Holocene environment. Sediments originate from a drowned coastal plain, while the estuaries are meso-tidal having a composite delta development with a landward side (flood-tide delta) and a seaward side (ebb-tide delta) with salt marshes and tidal flats at the boundaries next to strongly meandering upstream reaches partially mixed with seasonal stratification variations. It should be noted that spring tide currents favor turbulent flow exchanges and a lowering of the mixing stratification ratio with a less sharp interface. However, a zone of high salinity gradient is established at about mid-depth, while the surface and bottom layers are quite homogeneous (Hume and Herdendorf 1988; Dyer 1997; Seminara et al. 2001; Defra 2002).

Bonny River Estuary

The Bonny river system has the largest tidal volume within the Niger Delta (NEDECO 1961), composed of an estuarine and highly saline seawater located seaward of the river mouth (typical of the Niger Delta coastal region), and influenced by tide- and wind-driven surface currents. The Bonny River estuary lies in the strand coast, between longitudes 6°58′ and 7°14′ East, and latitudes 4°19′ and 4°34′ North. It has an estimated area of 206 km² and extends 7 km offshore to a depth of about 7.5 m (Irving 1962; Scott 1966). It is a partially mixed estuary with a typical tidal water zone and little freshwater input but extensive mangrove swamps and intertidal mud flats, which are being influenced by the semidiurnal tidal regime. In the Bonny River estuary, the salinity fluctuates with the season and tide regime and is influenced by the Atlantic Ocean (Dangana 1985) with a tidal range of about 0.8 m at neap tides and 2.20 m during spring tides (NEDECO 1961). The estuary width is about 3,000 m and narrows down to 200 m upstream with water depth decreasing upstream except in the relatively shallow expanse where the Opobo channel

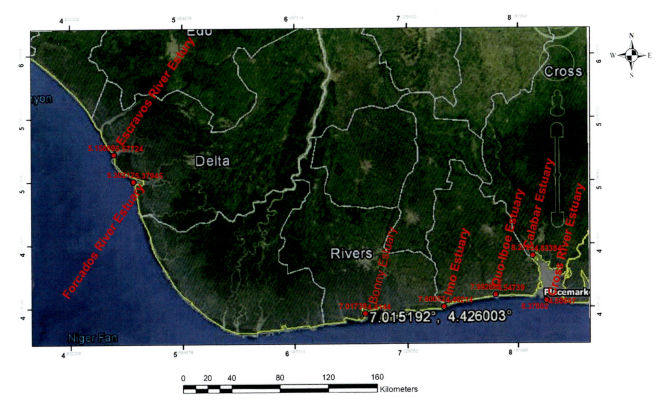

Fig. 6 Niger Delta estuaries

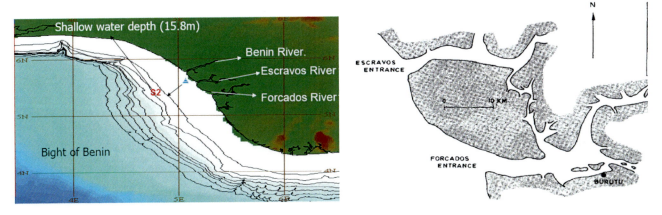

Fig. 7 The Escavos and Forcardos River estuaries

and the Hughes channel branch off. The river is tidal over almost the entire channel, with salt intrusion reaching New Calabar junction, a distance of 70 km. Freshwater discharge into the river system is relatively insignificant. Bonny River tidal range varies from 1.96 m on mean spring tides to 0.91 m on mean neap tides. Peak ebb and flood current velocities are 1.6 and 1.0 m/s, respectively, on spring tides and 1.5 and 0.8 m/s under average tides (Plate 3).

Cross River Estuary

The Cross River Estuary (Plate 4) encompasses rich habitats such as the mangroves, wetlands, lagoons, beaches, and mudflat ecosystems around it, as well as maintaining an exceptionally high level of biological productivity. The strength of the Cross River deltaic sedimentation is derived from the Cameroun Mountain. The Cross River estuary is

Plate 1 Overview of an estuary in Nigeria's coast

the largest in the West African subregion and is approximately 25 km wide at the mouth and more than 440 km long, with a tidal flushing of 1.83 billion cm^3 per day (Enyenihi 1991) and tidal amplitude of 3 m (Asuquo et al. 1998). It is the largest estuary along the Gulf of Guinea (Enyenihi 1991) covering an estimated area of 54,000 km^2 with about 74 % lying in Nigeria with the remaining 26 % in Cameroon (Enyenihi 1991). The estuary has a long coastline with fringing mangrove and a characteristic muddy bottom. Also, the estuary has no sandbar blockage but only a moderate tidal range of 3 m at Calabar. The beaches around the estuary are a potential attraction for tourists and perform numerous functions ranging from species abundance, research, and educational purposes to the protection and stabilization of the coastline.

Calabar Estuary

The Calabar River is a NE–SW-trending tidal river and a major tributary of the Cross River which has the largest estuary in West Africa (Plate 5). The Calabar estuary is mainly formed by the Cross River, but also receives waters from the Calabar and other streams. The Rio del Rey creek at the eastern end of the estuary marks the boundary between Nigeria and Cameroon. The estuary is 10–12 m broad at its mouth and maintains the same breadth for about 30 km. The upper facies of the Calabar River has a thickness of 8 m, while the lower facies is 10 m thick. The grain size distribution reflects that the facies has medium sands and negatively skewed sediments. Also, coarse sands, poorly to moderately sorted, and leptokurtic sediments characterize the upper facies.

Qua Iboe River Estuary

The Qua Iboe River estuary is a meso-tidal estuary with a N–S-trending estuary which opens into the Atlantic Ocean. The deepest part of the channel lies within the central portion and is 10 m deep (Plate 6).

Although sandy beaches are known to develop in some portions of the estuary, most of them are fringed with tidal mudflats and oligotrophic mangrove swamp (Essien et al. 2003). In the Qua Iboe River estuary, the upper facies has a thickness of 5 m, while the lower facies is 10 m thick. Both its upper and lower facies are characterized by moderately sorted, negatively skewed sediments. Very fine-grained, leptokurtic to very leptokurtic sediments typify its upper Facies, and medium sands and mesokurtic sediments typify its lower facies. Table 1, provides the details of tidal current velocities for the Qua Iboe River estuary and Calabar River. Tidal current velocity for other estuaries is not readily available.

Imo River Estuary

The Imo River is located in southeastern Nigeria and flows 150 miles into the Atlantic Ocean. The estuaries of the Imo

Plate 2 Niger Delta landscape

Plate 3 Liquified natural gas terminal within Bonny landscape (www.indypendent.org, 2010)

and Qua Iboe Rivers are underlain by sedimentary formations of Late Tertiary and Holocene ages with deposits of recent alluvium and beach ridge sands occurring along the coast (Plate 6). The Imo estuary is around 40 km wide, and the river has an annual discharge of 4 km^3 with 26,000 hectares of wetland and is characterized by fluvio-lagoonal deposits and littoral sands of the beach ridge complex, including organic silts, clays, and sand. The estuary comprises tidal creeks, small brackish water lagoons, and fringing mangrove swamps. The estuaries are shallow and have tidal amplitudes of about 1.0 m and receive seawater from the Atlantic Ocean (Plate 7).

Escravos and Forcados Estuaries

Escravos and Forcados estuaries (Fig. 7) are river estuaries which are tributaries of the Niger River located in the western Niger Delta flank. The river estuaries are a complex system where seawater is diluted with freshwater and the dilution process takes place at different mixing levels. This results from mixing due to many forcing mechanisms such as tides and tidal currents, waves and tidal waves, and induced motion and runoff (Dyer 1997). The Escravos River is a distributary of the Niger River; it flows through a westerly course and traverses zones of mangrove swamps and coastal sand ridges for 57 km before ending at the Bight of Benin on the Gulf of Guinea, where it flows into the Atlantic Ocean. At the Escravos Bar, the natural passageway is about 4 m deep with the sedimentation being of the order of 2 million m^3 per year in the lee of the breakwaters. The Escravos is linked by a maze of interconnected waterways to the Forcados, Warri, Benin, and Ethiope Rivers.

The Forcados River is a channel in the Niger Delta, in Southern Nigeria. It flows for approximately 198 km and meets the sea at the Bight of Benin. The Forcados River splits from the Niger River at the same point as the Nun River. At Forcados Entrance, the natural depth of the bar at the threshold is about 4.8 m.

> Box 4: Niger Delta Estuaries
> - Niger Delta is a tidal river estuary of fluvial erosion origin and a fully developed Holocene environment covering an area of about 680 km^2.
> - Bonny river system has the largest tidal volume within the Niger Delta and lies in the strand coast.
> - Cross River estuary is the largest along the Gulf of Guinea.
> - Calabar Estuary is a NE–SW-trending tidal river and a major tributary of the Cross River.
> - Qua Iboe River estuary is a meso-tidal estuary with N–S-trending.
> - Estuaries of the Imo and Qua Iboe Rivers are underlain by sedimentary formations of Late Tertiary and Holocene ages.
> - Escravos and Forcados estuaries are river estuaries which are tributaries of the Niger River located in the western Niger Delta flank.

Plate 4 Cross River estuary landscape

Plate 5 Calabar River landscape

Why Ecotourism a Lucrative Option in Niger Delta Region

The Niger Delta could be a major tourist attraction because of its rich cultural, historic, landscape and ecological interest, and vast areas that could take advantage of the growing worldwide demand for sustainable ecotourism at a communal level (Plate 8). The need to sustainably develop and conserve the region's natural resources, with potential to contribute to local development and alleviate the plight of the local communities in the region, therefore lies in the development of a viable ecotourism industry.

The importance of ecotourism is evident, with global spending on ecotourism increasing by 20 % annually, the increase being about 6 times the rate of growth of the general tourism industry, while greater interest in outdoor activities has increased the scope of ecotourism. Ecotourism also minimizes the ecological impact of modern-day tourism since it travels with environmental sensitivity and is

Plate 6 Mobil Qua Iboe terminal (www.oil-spill-info.com)

Table 1 Tidal current velocities in Qua Iboe River estuary and Calabar River between August and September 2006

		Max. surface tidal current velocity (m/s)		Max. bottom tidal current velocity (m/s)	
		Qua Iboe River estuary	Calabar River	Qua Iboe River estuary	Calabar River
Flood Current	Spring	80	46.99	56	32.89
	Mean	50.99	29	35.7	20.3
	Neap	48.78	27	34.15	18.9
Ebb Current	Spring	119.05	72.99	83.33	51.09
	Mean	80	49	56	34.3
	Neap	73.8	42.99	51.66	30.09

Source Antia et al. (2012)

sustainable. It focuses on local culture, wilderness adventures, personal growth, and learning new ways of life (Shapley 2002; Ukpolo et al. 2008).

Ecotourism is an environment-friendly activity that inculcates environmental values and ethics. Ecotourism continues to be a popular option because of its claim to support conservation attempts through market-based mechanisms. It is sustainable tourism, which is nature based and incorporates a desire to minimize negative social and environmental impacts and embrace economic, environmental, social, community, and visitor benefits (Boo 1990; Herath 2002). Its purpose may be to educate the traveler, to provide funds for ecological conservation, to directly benefit the economic development and political empowerment of local communities, or to foster respect for different cultures and for human rights. Ecotourism combines environmental responsibility with the generation of local economic benefits that will have both a development impact and serve as conservation incentives.

Box 5: Factors favoring decision for ecotourism
- Ecotourism is intended as a low-impact and often small-scale alternative to standard commercial tourism.
- Commitment to ecotourism as a trade activity is increasing globally.
- Ecotourism is primarily nature based and immensely contribution to export earnings, poverty alleviation, and employment creation, among others, if well managed.
- Provision of limited development assistance to address microlevel issues such as enterprise development, increased benefit sharing among key actors, etc.
- Existence of complimentary development assistance both internal and external governmental and non-governmental sources, e.g., IUCN and World bank.

Plate 7 Part of Imo River beach landscape

Plate 8 An overview of part of Niger Delta luxurious landscape

- Favorable legislative and policy climate for environmental protection at both local and international level and conventions such as CBD and RAMSAR.

Environmental Assessment Impact of Ecotourism

One of the more fundamental issues surrounding ecotourism is the lack of standards regarding its practice. It should be noted that ecotourism is a multidimensional, complex practice being characterized by a higher degree of risk, novelty, and interaction with culture and natural fragile biological diversity. Hence, in practice, the fragile sites of ecological and cultural significance are being exposed to the threat of degradation by unregulated tourism development and overvisitation. Natural habitat is also being destroyed through the provision of ecotourism goods and services as well as infrastructures development like lodges, private reserve, roads, among others (Boo 1990; Hawkins and Mann 2007).

Non-adherent to the carrying capacity of the ecotourism area through unregulated, nature tourism can damage the environment and corrode local cultures due to pressure on the fragile environment, as a result of litter, pollution, and habitat disruption, among others.

Invasion of officially reserved nature conservation area as a nature's genetic reservoirs due to lack of sufficient fund to manage and protect them might result in loss of biological diversity. Under the pressures of hunting, logging, agriculture, and fishing, forest and marine habitats are being destroyed and some of the wildlife is being driven to extinction.

Solution

Making ecotourism a positive economic and environmental tool requires policies that foster responsible nature tourism development, broad-based and active local participation, and conservation of the natural heritage. This is necessary in order to properly regulate and manage as well as protect the environment and cultural heritage from overbuilding of tourist facilities and influx of populations around fragile ecosystems (Mowforth and Munt 2009).

Raising local awareness about the value of biological resources, increasing local participation in the benefits of biodiversity conservation (through new sources of jobs and incomes), and generating revenues toward conservation of biologically rich areas is also necessary. This will enhance total participatory involvement of all stakeholders in the sustainable ecotourism development of the region.

Identifying and mobilizing funding for potential private nature tourism investments and formulation of fiscal policies to promote nature tourism and to maximize its economic and environmental benefits is indispensable.

Encouraging international exchange of information and know-how about nature tourism opportunities and operations, through technical and management training, to meet the needs and interests of international and domestic nature tourists is necessary for good practices.

There is need to monitor and certify the performance of ecotourism activities toward promoting "green tourism". This entails promoting environmentally responsible tourist operations that conserve energy, recycle waste, and proper adherent to regulated rules and policies guiding the parks and protected areas usage.

In addition, sustainability principles must also be incorporated in all nature-based tourism activities with a detailed pricing policy designed for ecosystem-based services.

Expected Outcome of Ecotourism Strategy for the Niger Delta Region

- The strategy must be designed to suit the current national and international development agenda and priorities without destructive effects.

- There is also a need to transform and present acceptable nature-based tourism products and services to national and international markets.
- Ecotourism development strategy should fall within the new array of 'green' products and services.

Conclusion

The Niger Delta is characterized by semidiurnal tidal regimes with tidal amplitude of about 1.2 m and higher ebb flow velocities than flood flow. The geomorphic unit has 21 major river mouths/tidal inlets that intersect the coast, breaking it up into a series of barrier islands, 16 of which are within the arcuate delta region. This natural delta receives its sediments, which mainly comprise medium to coarse unconsolidated sands, silt, clay, shale, and peat, from the suspended and traction loads of the Niger and Benue Rivers and their tributaries. The salinity intrusion within the Niger Delta depends strongly on the spatial extent of the diurnal tidal range within the region, which diminishes inland (Abam 1999) with more inundation in the western and arcuate delta part of the delta. Niger Delta could be a major tourist attraction because of its rich cultural, historic, landscape, and ecological interest and vast areas that could take advantage of the growing worldwide demand for sustainable ecotourism at the communal level, if well managed. Ecotourism development in the Niger Delta region should therefore be plan to combine environmental responsibility with the generation of local economic benefits that will have both a development impact and serve as conservation incentives. Such plan should be people oriented, encompassing a range of activities such as community-based management of coastal resources and large-scale infrastructure development (ports, industrial and residential parks, etc.) and assist in organizing integrated coastal area management (ICAM) at regional level with detailed plan preparation, implementation, and management process without compromising the socioeconomic development.

References

Abam TKS (1999) Impact of dams on the hydrology of the Niger Delta. Bull Eng Geol Environ 57(3):239–251

Abam TKS (2001) Regional hydrological research perspectives in the Niger Delta. Hydrol Sci 46(1):13–25

Antia VI, Emeka NC, Ntekim EEU, Amah AE (2012) Grain size distribution and flow measurements in qua-Iboe river estuary and calabar tidal river S. E. Nigeria. Eur J Sci Res 67(2):223–239

Asuquo FE (1998) Physicochemical characteristics and anthropogenic pollution of the surface water of Calabar River, Nigeria. Glob J Pure Appl Sci 5:595–600

Awosika LF, Dublin-Green CO, Folorunsho R, (2000) Study of main drainage channels of Victoria and Ikoyi Islands in Lagos Nigeria and their response to tidal and sea level changes. A report for the Coast and Small Island (CSI) Division UNESCO Paris, p 108

Boo E (1990). Ecotourism: the potentials and pitfalls. World Wildlife Fund:Washington, DC

Braide SA, Izonfuo WAL, Adiukwu PU, Chindah AC Obunwo CC (2004) Water quality of miniweja stream a swamp forest stream receiving non-point source waste discharges in Eastern Niger Delta, Nigeria. Scientia Africana, 3(1):1–8

Dagana LB (1985) Hydro geomorphological controls of the mangrove environment in parts of the Rivers State. In: Wilcox BHR and Powell CB (eds) Proceedings of the mangrove ecosystem of the Niger Delta workshop, pp 6–23

Dalrymple RW, Zaitlin BA, Boyd R (1992) Estuarine facies models: conceptual basis and stratigraphic implications. J Sediment Petrol 62:1130–1146

Defra (2002) Future coast. Halcrow, Defra CD-Rom, vol 3

Dyer KR (1997) Estuaries: a physical introduction, 2nd edn. Wiley, Chichester, pp 1–195

Enyenihi UK (1991) The cross River Basin: soil characteristics, geology, climate, hydrology and pollution. International workshop on methodology and quantitative assessment of pollution load of coastal environment (FAO/ UNAP/IOC/WHO), pp 1–17

Essien JP, Itah AY, Edwok SI (2003) Influence of electrical conductivity on microorganisms and rate of crude oil mineralization in Niger Delta ultisol. Global J Pure Appl Sci 9:199–203

Fairbridge RW, (1980) The estuary: its definition and geodynamic cycle. In: Olausson E and Cato I (eds) Chemistry and biogeochemistry of estuaries. Wiley, Chichester, pp 1–35

French P (2006) Managed realignment—the developing story of a comparatively new approach to soft engineering. Estuar Coast Shelf Sci 67(3):409–423

Hawkins DE, Mann S (2007) The World Bank's role in tourism development. Ann Tourism Res 34(2):348–363

Herath G (2002) Research methodologies for planning eco-tourism and nature conservation. Tourism Econ. 8:77–101

Hume TM, Herdendorf CE (1988) A geomorphic classification of estuaries and its application to coastal resource management—a New Zealand example. Ocean Shoreline Manage 11:249–274

Hayes MO (1975) Morphology of sand accumulation in estuaries: an introduction to the symposium. In: Cronin LE (Ed) Estuarine Research vol II Academic Press, New York, pp 3–22

Irving EG (1962) Bonny River entrance (admiralty chart, 3287). The Admiralty, London, p 249

Milliman RE (1991) An experiment designed to maximize the profitability of customer directed post-transaction communications. J Tech Writ Commun (Fall) pp 22–26

Moelry CL (1990) What is tourism? definitions, concept and characteristics. J Tourism Stud 1(1):3–8

Mowforth M, Munt I (2009) Tourism and sustainability: development and new tourism in the Third World, 3rd edn. Routledge, Oxford

NEDECO (1961) The waters of the Niger Delta: Reports of an investigation by NEDECO (Netherlands Engineering consultants). The Hague, pp 1–317

Nwilo PC and Onuoha A (1993). Environmental impact of human activities on the coastal areas of Nigeria. In: Awosika LF, Ibe AC and Schroader P (eds) Coastlines of Western Africa, ASCE, pp 222–233

Ogba CO, Utang PB (2010) Geospatial evaluation of Niger Delta coastal susceptibility to climate change (4039), FIG congress 2010, facing the challenges—Building the Capacity, Sydney, Australia, 11–16 April 2010

Pethick J (1994) Estuaries and wetlands: function and form. Wetland Manage, Thomas Telford, London, pp 75–87

Pritchard DW (1967) What is an estuary: physical view point. In: Lauff GH (ed) Estuaries. American Association for the Advancement of Science, Washington DC, pp 3–5

Scott JS (1966) Report on the fisheries of the Niger Delta special area. Port Harcourt, Niger Delta Development Board, p 109

Seminara G, Zolezzi G, Tubino M, Zanardi D (2001) Downstream and upstream influence in river meandering. Part two planimetric development. J Fluid Mech 438:213–230

Sexton WJ, Murday M (1994) The morphology and sediment character of the coastline of Nigeria-the Niger Delta. J Coast Res 10(4):959–977 Fort Lauderdale (Florida), ISSN 0749-0208

Shapley R. (2002) Tourism: a vehicle for development? In: Sharpley R and Telfer DJ (eds) Tourism and development: concept and issues. Aspects of Tourism Series 5, pp 11–34

Ukpolo UA, Emeke EE, Dimlayi C (2008) Understanding tourism. Nsukka University Press, Nsukka, pp 1–47

Importance of Mangrove Litter Production in the Protection of Atlantic Coastal Forest of Cameroon and Ghana

Sylvie Carole Ondo Ntyam, A. Kojo Armah, Gordon N. Ajonina, Wiafe George, J. K. Adomako, Nyarko Elvis, and Benjamin O. Obiang

Abstract

For this study, litterfall and structural characteristics of mangrove forest in Ghana and Cameroon were monitored from November 2008 to November 2010. The annual fluctuation of litterfall mass and carbon stocks increased with increases in air temperature (Dry season). During the study period, mean annual total litterfall production, mean carbon litterfall stocks and density were, respectively, 3,035 g/m^2, 12,454.15 g/m^2 and 24,500 stems/ha in Ghana and 5,410 g/m^2, 21,441.61 g/m^2 and 32,275 stems/ha in Cameroon. Litterfall biomass in both countries was made up of more than 80 % leaves. It also appeared that the structural development of the mangrove forest was positively related to the production of litterfall in each country, indicating the importance of litterfall productivity in the general growth of mangrove forest.

Keywords

Litterfall • Productivity • Mangrove • Structural characteristics • Protected areas

S. C. O. Ntyam (✉)
Department of Marine and Fisheries Sciences,
University of Ghana, Legon, Ghana
e-mail: sylondocarlo@yahoo.fr; sylondocarlo@gmail.com

S. C. O. Ntyam
Centre for Coastal and Marine Research (CERECOMA/IRAD),
Kribi, Cameroon

A. Kojo Armah · W. George · N. Elvis
Department of Marine and Fisheries Sciences,
University of Ghana, PO BOX LG 99, Legon, Ghana
e-mail: akarmah@yahoo.com

W. George
e-mail: wiafe@gmail.com; wiafe@yahoo.com

N. Elvis
e-mail: enyarko@hotmail.com

G. N. Ajonina
CWCS Coastal Forests and Mangrove Conservation Programme,
BP 54, Mouanko, Littoral Region, Cameroon
e-mail: gnajonina@hotmail.com

J. K. Adomako
Department of Botany, University of Ghana, Legon, Ghana
e-mail: jadomako@ug.edu.gh

B. O. Obiang
CEPFILD Circle of Forest Promotion and Local Initiatives
Development, BP 532, Kribi, Cameroon
e-mail: ondobenjamin@gmail.com; benja_ond@yahoo.fr

Introduction

Litterfall provides a significant contribution towards the coastal food chain, keeping the coastal ecosystems in a dynamic state through the intense biological activity which accompanies its decomposition (Ochieng and Erftemeijer 2002; Raulerson 2004) Litterfall in mangrove ecosystems represents an essential component of the organic production–decomposition cycle and is, in many ways, a fundamental ecosystem process (Adriamalala 2007; Conchedda et al. 2011). The major process by which the nutrient pool of a mangrove ecosystem becomes enriched is the export of decomposable organic matter, mostly in the form of plant litter. Although data are fairly common on the biology and ecology of mangrove plants around the world, there has been no previous work on mangrove litterfall in the West Central African region (World Bank 2004; Lovelock et al. 2005; Spalding et al. 2010).

This investigation aims at sustaining the importance/role of mangrove litter production in the management and stabilization of coastal and marine areas of Ghana and Cameroon within the West and Central African Atlantic coast. More specifically, the objectives were as follows:

S. Diop et al. (eds.), *The Land/Ocean Interactions in the Coastal Zone of West and Central Africa*, Estuaries of the World, DOI: 10.1007/978-3-319-06388-1_11,
© Springer International Publishing Switzerland 2014

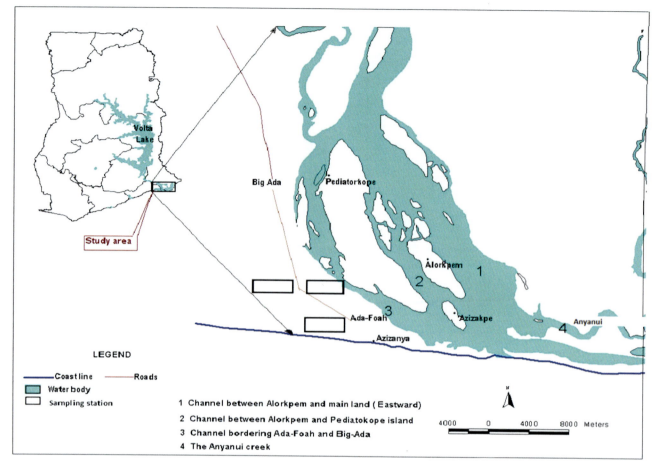

Fig. 1 Map showing the research area at Ada Songor Ramsar site in Volta estuary

1. To assess the distribution and structural characteristics of mangrove forest in the coastal areas of Ghana and Cameroon.
2. To determine mangrove litterfall production of mangrove forest in the coastal areas of Ghana and Cameroon.
3. To assess carbon stocks in mangrove litterfall in the coastal areas of Ghana and Cameroon.

Materials and Methods

Study Sites

Ada Songor Ramsar Site, Ghana

The study area, located at Ada-Foah, Songor Ramsar site (5°45′–6°00 N, 0°25′–0°35′), lies in the Dagme East District of the Greater Accra Region at about 79 km from the national capital, Accra (Fig. 1). It is the second largest Ramsar site along the coast of Ghana and covers an area of 53.33 ha and is the only natural point where the Volta River enters the sea. The mangrove in this area covers 28.740 ha and comprises mainly the Red mangrove (*Rhizophora* sp.) and the white mangrove (*Avicennia* sp.). The climate of the region is controlled by two air masses: the north-east trade winds and the south-west trade winds (UNEP 2007; Spalding et al. 2010); three types of climatic zones can be identified in the region: the humid south with two distinct rainy seasons, the tropical transition zone with two seasons of rainfall very close to each other, and the tropical climate north of lat 9°N, with one rainfall season that peaks in August. Average annual rainfall varies across the basin from approximately 1,600 to 300 mm in the south-eastern section of the basin in Ghana (Aheto et al. 2011).

Douala-Edea Reserve, Cameroon

The study area has been described (Fig. 2) by Ajonina and Usongo (2001). Douala-Edea Reserve (9°31′–10°05′E, 3°14′–3°53′N) is one of the largest and biologically rich mangrove reserve of Cameroon. It is situated within the Kribi-Douala Basin of the coastal Atlantic Ocean and covers a greater part of the coastal plains of the Cameroon Coast (160,000 ha). The area has a very dense hydrological network, being a meeting point of the estuaries of Cameroon's largest rivers (Rivers Sanaga, Nyong, Dibamba and Wouri).

Fig. 2 Map showing the sampling stations at Douala-Edea reserve in Cameroon estuary

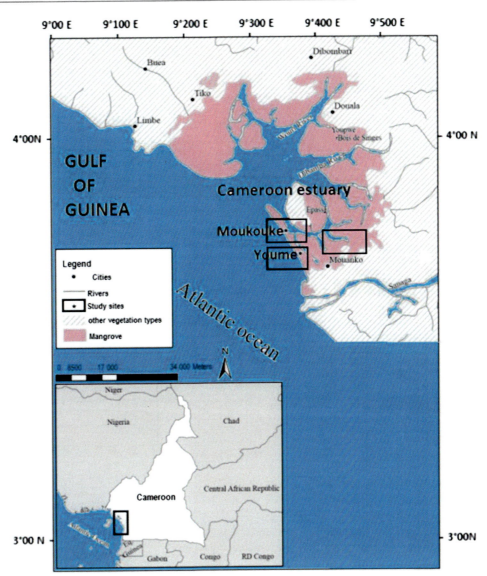

The Reserve is limited in the North by the rivers Wouri and Dibamba; the East by rivers Sanaga, Dipombe and Kwakwa; the South by river Nyong; and the West by the Atlantic Ocean for some 100 km coastline from river Nyong to the Cameroon Estuary. The climate is equatorial type characterized by abundant rain (3,000–4,000 mm) and generally high temperatures with monthly average of 24–29 °C with a dry season spanning from November to March.

Study Site Selection

A preliminary survey was conducted in all of the existing mangrove ecosystems in the Ada estuary complex in the Greater Accra region in Ghana and in the Cameroon estuary (Douala-Edea Reserve) in Cameroon. These zones in both countries are all protected areas and among the largest reserves, with a relatively high mangrove cover (Spalding et al. 2010; Ramsar-MAVA-UNEP 2012). The mangrove ecosystems were evaluated and identified based upon their main plant species (*Rhizophora racemosa*, *Avicennia germinans*) and their stability (very little degradation). Three study sites were then identified, based on the type of mangrove ecosystem vegetation: (1) pure red mangroves (*R. racemosa*); (2) pure white mangroves (*A. germinans*); and (3) mixed stand (red and white mangroves). These two reserves in both countries were selected for investigation because the mangrove ecosystems are relatively well conserved.

Selection and layout of plots

Combinations of sampling approaches were used to achieve a nested design. Targeted sampling (TS) method was used to select areas of species agglomerations where three study sites, representing the various species agglomerations (stands of pure *Rhizophora*, pure *Avicennia* and a mixture of them in equal proportion), were retained (Fig. 3). This was followed by a three-stage sampling approach to subdivide the plots

Fig. 3 Schematic sampling design showing the methodology used for the study (Target sampling followed by a point-centered quarter method one plot (20 × 20 m) subdivided in four subplots 4 (10 × 10 m) and each subplot in hundred quadrats 100 (1 × 1 m); inventory of seedlings in 100 (1 × 1 m) quadrats and saplings and trees in 4 subplots (10 × 10 m) in each mangrove stand (*Rhizophora*, *Avicennia*, mixed). After inventory, collection of litterfall has been made once a month in the morning in 4 subplots (10 × 10 m) in each mangrove stand, from November 2008 to November 2010 in both countries)

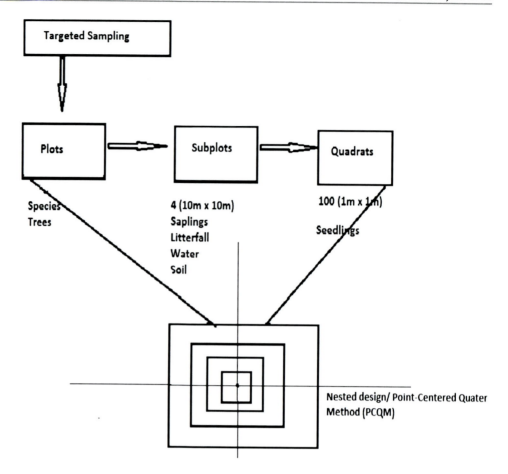

into a desired measurement level corresponding to the point-centred quarter method (PCQM) revised and described by Dahdouh-Guebas and Koedam (2006). At each site, each of the plots (20 × 20 m = 400 m²) corresponding to one species or mixed was marked out. Each plot was further divided into four 10 × 10 m (100 m²) subplots and each subplot into a hundred 1 × 1 m sampling quadrants, and forty of them were randomly selected for some measurements (Fig. 4). Therefore, for this research in each country, a total of 3 plots and 12 subplots covering an area of 0.24 ha were established. Plots and subplots were then used for some of the parameters as tree inventory, litterfall production and water and soil measurements (Dahdouh-Guebas and Koedam 2006; Armah et al. 2009; Spalding et al. 2010). After selecting and laying out the plots, they were monitored for 26 months, selecting ecological processes from November 2008 to November 2010 (Ghana and Cameroon).

Measurement of Tree and Stand Parameters for Forest Structure

Plots were designed to inventory trees, and subplots were used to inventory saplings and seedlings (Fig. 5). In each plot, all individual trees with height greater than 4 m were identified, counted and measured for diameter at breast height (DBH) and height. In the subplots, all saplings with height between 1 and 4 m and seedlings below 1 m were counted and measured for DBH and height (English et al. 1997; Dahdouh-Guebas 2011). During the inventory, red mangrove, the most abundant species, with prop-roots and multiple stems, had their DBH measured above the highest

Fig. 4 Materialization of the sampling design in the field at Douala-Edea reserve, Cameroon (photo Carole Ntyam 2010)

Fig. 5 Measurement of sapling in *Rhizophora* (a) stand in Ghana, Ada Songor Ramsar site and measurement of tree in *Rhizophora* stand in Cameroon, Douala-Edea reserve, Cameroon (b) (photo Carole Ntyam 2008 in Ghana and 2010 in Cameroon)

prop-root where the root no longer influences the diameter of the stem. The diameter and height of each tree were, respectively, recorded by measuring the circumference of the tree using a flexible tape measure and by using clinometers (Kathiresan 1997 ; English et al. 1997; Armah et al. 2009). The diameter of each sapling or seedling was recorded using calipers. Some sapling height was recorded using a clinometer. The abundance of seedlings and saplings was recorded in forty 2 × 2 m sampling plots, ten of them selected randomly inside each of the four 100 m² (10 × 10 m) vegetation subplots (Kathiresan; English et al. 1997). Within each subplot, the number of seedlings and saplings were taken as a measure of regeneration (Chen et al. 2009; Din et al. 2008; Nfotabong Atheull 2011; Aheto et al. 2011).

Litterfall Collection

Litterfall was collected in 1 m² litter traps constructed of fibreglass screening (1 mm mesh) and installed at about 1 m, above ground to prevent loss from flooding (Fig. 6). At each study site (20 × 20 m), four traps were placed, one in the centre of each of the four 10 × 10 m vegetation subplots. Litter collection was carried out on a monthly basis for a period of 26 months starting from November 2008 to November 2010, in Ghana and Cameroon using the method described by Odum and Heald (1975). At each period, all loose material of recognizable identification was collected within the traps (Nfotabong Atheull 2008; Kairo and Bosire 2009; Conchedda et al. 2011; Twilley and Day 1999).

Fig. 6 Collection of litterfall in pure *Rhizophora* stands (photograph Carole Ntyam 2010 Cameroon)

Data Analysis

Laboratory Analysis of Litterfall Components

Field sample analyses in Ghana were mainly done in Ghana Atomic Energy Commission (GAEC) at the Chemistry Department and in the University of Ghana (Soil Department and Ecology Laboratory Centre). In Cameroon, laboratory facilities of the IRAD institute and University of Dschang were used for analysing the litter samples.

Litterfall collected from each subplot was set in a 60 °C oven to dry for at least 24 h until a constant mass was reached and then separated into fractions (leaves, flowers, fruits, seeds and twigs) and weighed to determine the mass remaining (Bosire et al. 2005; Arreola-Liźarraga 2004; Snowdon and Raison 2005). Litter samples were ground into fine powder and sub-samples collected for chemical analysis after thorough mixing (Chiambeng 1989; Arreola-Liźarraga 2004; Raulerson 2004; Spalding et al. 2010). The leaf litter was analysed for extractable N, P, K, Mg and Ca and total N, P and C. The determination of nutrients was done mainly using a spectrophotometer (Din et al. 2008; Armah et al. 2009).

To determine the carbon pool of litterfall (above-ground components), the method determined by Kauffman and Donato (2012) was used. Carbon pools of above-ground biomass were then assessed by multiplying the biomass of individual components by their specific carbon concentration (percentage). Carbon concentrations of above-ground biomass ranging from 0.46 to 0.5 were considered (Kauffman et al. 2011).

Stand Structural Analysis

Forest structural classification was done based on tree diameter measurements at breast height (DBH) into eight tree size classes as: <1 cm seedlings, ≥1 to <3 cm small saplings, ≥3 to <5 cm medium-sized saplings, ≥5 to <7 cm large saplings, ≥7 to <10 cm small trees (posts), ≥10 to <30 cm medium-size trees (poles), ≥30 to <50 cm large trees (standards), and ≥50+ cm giant trees (veterans) (Ajonina 2008). Secondary tree and stand parameters were estimated using standard forest inventory and mensuration procedures (Loetsch et al. 1973; Hellier 1988; Husch et al. 2003) also with adaptations to mangrove forests (Ajonina 2008; Alongi 2011; Aheto et al. 2011).

The tree basal area is:

$$g = \pi/4D^2 \qquad (1)$$

and the tree volume:

$$V = 0.6\,gh. \qquad (2)$$

Further data processing was done of mangrove vegetation and structure, frequency, density, basal area, average diameter, average height and importance values for each mangrove species (Arreola-Liźarraga et al. 2004), and the complexity index was calculated for each site (Hossain et al. 2008; Day and Machado 1986; Arreola-Liźarraga et al. 2004). Stand parameters were obtained by summation and conversion of tree parameters to hectare estimates. These characteristics were calculated using the methods and formula worked out by Kathiresan (1997) to study mangroves:

Density is measured species-wise and total in each plot as follows:
- Density of each species (no/ha) = no. × 10,000 m^2/area of plot in m^2.

- Total density of all species = sum of all species densities.
- Basal area is measured species-wise and total in each plot as follows:
  - Basal area (m^2) of each species = 0.005 × DBH.
  - Total basal area of all species (m^2/ha) = sum of all species basal area/area of plot in m^2 × 10,000 m^2 individuals of all species × 100.

All statistical tests were performed using SPSS version 18 (Statistical Package for Social Sciences).

Results and Discussion

Environmental/Climatic Characteristics

The mean air temperature varied between 26 and 30.1 °C in Ghana and 26.80 and 29.5 °C in Cameroon, with higher values from January to March. In general, most rainfall occurred between July and October for Cameroon and May and June for Ghana. The lower values were between December and February in Ghana and Cameroon. The mean rainfall varied between 50 and 300 mm in Ghana and 100 and 620 mm in Cameroon, with higher values in August and September in Cameroon and May and June in Ghana (Fig. 7).

Mangrove Stand Structure

Distribution of Mangrove Species

Pure *Avicennia* Stand in Ghana
Analysis of the size class distribution of the trees showed very high density of seedlings in Ghana and none in Cameroon. Saplings on the other hand were found in Cameroon, but not in Ghana. There were no small trees (class 5) in either country; from medium to giant trees, the density was relatively low in both countries.

It appeared that there had been high natural recruitment into the lower diameter classes since these stands of *Avicennia* (Fig. 8). The relatively high sapling and seedling density under the canopy, respectively, in Cameroon and Ghana implies great natural regeneration capacity of the stands.

Pure *Rhizophora* Stand
In Ghana and Cameroon, analysis of the size class distribution of the trees showed very few individuals in the upper diameter classes (from class 5 to 8) and a preponderance of individuals in the lower classes (Fig. 9). It appeared that pure *Rhizophora* stands had a lot more saplings (small, medium and large), followed by seedlings.

Fig. 7 Climatic variables in Cameroon and Ghana studied sites

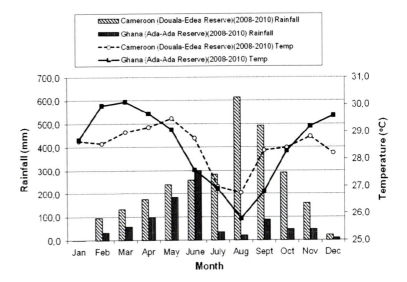

Fig. 8 Distribution of *Avicennia* species by class of diameter.
1 = <1 cm; 2 = ≥1 to <3 cm;
3 = ≥3 to <5 cm;
4 = ≥5 to <7 cm;
5 = ≥7 <10 cm;
6 = ≥10 to <30 cm;
7 = ≥30 to <50 cm;
8 = ≥50+ cm

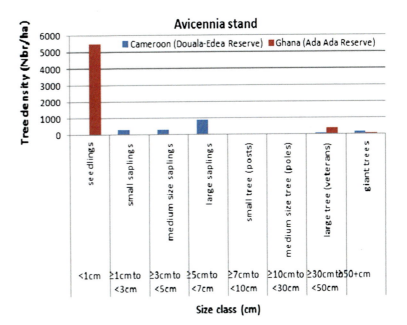

It is evident that there has been high natural recruitment into the lower diameter classes since the establishment of these stands of *Rhizophora*. The relatively high sapling and seedling density under the canopy implied great natural regeneration capacity of the stands. In both countries, we also observed that the densities of species are very high in Cameroon, compared with Ghana.

Rhizophora and *Avicennia* in Balanced Mixed Stand

Analysis of the size class distribution of the trees showed that Cameroon has the highest value of seedling density, compare with Ghana. Ghana on the other hand has a very high density of saplings, compared with Cameroon, with only a very low density of small saplings. Most of the trees (small to giant), were observed to be sparsely distributed in the upper diameter classes (class 5–8) in Cameroon, while none of them were seen in Ghana (Fig. 10). It was shown that there had been high natural recruitment into the lower diameter classes in these mixed stands in both countries (Fig. 10). The relatively high sapling and seedling density under the canopy, respectively, in Ghana, and Cameroon implied great natural regeneration capacity of the stands, then relatively well conserved.

In the mangroves of both countries, the number of saplings, seedlings and trees ranged, respectively, from 200 to 17,100, 5,500–7,100 and 175–400 stems/ha in Cameroon, and 7,600–23,600, 1,900–5,500 and 225–575 stems /ha in Ghana (Figs. 11, 12, 13). The relatively highest number of

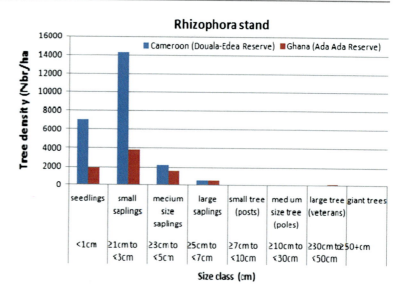

Fig. 9 Distribution of Rhizophora species by class of diameter. 1 = <1 cm; 2 = ≥1 to <3 cm; 3 = ≥3 to <5 cm; 4 = ≥5 to <7 cm; 5 = ≥7 to <10 cm; 6 = ≥10 to <30 cm; 7 = ≥30 to <50 cm; 8 = ≥50+ cm

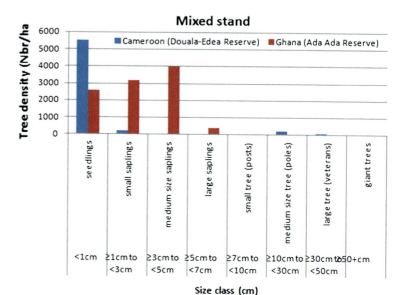

Fig. 10 Distribution of mixed stands species by class of diameter. 1 = <1 cm; 2 = ≥1 to <3 cm; 3 = ≥3 to <5 cm; 4 = ≥5 to <7 cm; 5 = ≥7 to <10 cm; 6 = ≥10 to <30 cm; 7 = ≥30 to <50 cm; 8 = ≥50+ cm

seedlings and saplings in Cameroon and Ghana can attributed to the lower canopy height of mangrove.

Structural Characteristics of Mangrove Forest

Pure *Rhizophora* Stand

The data about structural attributes of *Rhizophora* stands indicated that the forest structure, shown mainly by the mean values of density, basal area, volume and dominant height in both countries, was, respectively, 46.7 m, 24 375 stems/ha, 37.35 m²/ha and 836.18 m³ in Cameroon and 14 m, 8,225 stems/ha, 44.21 m²/ha and 279.66 m³ in Ghana. It appeared that, in Fig. 11, the results confirm that *Rhizophora* in Cameroon is significantly larger in terms of mean height (46.7 m), mean density (24,375 stems/ha) and mean volume (836.18 m³). *Rhizophora* in Ghana is significantly larger in terms of mean basal area (44.21 m²/ha).

From this study, it has been shown that Cameroon has the mean highest *Rhizophora* density with 24,375 stems/ha. Before this zone had been designated as protected area, large mangrove trees were cut as construction materials and firewood for bakeries that resulted in denser trees of smaller size (BA = 37.35 m²/ha). Ghana has a lower density (8,225 stems/ha) of smaller size (BA = 44.21 m²/ha). This may be is due to less utilization of large mangrove trees since the economic activity in this area was charcoal making, which utilizes the smaller sizes of mangrove trees (Meentemeyer 1982; Nfotabong Atheull 2008, 2011; Spalding et al. 2010; Aheto 2011). The highest average canopy height was in Cameroon (46.7 m), while the lowest was in Ghana (14 m).

Importance of Mangrove Litter Production

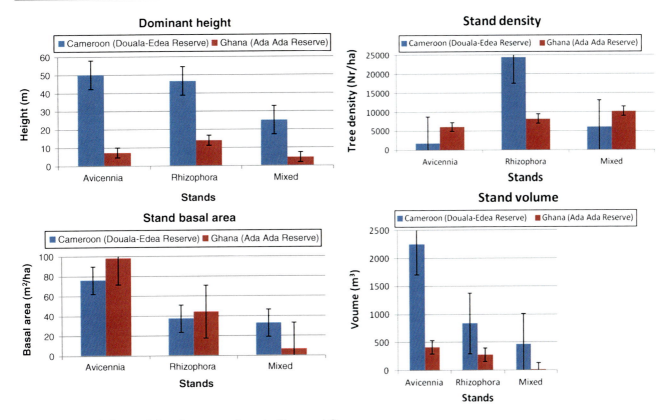

Fig. 11 Structural characteristics of mangrove forest in Ghana and Cameroon

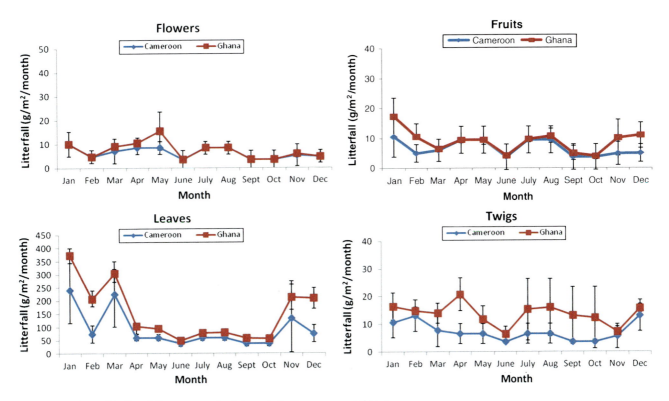

Fig. 12 Total monthly litterfall component in *Avicennia* in Cameroon and Ghana

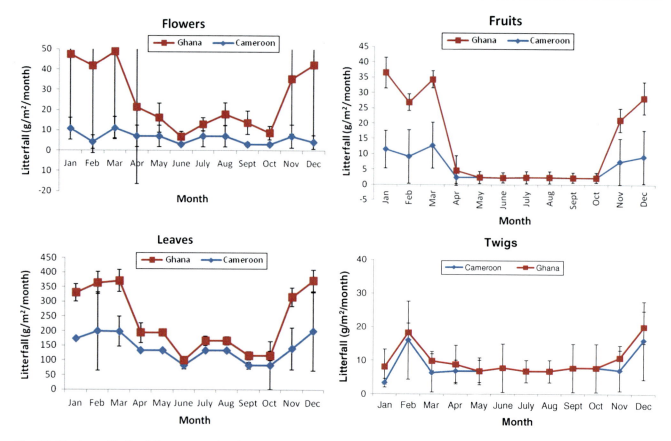

Fig. 13 Total monthly litterfall component in *Rhizophora* in Cameroon and Ghana

This can be attributed to less pressure to utilize the trees due to the vast mangrove area and relative small human population compared with the mangrove area in Cameroon. Although *Rhizophora* stands in Ghana had the largest basal area (44.21 m^2/ha), it was noted that the average height was only 14 m. This may be due to cutting branches of large trees for charcoal making or drying of fish, rather than cutting down the trees (Wattayakorn et al. 1990; Hegazy 1998; Nfotabong Atheull 2011; Worlanyo Aheto 2011).

Pure *Avicennia* Stand

Considering the structural characteristics of *Avicennia* stands showed that the forest structure, presented mainly by the mean values of density, basal area, volume and dominant height in both countries, was, respectively, 50.1 m, 1,800 stems/ha, 76.07 m^2/ha and 2,246.87 m^3 in Cameroon, and 7.3 m, 6,075 stems/ha, 98.11 m^2/ha and 412.82 m^3 in Ghana (Fig. 11).

From this study, it has been shown that Ghana has the mean highest *Avicennia* density, with 6,075 stems/ha and Cameroon has lower density (1,800 stems/ha) of smaller size (Ba = 44.21 m^2/ha). Before this zone has been designated as protected area, large mangrove trees were cut as construction materials and firewood for bakeries that resulted in denser trees of smaller size (BA = 76.07 m^2/ha). This is maybe due to less utilization of large mangrove trees since the economic activity in this area was charcoal making and firewood selling, which utilized the smaller size of mangrove trees. The highest average canopy height was in Cameroon (50.1 m), while the lowest was in Ghana (7.3 m).

This can be attributed to less pressure to utilize the trees due to the vast mangrove area and relatively small human population compared with the mangrove area in Cameroon. Although *Avicennia* stands in Ghana had the highest basal area (98.11 m^2/ha), it was noted that the average height was only 7.3 m. This may be due to cutting branches of large trees for charcoal making or drying of fish, rather than cutting down the trees (Cintrón and Schaeffer-Novelli 1984; Nfotabong Atheull 2008, 2011; Spalding et al. 2010; Worlanyo Aheto 2011).

Rhizophora and *Avicennia* in Balanced Mixed Stands

In mixed stands, the mean values of density, basal area, volume and dominant height in the countries was, respectively, 25.1 m, 6,100 stems/ha, 32.76 m^2/ha and 466.22 m^3 in Cameroon and 4.7 m, 10,200 stems/ha, 6.62 m^2/ha and 14.78 m^3 in Ghana (Fig. 14). It appeared (Figs. 11, 12, and

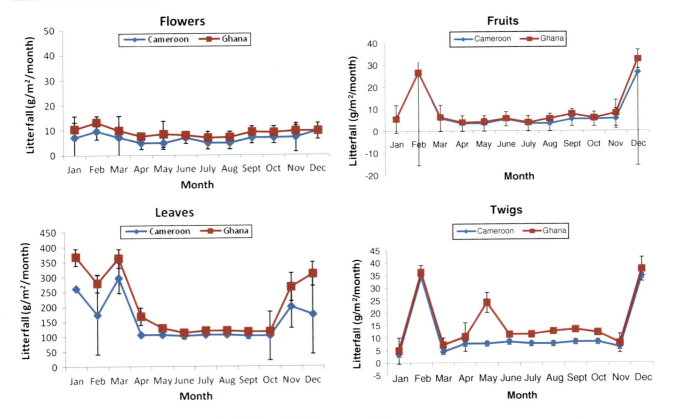

Fig. 14 Total monthly litterfall component in mixed stands in Cameroon and Ghana

13) that mixed stands is significantly larger in terms of mean height (25.1 m), average basal area (32.76 m^2/ha) and mean volume (466.22 m^3) in Cameroon and density (10,200 stems/ha) in Ghana.

From this research, it has been observed that Ghana has the mean highest species density, with 10,200 stems/ha (Figs. 11, 12, and 13). Cameroon has lower density (6,100 stems/ha) of smaller size with a larger basal area (BA = 32.76 m^2/ha). Before this zone has been designated as protected area, large mangrove trees were cut as construction materials and firewood for bakeries that resulted in denser trees of smaller size (BA = 76.07 m^2/ha). This may be due to less utilization of large mangrove trees since the economic activities in this area were charcoal making and firewood selling, which utilized the smaller size of mangrove trees (Adjonina 2008; Chen et al. 2009; Nfotabong Atheull 2008, 2011; Aheto 2011; Dahdouh-Guebas 2011).

The highest average canopy height was in Cameroon (25.1 m), the lowest in Ghana (4.7 m) (Figs. 11, 12, and 13). This can be attributed to less pressure to utilize the trees due to the vast mangrove area and relatively small human population compared with the mangrove area in Cameroon. The mixed stands in Cameroon had the highest basal area (32.76 m^2/ha), with a high average height of only 25.1 m showing that mangrove species in mixed stands were more or less well conserved.

Mangrove Litterfall Production

Litter Composition

A substantial portion of mangrove products returns to the environment in the form of litterfall. This is an important source of organic detritus, which supports important detrital marine food webs (Odum and Heald 1975; FAO 2007; Conchedda et al. 2011). It is a direct food source for various herbivore crustacean and molluscs. Table 1 and Fig. 14 show the monthly dry weight of litterfall in the mangrove forests of Cameroon and Ghana. Litterfall rate is seasonal, generally being low in the rainy season and high in the dry season (Nfotabong Atheull 2011; Worlanyo Aheto 2011). In Ghana, the total dry weight of litterfall was 278 g/m^2/year (9.15 %) of flowers, 143 g/m^2/year (4.71 %) of fruits/seeds, 2,467 g/m^2/year (81.28 %) of leaves and 147 g/m^2/year (4.84 %) of twigs. In Cameroon, the total dry weights for flowers, fruits/seeds, leaves and twigs were, respectively, 231 g/m^2/year (4.26 %), 247 g/m^2/year (4.56 %), 4,608 g/m^2/year (85.17 %) and 324 g/m^2/year (5.98 %).

Total Annual Litterfall

For Ghana sites, the total annual litterfall was 3,035 g/m^2/year or 30.3 t/ha (average value: 2.5 t/ha), consisting of *Avicennia* (27.47 %), *Rhizophora* (48.66 %) and mixed (23.85 %). Leaf litterfall comprised the largest component

Table 1 Total annual litterfall (g/m^2)

Mangrove forest stand composition	Litter component	Total annual production (g/m^2)	Mean monthly production (g/m^2)	SE	CV
Avicennia	Flowers	76	6.3	0.7	37.7
	Fruits/seeds	80	6.7	0.8	40.7
	Leaves	1,097	91.4	20.3	77.1
	Twigs	86	7.2	1.0	46.7
	Total	1,339	111.6	21.3	66.2
Rhizophora	Flowers	75	6.2	0.8	45.8
	Fruits/seeds	67	5.5	1.2	74.8
	Leaves	1,685	140.4	12.8	31.6
	Twigs	99	8.2	1.1	46.1
	Total	1,926	16.5	14.7	31.8
Mixed	Flowers	80	6.7	0.5	24.7
	Fruits/seeds	100	8.3	2.5	102.9
	Leaves	1,826	152.2	20.0	45.4
	Twigs	138	11.5	3.2	94.8
	Total	2,145	178.8	21.5	41.7
Avicennia	Flowers	12	1.0	0.6	213.9
	Fruits/seeds	28	2.3	0.8	117.8
	Leaves	718	59.8	14.4	83.2
	Twigs	77	6.4	1.1	60.4
	Total	834	69.5	14.4	71.8
Rhizophora	Flowers	238	19.8	4.2	72.6
	Fruits/seeds	99	8.3	2.9	123.0
	Leaves	1,119	93.3	19.7	73.0
	Twigs	20	1.7	0.5	112.7
	Total	1,477	123.1	27.0	76.1
Mixed	Flowers	28	2.3	0.3	44.5
	Fruits/seeds	16	1.3	0.5	129.2
	Leaves	630	52.5	12.6	83.1
	Twigs	50	4.1	1.2	98.3
	Total	724	60.3	12.4	71.0

(81.28 %) (Table 1). For Cameroon sites, the total annual litterfall was 5,410 g/m^2/year or 54.1 t/ha (average value: 4.5 t/ha) made up by *Avicennia* (24.75 %), *Rhizophora* (35.60 %) and mixed (39.64 %) (Fig. 14). Leaf litterfall comprised the largest component (85.17 %) (Table 1). It is apparent from Figs. 12, 13 and 14 that the total litterfall of those two countries within the West and Central African ecoregion was 8,445 g/m^2/year or 84.4 t/ha (average value: 7 t/ha), with a high percentage of leaf litterfall (83.77 %).

Litterfall values for mangrove forests worldwide range from 2 to 16 t/ha/year (Day and Machado 1986; Spalding et al. 2010); thus the values of litterfall from the mangrove sites of Ghana (average annual production: 2.5 t/ha/year) and those of Cameroon (average annual production: 4.5 t/ha/

year), actually fall below the minimum of this range and are a little bit higher than those of a riverine mangrove wetland (1.7 t/ha/year) reported by Arreola-Lizárraga et al. (2004).

Pattern of Total Monthly Litterfall

Figure 14 show that the total value of mangrove leaves in Cameroon was higher than that in Ghana. However, the response to environmental conditions was similar in both countries, since all tend to shed more litter during dry season than during the rainy season (Conchedda et al. 2011). Although data were not presented, it was appeared that low values of total leaf litter of studied mangroves in Ghana were observed in June, and in Cameroon from June to September and October for *Avicennia* and *Rhizophora*,

Table 2 Carbon concentration of litter in mangrove protected areas of Cameroon and Ghana

Country	Mangrove forest stand composition	Jan	Feb	Mar	Apr	May	June	July	Aug	Sept	Oct	Nov	Dec	Mean	SE	CV (%)
Cameroon	*Avicennia*	47.7	47.8	47.8	47.7	47.7	47.8	47.7	47.7	47.8	47.8	47.8	47.8	47.8	0.0	0.1
	Rhizophora	47.3	47.3	47.3	47.2	47.2	47.2	47.2	47.2	47.2	47.2	47.3	47.3	47.2	0.0	0.1
	Mixed	47.7	47.7	47.7	47.5	47.5	48.0	47.5	47.5	48.0	48.0	47.8	47.7	47.7	0.1	0.4
Ghana	*Avicennia*	49.1	49.0	49.5	49.6	48.7	48.7	49.4	49.4	48.6	48.6	49.0	49.1	49.0	0.1	0.7
	Rhizophora	49.3	49.3	49.1	48.9	49.2	49.2	49.7	49.7	49.6	49.6	49.5	49.2	49.4	0.1	0.5
	Mixed	49.5	49.0	48.9	49.8	49.2	48.9	48.5	48.5	49.2	49.2	49.4	49.8	49.2	0.1	0.8

and from April to October in mixed mangrove. This may be attributed to the change from high-shedding rates to high-leafing rate. This is further exemplified by the lower quantity of stipules in this period (Spalding et al. 2010; Dahdouh-Guebas 2011).

According to the rainfall variations in Ghana and Cameroon, it appeared that the leaves and twigs/branches litterfall of *Avicennia*, *Rhizophora* and mixed stands increased with the decrease in rainfall (Fig. 14), and obviously in the dry season (Kairo and Bosire 2009; Spalding et al. 2010). This can be explained by the fact that interstitial water salinity, which is relatively high in the dry season, increases stress in mangroves, resulting in increased leaf and branch loss by the mangrove trees, indicative of their adaptive measures to reduce water loss (Spalding et al. 2010; Dahdouh-Guebas 2011; Ramsar-Mava-Unep 2012).

Leaf production was found to be continuous throughout the study period, which suggests that environmental conditions are favourable for leaf emergence all year round, and the stress does not appear to limit leaf production. Similar results have been reported by Aheto (2011), Spalding et al. (2010), Crona et al. (2009), and Dahdouh-Guebas (2011). Seasonal fluctuations have been found in the litterfall of several mangrove species, notably of the genera *Avicennia* and *Rhizophora* (Spalding et al. 2010). Litterfall can also be observed throughout the year with little (Lovelock et al. 2005; Conchedda et al. 2011) or marked (Day and Machados 1986; Wattayakorn et al. 1990; Bosire et al. 2006; Andriamalala 2007; Egnankou Wadja 2009) seasonal variations. The major flowering and fruiting seasons of *R. racemosa* and *A. germinans* in Ghana and Cameroon were mainly in the dry season (except for fruits in *Avicennia* Ghana and Cameroon, and flowers in mixed mangroves in Cameroon were highest in the dry season). The trend of total litterfall products followed the changes of flower and fruit litterfall products (Table 1). It was observed that total litterfall decreased in the rainy season. A general trend of litterfall peaks occurring during the dry season has been reported in a number of mangrove studies (Ochieng and Erftemeijer 2002; Lovelock et al. 2005).

Carbon Stocks in Total Litterfall in Cameroon and Ghana

Carbon pools in litterfall were determined mainly using the Kauffman and Donato (2012) method, where carbon concentrations (ranging from 46 to 50 %) are multiplied by the biomass of litterfall in both countries (Table 2). It appeared that in Ghana sites the total mean carbon stock was 3,410.98 g/m^2 in *Avicennia*, 6,067.33 g/m^2 in *Rhizophora* and 2,975.84 g/m^2 in mixed stands; while in Cameroon the carbon stocks were 5,329.73 g/m^2 in Avicennia, 7,582.52 g/m^2 in *Rhizophora* and 8,529.36 g/m^2 in mixed stands. It was also clearly shown in Fig. 15 that the highest peak of carbon stock mainly appeared in the dry season. From the results above, it was shown that the period (dry season) or country (Cameroon) with high total mean biomass of litterfall also showed high-carbon content.

Globally, according to the rainfall variation in Ghana and Cameroon, it appeared that the leaves and twigs/branches litterfall of *Avicennia*, *Rhizophora* and mixed stands increased with decreasing rainfall, then obviously in the dry season (Figs. 3, 7). This can be explained by the fact that interstitial water salinity, which is relatively high in the dry season, increases stress in mangroves, resulting in increased leaf and branches loss by the mangrove trees, which is indicative of their adaptive measures to reduce water loss. An analysis of variance (ANOVA) of vegetative and reproductive litterfall showed very significant differences ($p \leq 0.0001$) in twigs/branches, leaves and flowers in Ghana and Cameroon between sites (Lovelock et al. 2005; Spalding et al. 2010; Alongi 2011; Dahdouh-Guebas 2011).

The present study also present higher total mean litterfall, trees densities and carbon stocks in Cameroon sites compare with those from Ghana with relatively lower values. The structural development of the mangrove forest was then found to be positively related to the production of litterfall in each country. Similar results were found in the West and Central African Ecoregion (Baba et al. 2004; Bosire et al. 2006; Ajonina 2008; Nfotabong Atheull 2008, 2011; Worlanyo Aheto 2011; Kauffman and Donato 2012).

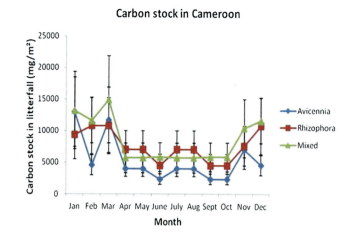
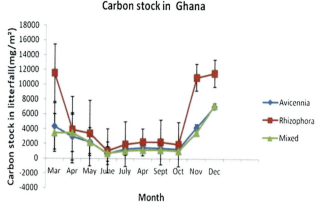

Fig. 15 Carbon stocks in litterfall in Cameroon and Ghana

Conclusion

Mangrove forests form the interface between marine and terrestrial environments. They are also recognized as essential nursery habitat for a diverse community of fish, which find protection and abundant food in these environments, especially during their early stages (FAO 2007). Mangroves litterfall are useful contributors of nutrient mass in a mangrove environment and contain sufficient amounts of minerals, vitamins and amino acids, which are essential for the growth and nourishment of marine organisms and livestock (Ajonina 2008; Egnankou Wadja 2009; Spalding et al. 2010).

The quantitative findings from the present study indicate in both countries that: (1) the major leaf litterfall, flowering and fruiting of *R. racemosa* and *A. germinans* and mixed stands were mainly in the dry season. (2) Leaf production was continuous throughout the study period. (3) The mean annual total litterfall and the carbon stocks of mean total litterfall were, respectively, higher in Cameroon (5,410 and 21,441.61 g/m^2) than in Ghana (3,035 and 12,454.15 g/m^2). (4) The mangrove forest in Cameroon had a higher tree density (32,275 stems/ha) than in Ghana (24,500 stems/ha). (5) The highest peak of carbon stocks mainly appeared in the dry season in both countries. It was also shown that, Cameroon with high total mean biomass of litterfall showed also high-carbon content. (6) The mangrove forest in Cameroon seemed to be more developed and productive.

From the results it appeared that, the litter fall production was strongly correlated with forest structure parameters, such as DBH, tree height, density and basal area. As forest structural characteristics decreased, litterfall production and carbon stocks also decreased.

The findings of this study demonstrate as well, that relative high production values of mangrove litterfall production in the coastal zones of Ghana and Cameroon continue to increase, as well as structural parameters and carbon stock, coastal and marine resources will continue to have an essential nursery habitat. The coastline will also be more protected and stabilized. Therefore, there is a need to step up sustainable management of this vital mangrove ecosystem. This could be achieved by participatory management approach where all stakeholders, especially the local communities are involved in the coastal and marine areas of Ghana and Cameroon.

References

Aheto DW Ama A, Aduomih O, Obodai EA (2011) Structural parameters and above-ground biomass of mangrove tree species around the Kakum river estuary of Ghana. Department of Fisheries and Aquatic Sciences, School of Biological Sciences University of Cape Coast (UCC), Ghana, 11 p

Ajonina GN (2008) Inventory and modelling mangrove forest stand dynamics following different levels of wood exploitation pressures in the Douala-Edea Atlantic coast of Cameroon, Central Africa. Mitteilungen der Abteilungen für Forstliche Biometrie, Albert-Ludwigs-Universität Freiburg, 2, 215 p

Ajonina G, Usongo L (2001) Preliminary quantitative impact assessment of wood extraction on the mangroves of Douala-Edea forest reserve, Cameroon. Trop Biodivers 7(2–3):137–149

Alongi DM (2011) Carbon payments for mangrove conservation: ecosystem constraints and uncertainties of sequestration potential. Environ Sci Policy 14:462–470

Andriamalala CAJ (2007) Etude écologique pour la gestion des mangroves à Madagascar: comparaison d'une mangrove littorale et d'estuaire à l'aide de la télédétection. Thèse de doctorat PHD, Université de Bâle, Suisse, 280 p

Armah AK, Adomako JK, Agyeman DY, Agyekumhene A (2009) Reforested mangrove evaluation: a case study from Ada, Ghana. Resource and Environmental Development Organisation (REDDR 17O)

Arreola-Liźarraga JA, Flores-Verdugo FJ, Ortega-Rubio A (2004) Structure and litterfall of an arid mangrove stand on the Gulf of California, Mexico. Aquat Bot 79:137–143

Baba S, Gordon C, Kainuma M, Aviyor JS, Dahdouh-Guebas F (2004) The global mangrove database and information system (GLOMIS): present status and future trends. In: Proceedings of 'the colour of ocean data' symposium, Bruxelles, 25–27 Nov 2002 Edited by: Vanden Berghe E, Brown M, Costello M, Heip C, Levitus S, Pissierssens P. IOC Workshop Reports 188, UNESCO/IOC, Paris

Bosire JO, Dahdouh-Guebas F, Kairo JG, Kazungu J, Dehairs F, Koedam N (2005) Litter degradation and CN dynamics in reforested mangrove plantations at Gazi Bay, Kenya. Biol Conserv 126:287–296

Bosire JO, Dahdouh-Guebas F, Kairo JG, Wartel S, Kazungu J, Koedam N (2006) Success rates and recruited tree species and their contribution to the structural development of reforested mangrove stands. Mar Ecol Prog Ser 325:85–91

Chen L, Wang W, Zhang Y, Lin G (2009) Recent progresses in mangrove conservation, restoration and research in China. J Plant Ecol 2:45–54

Cintrón G, Schaeffer-Novelli Y (1984) Methods for studying mangrove structure of latin America. In: Snedaker SC, Snedaker JG (eds) The Mangrove ecosystem: research methods. Kluwer Academic Publishers

Conchedda G, Lambin EF, Mayaux P (2011) Between land and sea: livelihoods and environmental changes in mangrove ecosystems of Senegal. Ann Assoc Am Geogr 101:1259–1284 10.1080/00045608.2011.579534

Crona BI, Rönnbäck P, Jiddawi N, Ochiewo J, Maghimbi S, Bandeira S (2009) Murky water: analyzing risk perception and stakeholder vulnerability related to sewage impacts in mangroves of East Africa. Glob Environ Change 19:227–239

Dahdouh-Guebas F (2011) World atlas of mangroves: Mark Spalding, Mami Kainuma and Lorna Collins (eds). Hum Ecol 39:107–109

Dahdouh-Guebas F, Koedam N (2006) Empirical estimate of the reliability of the use of point-centred quarter method (PCQM): solution to ambiguous field situation and description of the PCQM+ protocol. For Ecol Manage 228:1–18

Day JW, Machado N (1986) The productivity and composition of mangrove forest, Laguna De Ti~Rminos, Mexico. Aquat Bot 27:267–284 Elsevier Science Publishers B.V., Amsterdam

Din N, Saenger P, Priso RJ, Dibong DS, Blasco F (2008) Logging activities in mangrove forests: a case study of Douala Cameroon. Af J Environ Sci Technol 2(2):22–30

Egnankou Wadja M (2009) Rehabilitation of mangroves between Fresco and Grand-Lahou (Cote D'Ivoire): important fishing areas. Nat Faune Mag, FAO 24(1):81–88

English S, Wilkinson C, Basker V (1997) Survey manual for tropical marine resources, 2nd edn. Australian Institute of Mar. Sci., Townsville, pp 119–195

FAO (2007) The world's mangroves 1980–2005. Food and Agriculture Organisation of the United Nations Forestry paper, Rome

Hegazy AK (1998) Perspectives on survival, phenology, litterfall and decomposition, and caloric content of Avicennia marina in the Arabian Gulf region. J Arid Environ 40:417–429

Hellier C (1988) The mangrove wastelands, mangroves of Vietnam, Phan Nguyen, Hoang Thi San IUCN wetlands program. J Ecol 18(2), Hong

Hossain M, Othman S, Bujang JS, Kusnan M (2008) Net primary productivity of Bruguiera parviflora (Wight & Arn.) dominated mangrove forest at Kuala Selangor, Malaysia. For Ecol Manag 255:179–182

Husch B, Beers TW, Keuhaw Jr JA (2003) Forest mensuration, 4th edn. Wiley, 443 pp

Kairo JG, Bosire JO (2009) Ecology and restoration of Mangrove Forests in Kenya. Nat Faune Mag, FAO 24(1):41–48

Kathiresan K (1997) Methods of studying mangroves centre of advanced study in marine biology Annamalai University, 116–124 pp

Kaufman JB, Donato DC (2012) Protocols for the measurement, monitoring and reporting of structure, biomass and carbon stocks in mangrove forests. Working paper 86, CIFOR, Bogor, Indonesia. Version cite dans le doc

Kaufman JB, Heider C, Cole T, Dwire KA, Donato DC (2011) Ecosystem C pools of Micronesian mangrove forests: implications of land use and climate change. Wetlands 31:343–352

Loetsch I, Zohier F, Haller K (1973) Forest inventory, vol 2, 2nd edn. BIV, Germany, 469 pp

Lovelock CE, Feller IC, Mckee KL, Thompson R (2005) Variation in mangrove forest structure and sediment characteristics. Bocas del Caribbean J Sci 41(3):456–464 College of Arts and Sciences University of Puerto Rico, Toro, Panama

Meentemeyer V, Box EO, Thompson R (1982) World patterns and amounts of terrestrial plant litter production. Bioscience 32:125

Nfotabong Atheull A (2008) Utilisation des mangroves par les habitants des zones côtières près de Kribi, du Nyong et de l'estuaire du Cameroun. Mémoire de DEA, Université Libre de Bruxelles-ULB, Bruxelles, Belgique

Nfotabong Atheull A (2011) Impact of anthropogenic activities on the vegetation structure of mangrove forests in Kribi, the Nyong river mouth and Cameroon estuary. Ph. D Thesis, Université Libre de Bruxelles-ULB, Brussels, Belgium, The University of Douala, Cameroon, 196 p+appendices

Ngeh P (1989) Effects of land clearing methods on a tropical forest ecosystem and the growth of Terminalia ivorensis. Thesis submitted to the University of Edinburg for the degree of Doctor of Philosophy, 194 p

Ochieng CA, Erftemeijer PLA (2002) Phenology, litterfall and nutrient resorption in Avicennia marina (Forssk.) Vierh in Gazi Bay, Kenya. Trees 16:167–171

Odum EP, Heald EJ (1975) The detritus bases food web of an estuarine mangrove community. In: Cronin LE (ed) Estuarine research. Academic Press, New York, pp 265–286

Ramsar-MAVA-UNEP (2012) Integrated critical mangroves conservation and sustainable use program framework document

Raulerson GE (2004) Leaf litter processing by macrodetritivores in natural and restored Neotropical Mangrove Forest. PhD thesis, graduate faculty of the Louisana State University and Agricultural and Mechanical College, Department of Oceanography and Coastal sciences, 142 p

Snowdon P, Ryan and Raison J (2005) Review of C: N ratios in vegetation, litter and soil under Australian native forests and plantations. Technical report no. 45, CSIRO forestry and forest products national carbon accounting system. 60 p

Spalding M, Kainuma M, Collins L (2010) World atlas of mangroves. The International Society for Mangrove Ecosystems, Okinawa

Twilley RR, Day Jr., JW (1999) The productivity and nutrient cycling of mangrove ecosystem. In: Yañez-Arancibia A, Lara-Dominguez AL (eds) Ecosistemas de manglar en América Tropical. Instituto de Ecologia, A.C. México, UICN/HORMA, Costa Rica, NOAA/NMFS Silver Spring, MD, USA, pp 127–152

UNEP (2007) Mangroves of Western and Central Africa UNEP-regional seas programme/UNEP-WCMC

Wattayakorn G, Wolanski E, Kjerfve B (1990) Mixing, trapping and outwelling in the Klong Ngao mangrove swamp, Thailand. Estuar Coast Shelf Sci 31:667–688

World Bank/ISME/CenTER (2004) Principles for a code of conduct for the management and sustainable use of mangrove ecosystems, annual report

Carbon Budget as a Tool for Assessing Mangrove Forests Degradation in the Western, Coastal Wetlands Complex (Ramsar Site 1017) of Southern Benin, West Africa

Gordon N. Ajonina, Expedit Evariste Ago, Gautier Amoussou, Eugene Diyouke Mibog, Is Deen Akambi, and Eunice Dossa

Abstract

Carbon budgets were established for pockets of mangrove forests within the western coastal wetland complex of southern Benin (Ramsar site 1017) using standard methods. We assessed carbon stocks in four (0.1 ha) transects in sites with different levels of degradation of mangrove vegetation from non-forested, very degraded, degraded, to non-degraded mangroves in Kpétou, Djègbadji, Adounko (1), and Adounko (2) sites, respectively. Species encountered were typical of the west-central African region: *Avicennia germinans, Rhizophora racemosa, Acrosticum aureum* (fern), and *Paspalum vaginatum.* The maximum diameter was 24.8 and 20.3 cm for *Avicennia* and *Rhizophora,* respectively, in non-degraded mangrove sites, at a maximum height of 21 m. Standing volumes varied from 6.55 m^3/ha in degraded mangroves to 50.42 m^3/ha in the non-degraded mangrove sites. The total annual carbon stock, to a soil depth of one meter, varied from 179.39 t/ha in the non-forested site, to 288.77 t/ha in degraded mangroves and 184.01 t/ha for non-degraded mangroves. Aboveground biomass carbon in non-degraded sites was on average 35.07 t/ha, more than five times that in degraded mangrove sites. The distribution of total ecosystem carbon stocks for the aboveground component for the mangrove sites ranged from 0.82 % for non-forested sites, 2.94 % for degraded to 16.01 % for non-degraded mangrove forests. The belowground carbon stocks component ranged from 179.39 to 288.77 t/ha, being the highest in degraded mangrove sites due to leached organic matter following degradation. Potential annual net CO_2e emissions from up to 30 cm soil depth (compared to non-wooded mangrove vegetation) is estimated to increase from 10.28 to 99.64 t CO_2e (22 % of non-wooded

G. N. Ajonina (✉) · E. D. Mibog
CWCS Coastal Forests and Mangrove Programme,
BP 54 Mouanko, Littoral Region, Cameroon
e-mail: gnajonina@hotmail.com

E. E. Ago
Unit of Biosystem Physics, University of Liege Gembloux Agro-
Bio Tech (GxABT), 8 Avenue de la Faculté, 5030, Gembloux,
Belgium

G. Amoussou · I. D. Akambi · E. Dossa
Benin Ecotourism Concern (Eco-Benin), Zogbadjè, Rue début
Clôture IITA, Calavi 03, BP 1667, Jéricho, Benin

G. N. Ajonina
Institute of Fisheries and Aquatic Sciences, University of Douala
(Yabassi), Douala, Cameroon

S. Diop et al. (eds.), *The Land/Ocean Interactions in the Coastal Zone of West and Central Africa,* Estuaries of the World, DOI: 10.1007/978-3-319-06388-1_12,
© Springer International Publishing Switzerland 2014

mangroves) following mangrove degradation and 43 t CO_2e (10 %) avoided CO_2e emissions from undisturbed mangrove forests.

Keywords

Mangroves • Degradation • Carbon budget • CO_2 emission • Benin

Introduction

Mangroves are among the carbon-richest forests in the tropics, containing an average of 1,023 t carbon per ha. Mangrove deforestation generates global emissions of 0.02–0.12 Pg carbon per year, while storing up to 20 Pg C annually (Donato et al. 2011), roughly equivalent to 2.5 times annual global carbon dioxide (CO_2) emissions despite accounting for only around 0.7 % (around 140,000 km^2) of global tropical forests (Giri et al. 2010). Apart from high carbon storage potentials, mangroves provide other valuable ecosystem services like protection from storms, floods, and erosion; provision of timber and non-timber forest products (Daïnou et al. 2008); processing of waste and nutrient pollution; aquaculture and agriculture support; and habitat for aquatic and terrestrial species.

In Benin, the total mangrove cover is about 66 km^2 within its 120 km littoral zone consisting largely of a sandy coast without developed estuaries and deltas. The mangroves of Benin tend to be limited to the edge of the extensive network of brackish coastal lagoons in the south of the country. Despite replanting efforts, these mangroves are in an advanced state of degradation (UNEP 2007). Because of their large ecosystem, carbon stocks, their vulnerabilities to land use, and the numerous other ecosystem services they provide, mangroves are increasingly considered as prime ecosystems participating in climate change mitigation strategies through reforestation and restoration (Alongi 2002; Donato 2012).

To participate in global climate change-related processes, such as reduced emissions from deforestation and degradation (REDD+), and Clean Development Mechanism (CDM), it is necessary to determine carbon stocks and emissions baselines. Many carbon stock assessments in mangrove ecosystems have been carried out recently around the world (Donato et al. 2011; Kauffman and Donato 2012; Adame et al. 2013) mainly assessing mangroves from more or less temporal sample plots in mostly undisturbed states. Until recently in Central Africa (Ajonina et al. 2013), lack of permanent mangrove sample plots made it difficult to gauge mangrove ecosystem recovery dynamics from increasing degradation pressures for its vital ecosystem services that sustain coastal rural economies, especially in the west–central African region.

The present study was carried out to assess carbon stocks in mangrove ecosystems subjected to various level of degradation as part of a larger initiative to set up a CDM tree planting project to restore degraded mangrove areas of southern Benin.

Materials and Methods

Site Description

Biophysical Characteristics

Mangroves in southern Benin are concentrated within a wetland complex area that includes Ahémé Lake, the Aho Channel coastal lagoon mouth, and Roy Ramsar site 1017 (latitude 7°N and longitudes 1°35′ and 2°30′E) (see Fig. 1) (UNEP 2007). The climate is equatorial with a mean total rainfall of 1,200 mm/year mostly within two seasons (March–late July and September–November) and two dry seasons (late July–early September and late November–early March). Average daily temperatures are around 27 °C, relative humidity ranges from a low of 78 % in January–February to a maximum of 95 % in September. The western wetland complex is located in the western part of the coastal sedimentary basin whose main series are the Upper Cretaceous, Eocene, the Continental Terminal, and recent formations. The region has a dense hydrological network with three major rivers: the Couffo, Mono, and Sazué. The Couffo, 190 km long, is a small river in the Guinean regime with two seasons of flood. It has its source in Togo country at 240 m altitude, near the border village of Tchetti, and empties into Lake Ahémé. The Mono, with 148 km of its course in Benin, has a bed slope within the watershed of southern Benin that is very low (0.06 to 0.4 m/km). The Sazué, 63 km long, is formed by the meeting of two rivers: the Dévédon (22 km) and Salédo (40 km). The slopes of the bed are extremely low. Apart from these major rivers, there are others with various physiognomies:

Zoko lakes, Dévé, Togbadji, Egbo, Doukon, Toho Godogba, Wozo, and DatchiDofe; Large marshy depressions south of Athiémé and west of Comè districts, where several vegetation formations can be encountered especially pockets of mangrove, riparian forests, periodically flooded swamp forests, marshes, grasslands, and floating aquatic

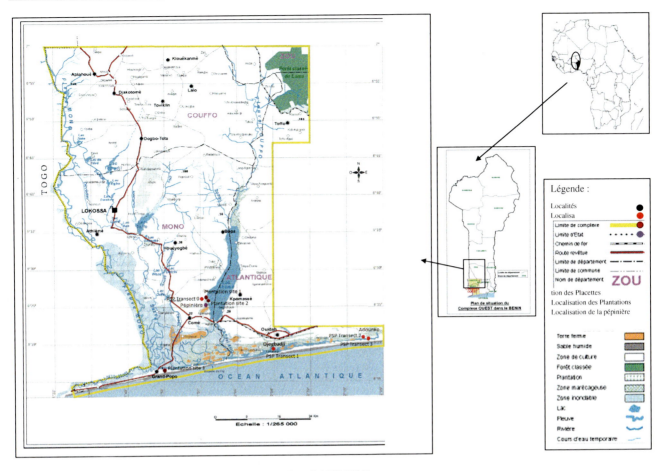

Fig. 1 Map showing the location of mangrove assessment sites (PAZH 2001)

vegetation. Physiognomy and the recovery of each of these courses depend on the floristic composition of plant communities that are themselves based on spatiotemporal variations of environmental and climatic conditions. The vegetation of dry lands is less diversified. These are the rainforest's semi-deciduous tree and shrub savannah. Natural formations have undergone or are under increasing pressure and have mostly been replaced by various farmlands and plantations, and the remainder is in a state of more or less advanced degradation. The specificity of this area is that it consists of a natural habitat dominated by mangroves species especially *R. racemosa* and *A. germinans*. The region houses exceptionally high biodiversity with rich wetland flora of 364 species in 100 families of which the distributed and most represented are Poaceae (34 species), Rubiaceae (29 species), Cyperaceae (28 species), Fabaceae (20 species), Euphorbiaceae (16 species), Moraceae (14 species), and Apocynaceae (11 species). Mangrove species include *R. racemosa*, *A. germinans*, *Conocarpus erectus* and *Laguncularia racemosa*, and *Acrostichum aureum* (mangrove fern); associate mangrove species frequently encountered are as follows: *Dalbergia ecastaphyllum*, *Drepanocarpus lunatus*, *Phoenix reclinata*, more herbaceous ones such as *P. vaginatum, Sesuvium portulacastrum, Philoxerus vermicularis,* and more or less *Fimbristylis ferruginea, Crotalaria retusa, Hibiscus tiliaceus, Annona senegalensis, Chrysobalanus orbicularis, Elaeis guineensis,* and *Cocos nucifera.* The wetland system provides not only migration corridors for marine and inland species, but also a wintering habitat for birds including terns and other Palearctic species. Protected species including four species of marine turtles frequent the coast between September and March: olive ridley (*Lepidochelys olivacea*), leatherback (*Dermochelys coriacea*), green turtles (*Cheloniamydas*), and hawksbill (*Eretmochelys ecosystem*). Fish species are mostly Cichlides consisting of 7 species dominated by *Sarotherodon melanotheron* and *Tilapia guineensis*. Over 43 species of pelagic species occur in 14 families especially the mostly exploited Carcharinidae, Sphyrnidae, Pristidae, Ginglimo stomatidae, Clupeidae, and Carangidae et Scombridae. Demersale species consisting of 51 species was divided into 24 families the most importance ones include: Dasyatidae, Mobulidae, Zanobatidae, Rajidae, Rhinobatidae, Rhinopteridae, Triakidae, Torpedinidae, Ariidae, Haemulidae, Polynemidae, Serranidae, Sciaenidae, Sparidae, and Lutjanidaeet Cynoglossidae. Some large aquatic mammals also frequent the sites including *Hippopotamus amphibious* and *Trichechus senegalensis* (Manatees) especially in valleys of River Monoor lakes and terrestrial mammals especially red-bellied monkeys (*Cercopithecus erythrogaster*).

Socioeconomic Environment and Human Activities

This complex is full of unique assets: natural, landscape, religious, tourist, economic, etc. Indeed, the area is subject to high anthropogenic pressure manifested by uncontrolled exploitation of its resources, which have consequences of increased erosion, ecological change affecting biodiversity, etc. The region is characterized by a wide range of ethnic diversity with a population of over 1 million people mostly made up of two (02) groups:

- On the coast: the Pla between Pahou and Avlékété, the Pédah between Ouidah and Djègbadji, the Aizo between Cococodji and Godomey, the Fon and the Keta were found along the coast and everywhere, Gengroup and Ewe, late comers occupy the entire coast from Ghana (origin) to Nigeria.
- At North coast, mostly the Adja, Fon, Nago, and Mina were found.

The most prominent human activities in the area are fishing in nearby lakes and lagoonal systems. Agriculture includes vegetable and arable crop farming, especially tomatoes, beans, onions, pepper, maize, and cassava. Small livestock production systems (domestic birds, rabbits, goats, and sheep) are mostly based on free-ranging systems around homesteads. Salt is also produced around the mangrove areas and conversion of palm wine into alcohol. The large amounts of clay in the mangrove area are used in local pottery industry.

Establishment of Permanent Sample Plots and Quantification of Carbon Pools

The method used was that of permanent sample plots described and applied to the rainforest by Condit (1998), adapted to the mangrove forests in Cameroon by Cameroon Wildlife Conservation Society (CWCS) (see Ajonina 2008); it should be noted that the same method was used in Ghana (Ajonina 2011), and most recently in the mangrove countries of the Congo Basin under the UNEP REDD + Central African mangroves evaluation of multiple services mangrove ecosystem in the Congo Basin (Cameroon, Congo, DRC, and Gabon) (Ajonina et al. 2013). Transects of 100×10 m (0.1 ha) were established following a random azimuth from a random fixed point. Each transect was divided into three plots of 0.02 ha (10×20 m) separated by an interval of 10 m along the transect with further division into six square subplots of 10×10 m. The angle of each

subplot is fixed with painted PVC plastic pipe with boundaries demarcated by strong ropes. All plots were well mapped from their GPS coordinates. Four transects were established in sites with different levels of degradation of mangrove vegetation from non-woody, very degraded, medium or less degraded to non-degraded mangroves in Kpétou, Djègbadji, Adounko (1), and Adounko (2) sites, respectively (Table 1; Figs. 2, 3). Pre-qualitative criteria for the characterization of different degradation regimes were based on the characteristics of the canopy (closure, open), trees (diameter, height, density) and undergrowth (density), and the frequency of indicator species later compared with a unique post-quantitative criteria of basal surface (m^2) per ha. The monitoring protocol consisted of species identification, mapping, tagging, and measurements of all trees inside the plot using modified forestry techniques for mangroves (Pool et al. 1977; Cintron and Novelli 1984; Kauffman and Donato 2012). Four carbon pools were considered in the present study, including vegetation carbon pools (both above- and belowground), litter, coarse deadwood, and soil.

Measurement of Vegetation Carbon

The most commonly assessed carbon stock in forestry is usually in the aboveground component because trees dominate the aboveground carbon pools and serve as indicator of the ecological conditions of most forests. Inside the plots, all trees with diameter of the stem at breast height (dbh_{130}) $\geq$ 1.0 cm were identified, measured, tagged, and mapped (Fig. 4). Data on species, dbh, live/dead, and height were recorded for all individuals. In *Rhizophora*, dbh was taken 30 cm above highest stilt root. Aboveground roots and saplings (dbh $\leq$ 1 cm) were sampled inside five 1-m^2 quadrants placed systematically at 1-m intervals along the 10×10 m plot (Fig. 2b).

Dead and Downed Wood

Deadwood was estimated using the transect method whose application is given in Kauffman and Donato (2012). The line intersect technique involves counting intersections of woody pieces along a vertical sampling transect. The diameters of deadwood (usually more than 0.5 cm in diameter) lying within 2 m of the ground surface were measured at their points of intersection with the main transect axis. Each deadwood measured was given a decomposition ranking: rotten, intermediate, or sound.

Soil Samples

Mangrove soils have been found to be major reservoirs of organic carbon (Donato et al. 2011). Soil carbon is mostly concentrated in the upper 1.0 m of the soil profile. This layer is also the most vulnerable to land-use change, thus

Table 1 Selected sites for mangrove permanent plot establishment

Site		GPS fixes		Type of plot	Vegetation condition
No	Name	N	E		
1	Kpétou	06°25,945′	001°55,557′	Blank area where mangrove stands do not exist in the natural state	Area in Kpétou village (Agatogbo) on the shores of Lake Ahémé. This area is periodically flooded by tides and also influenced through communication of the lake with the Atlantic sea through the Mono River. Presence of pockets of mangroves (*Rizophora racemosa* and *A. germinans*). The dominant plants are grasses (*P. vaginatum*). Site for mangrove reforestation efforts of the NGO Eco-Benin. Heavy fishing by dugout canoes along the banks of the lake
2	Djègbadji	06°19,901′	002°05,034′	Very disturbed mangroves	Situated within the Djègbadji village in Ouidah town is a branch of the coastal lagoon, periodically flooded and tidally influenced through communication with the seawater. Presence of mangrove species as *Rizophora racemosa* and *A. germinans* scattered in the various islands and shores. Also presence of sea grasses (*P. vaginatum*) lining the floor. Mangroves in this area are highly disturbed by human activities, especially women who collect wood to prepare salt, high activity in this area, but also for firewood. Floods also disrupt natural regeneration
3	Adounko	06°21,164′	002°16,998′	Medium mangrove degradation	Area in Adounko village in the coastal lagoon. It is the part of the Marine Protected Area that the Benin Agency for the Environment is setting up of about (54 ha) area periodically flooded by tides and influenced through communication with the seawater. Presence of mangroves (*R. racemosa* and *A. germinans*) scattered on the shore. Also presence of grasses (*P. vaginatum*) and *Acrosticum* and also other species that are at the interface between the land and sea mangrove zone. Area also disturbed by human activities, especially women who collect firewood and other food plants
4	Adounko	06°21,056′	002°17,456′	Non-degraded mangroves	Area in the Adounko village near Togbin town within the coastal lagoon periodically flooded by tides and influenced through communication with the seawater. Presence of large stands of mangroves (*R. racemosa* and *A. germinans*). Presence of other species that are at the interface between land and sea. The mangrove zone is protected by the state because it is the part of the Marine Protected Area that the Benin Agency for the Environment is currently implementing. It is also sacred forest with spiritual worship of some deities

contributing most to emissions when mangroves are degraded. Soil cores were extracted from each of the 20×10 m plots using a core of 5.0 cm diameter and systematically divided into different depth intervals (0–15, 15–30, 30–50, and 50–100 cm), following the protocol recommended by Kauffman and Donato (2012). A sample of 5 cm length was extracted from the central portion of each depth interval to obtain a standard volume for all subsamples (Fig. 5).

A total of 48 soil samples were collected and placed in pre-labeled plastic bags. In the laboratory (Laboratory of Soil Science, Water and Environment of the Centre for Agricultural Research Agonkanme'-Benin.), samples were

weighed and oven-dried to constant mass at 70 °C for 48 h to obtain wet: dry ratios (Kauffman and Donato 2012). Bulk density was calculated as follows:

$$\text{Soil bulk density} \left(\text{gm}^{-3} \right) = \frac{\text{Oven} - \text{dry sample mass (g)}}{\text{Sample volume} \left(\text{m}^3 \right)}$$

(1)

where volume = cross-sectional area of the corer × the height of the sample subsection.

Of the dried soil samples, 5–10 g subsamples were weighed out into crucibles and set in a muffle furnace for combustion at 550 °C for 8 h through the process of loss on

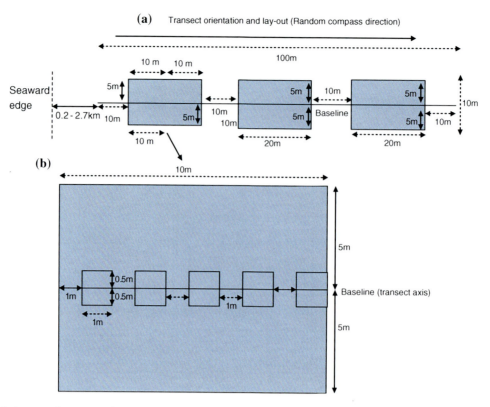

Fig. 2 a Schematic layouts of mangrove forest stands permanent sample plots and, **b** roots and sapling inventories (After Ajonina 2008)

Fig. 3 Permanent transect layout: Kpétou village site (transect 1) with non-mangrove woody vegetation (*left*) and degraded mangrove vegetation (*Photos* Diyouke Eugene and Gordon Ajonina 2012)

ignition (LOI) and cooled in desiccators before reweighing. The weight of each ashed sample was recorded and used to calculate organic concentration (OC). Total soil carbon was calculated as

$$\text{Soil C (tonnes/ha}^1) = \text{bulk density } (g/cm^3) \\ * \text{ soil depth interval (cm)} * \%\, C \quad (2)$$

The total soil carbon pool was then determined by summing the carbon mass of each of the sampled soil depths.

Data Analysis and Algometric Computations

Field data were organized into various filing systems for ease of analysis and presentation. Both structural and

Fig. 4 Diameter measurement and tagging trees (*Photo* Gordon Ajonina 2012)

Fig. 5 Collecting soil samples (*Photo* Gordon Ajonina 2012)

biophysical data were entered into prepared data sheets. Standing volume was determined using locally derived algometric relations from sample data with dbh as the independent variable following Eq. (3):

$$V = a * D^b \quad (3)$$

where

- V Volume (m³/ha)
- D diameter: 1 cm ≤ D < 30 cm
- a and b are coefficients.

Biomass conversion/expansion factor (BC/EF), which is the ratio of total aboveground biomass to stand volume, and shoot/root ratio (SRR) developed by Ajonina (2008), were used for the estimation of total tree biomass and carbon densities. The BC/EF used in the study was 1.18 (Ajonina 2008) which is comparable to that reported for humid tropical forests by Brown (1997).

Belowground root biomass per tree was determined using the equation (Ajonina 2008):

$$\text{Root biomass per tree} = 1.385 * \text{Stem Diam}^{-0.4331} * \text{Stem volume}. \quad (4)$$

The algometric equation derived for estimating biomass samples after drying (temperature constant at 105 °C) of 52 mangrove fern (*Acrosticum aurerum*) is

$$y = b + a * H * D^2 \quad (5)$$

where D and H are the diameter and height of mangrove fern.

Biomass and carbon in grasses was estimated from a composite sample of grass collected in 15 quadrants of 1 m² systematically placed along the first subplot (10 × 10 m) in each plot (20 × 10 m), and their weight was measured in the field with a balance. Dry weight and the carbon concentration were determined by the Laboratory of Soil Science, Water and Environment of the Centre for Agricultural Research Agonkammey-Benin.

Deadwood

Deadwood volume was estimated using the protocol of Kauffman and Donato (2012):

$$\text{Volume } (\text{m}^3/\text{ha}) = \Pi^2 * \frac{\sum_{i=1}^{n} d_i^2}{8L} \quad (6)$$

where $d_i = d_1, d_2 \ldots d_n$ are diameters of intersecting pieces of deadwood (cm) and L = the length of the intersecting line (transect axis of the plot) generally $L = 20$ m being the length of each plot or 100 m being the length of transects. Deadwood volumes were converted to carbon density estimates by using the different size-specific gravities provided by Kauffman and Donato (2012).

Total Ecosystem Carbon Stock and CO₂e Emission Factor in Degraded Mangrove Forests

The total ecosystem carbon stock to soil depth of one meter was obtained by adding the aboveground and belowground carbon densities. The carbon dioxide equivalent was calculated by multiplying the total ecosystem carbon stock by the carbon dioxide emission factor 3.67. Carbon dioxide emission factors for top 30 cm soil in the different mangrove degradation regimes were expressed as excess over the non-wooded mangrove site.

Table 2 Structural characteristics of mangrove woody species within the sites inventoried in southern Benin

Condition	Species	Stem density (Stems/ha)	Max diameter (cm)	Mean diameter (cm)	Max height (m)	Mean height (m)	Stand basal area (m²/ha) G	±SE	Stand volume (m³/ha) Vol	±SE
Degraded	A. germinans	733.5	16.8	5.7	4.8	3.7	1.51	0.01	4.02	0.03
	Acrosticum aurerum	300	10.4	3.1	3.9	3.1	0.34	0.01	0.58	0.02
	R. racemosa	625	8.75	6.3	8.8	7.8	0.40	0.03	2.23	0.23
	Total (mean)	1,508.5	19.35	3.3	9.3	3.7	2.09	0.01	6.55	0.06
Non-degraded	A. germinans	183	21.1	10.2	13.5	10.5	1.89	0.05	12.71	0.45
	R. racemosa	1,233	20.3	8.2	21.1	10.8	8.24	0.01	37.72	0.13
	Total (mean)	1,416	21.1	8.5	18.6	10.7	10.14	0.01	50.42	0.13
All sites	A. germinans	550	24.8	7.2	5.1	5.9	1.64	0.01	6.92	0.15
	Acrosticum aurerum	300	10.4	3.1	3.9	3.1	0.34	0.01	0.58	0.02
	R. racemosa	828	20.3	6.9	21.1	8.8	3.02	0.02	14.06	0.01
	Mean	559	24.8	5.8	21.1	5.9	1.66	0.01	7.19	0.02

Results and Discussions

Floristic Composition, Structure, and Biometric Characteristics of Mangrove Stands in the Mangrove Sites Inventoried

The details of floristic composition, structure, and biometric characteristics are presented in Table 2 and Fig. 6. Species frequently encountered in the sites surveyed are typical species of mangroves in West Africa, Central, and Western bloc (Tomlinson 1986). They are as follows: *A. germinans*, *R. racemosa* as woody species, *A. aurerum* (mangrove fern), and *Paspalum vaginatus* (herbaceous). The maximum diameter was 24.8 and 20.3 cm for *Avicennia* and *Rhizophora*, respectively, in non-degraded mangrove sites with maximum height of 21 m. Standing volumes varied from 6.55 m³/ha in degraded mangroves to 50.42 m³/ha in the non-degraded mangrove sites. Stem diameter distributions depart from the exponential law obeyed by pristine stands (Husch et al. 2003) showing the disturbed nature of the sites as reported by Daïnou et al. (2008). The highest stem density of diameter between 0 and 5 cm in degraded mangroves (Fig. 6) compared to non-degraded indicates that the latter were characterized by more regeneration of species due to canopy opening. The stem density of diameter 5–30 cm in degraded mangroves was lower than that in non-degraded, indicating high human pressure on young and adult stems in the region.

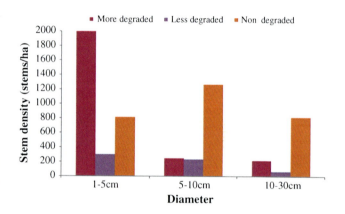

Fig. 6 Stem density–diameter distribution in mangroves of southern Benin

Carbon Budget Within the Inventoried Sites

The carbon budget at the different sites surveyed is presented in Table 3 and Fig. 7. The total annual carbon stock to soil depth of one meter varied from 179.39 t/ha in non-forested sites to 288.77 t/ha in degraded mangroves and 184.01 t/ha for non-degraded mangroves. Aboveground biomass carbon in non-degraded sites was 35.07 t/ha, more than five times that in degraded mangrove sites. The distribution of total ecosystem carbon stocks for the aboveground component for the mangrove sites varied from 0.82 % for non-forested sites, 2.94 % for degraded, and 16.01 % for non-degraded sites (Table 3). The

Table 3 Carbon budget in the different degraded mangrove sites in southern Benin

Component	Carbon density (t/ha)							
	Site 1: Kpétou		Site 2: Djègbadji		Site 3: Adounko		Site 4: Adounko	
	Non-wooded mangroves		Degraded mangroves		Degraded mangroves		Non-degraded mangroves	
	Mean	±SE	Mean	±SE	Mean	±SE	Mean	±SE
Basal area (m²/ha)	0.00	0.00	1.16	0.01	4.12	0.17	10.14	0.03
Aboveground								
Live								
Trees and stems (woody vegetation)			3.03	1.79	0.15	0.05	0.68	0.11
Roots			0.22	0.12	1.52	0.56	0.10	0.04
Seedlings with herbs			3.52	1.77	0.01	0.01	22.70	11.66
Herbs	1.48	0.09						
Subtotal (live)	1.48	0.09	6.77	1.25	1.67	0.18	23.48	1.97
% total	0.82		2.28		0.70		10.72	
Dead								
Deadwood			0.74	0.19	1.18	0.13	3.56	0.35
Dead roots			0.00	0.00	2.87	1.32	0.02	0.02
Dead seedlings with herbs			1.12	1.12	0.57	0.53	8.01	2.38
Subtotal (dead)			1.86	0.28	4.62	0.38	11.59	0.89
% total	0.00		0.62		1.94		5.29	
Subtotal (aboveground)	1.48	0.09	8.63	0.78	6.29	0.21	35.07	1.25
% total	0.82		2.90		2.94		16.01	
Belowground								
Tree roots			0.32	0.12	0.08	0.02	0.33	0.05
Sediments								
0–15 cm depth	60.31	3.01	59.72	1.42	52.87	0.85	59.85	25.96
15–30 cm depth	63.22	12.12	53.54	15.66	38.62	1.93	41.56	7.29
30–50 cm depth	12.85	2.97	48.28	3.20	20.61	0.36	24.52	1.82
50 cm–100 m depth	43.01	2.40	126.92	23.31	119.46	56.97	57.75	1.63
Subtotal (sediments)	179.39	6.65	288.45	11.35	231.57	16.57	183.68	7.19
Subtotal (belowground)	179.39	6.65	288.77	10.49	231.64	12.73	184.01	5.93
% total	99.18		97.10		97.36		83.99	
Total (ecosystem)	180.87	5.12	297.40	5.00	237.94	5.51	219.08	2.67
Total CO_2e of the ecosystem	663.18	18.79	1,090.47	18.35	872.43	20.22	803.29	9.80
CO_2e up to 30 cm depth	458.79	55.86	448.51	65.99	359.15	11.05	502.09	126.80
CO_2e emission factor up to 30 cm depth	0.00	0.00	−10.28	10.13	−99.64	−44.81	43.31	70.94
CO_2e emission factor up to 30 cm depth (%)	0.00		−2.24		−21.72		9.44	

belowground carbon stock component ranged from 179.39 to 288.77 t/ha (Fig. 7), being highest in degraded mangrove sites due to organic matter accumulation and leaching. This difference in the distribution of carbon stock at belowground and aboveground from degraded to non-degraded mangrove ecosystems in southern Benin reveals the important role of vegetation in the carbon budget estimation. The trend in the carbon balance within the surveyed mangrove sites is also typical of a mangrove ecosystem that

stores most of its carbon as belowground (Kauffman and Donato 2012). These figures are close to those reported by Kauffman and Donato (2012) who obtained total stock of carbon of 278.0 t/ha for the large mangrove delta of the river Ganges in Bangladesh Sunderbans with an aboveground component of 83.7 t/ha (30.1 %) and underground component of 194.9 t/ha (69.9 %). However, the values are much lower than those obtained for the mangroves of Central Africa (Cameroon, Congo, Gabon, and DRC):

Fig. 7 Distribution of annual carbon stock in different mangrove sites in southern Benin

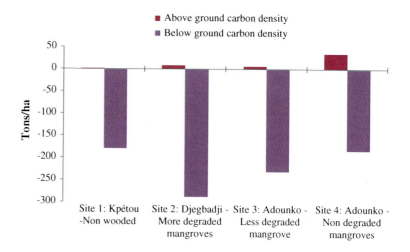

925.4 t/ha (85 % underground component) in degraded mangrove and 1,520 t/ha (35.1 %air component) for non-degraded mangroves (Ajonina et al. 2013). This difference could be due to variability in climatic and environmental conditions, especially temperature, amount of rainfall and its distribution, and solar radiation in the western part of Africa compared to the central. It was recognized that the carbon fluxes which are exchanged by the terrestrial ecosystems with the atmosphere in Africa are found to be strongly dependent on water availability, temperature, and solar radiation (Merlbold et al. 2009; Williams et al. 2008; Kutsch et al. 2008; Hanan et al. 1998). Potential net CO_2e emissions from up to 30 cm soil depth (compared to non-wooded mangrove vegetation) are expected to increase from 10.28 to 99.64 t CO_2e/ha (22 % of non-wooded mangroves) following mangrove degradation and 43 t CO_2e/ha (10 % of non-wooded mangroves) avoided CO_2e emissions from undisturbed mangrove forests (Table 3).The values tend to fall within the interval recorded elsewhere: For example, Lovelock et al. (2011) found 2,900 t CO_2/km^2/year (29.0 t/ha/year) emissions from degraded mangroves in Belize, Cahoon et al. (2003) found 1,500 t CO_2/km^2/year (15.0 t/ha/year) emissions from degraded mangroves in Honduras, and Burford and Longmore (2001) found 1,750 t CO_2/km^2/year (17.5 t/ha/year) emissions from degraded mangroves in Australia.

Different Allometric Relations and Coefficient Values Obtained at the Site

$$V = 0.000103 \, D *^{2.6207} \, (R^2 = 0.976 \; n = 233) \quad (7)$$

$$y = 0.112 \, D^2H + 6.462 (R^2 = 0.795 \; n = 52) \quad (8)$$

Conclusion

The assessment of carbon stocks in mangrove wetlands of the West Complex of southern Benin (Ramsar site 1017), using standard methods, presents a typical distribution between above- and belowground components similar to other mangrove ecosystems of the world. However, the carbon stocks are much lower than those in the Central Africa region. In future, it appears very important to continue to monitor the carbon stock of all mangrove ecosystems inventoried to better define strategies to diminish carbon emissions.

Acknowledgments This study has been prepared with the financial assistance of the IUCN National Committee of the Netherlands, Wetlands International and Both Ends, partners in the Ecosystem Alliance. The authors would like to thank all individuals and institutions that have contributed to the realization of this work. We also express gratitude to the leaders and authorities in the villages in which the sites were surveyed for their hospitality and especially to field assistants from the different site inventoried, as well as partners involved in laboratory analysis and particularly the officials of the Laboratory of Soil Science, Water and Environment of the National Institute of Agricultural Research of Benin (INRAB) and those who took part in the workshop to discuss the preliminary results of this study. The administrative staff of NGO ECO-BENIN is heartily thanked for their sense of hospitality and work beyond working hours during the study period.

References

Adame MF, Kauffman JB, Medina I, Gamboa JN, Torres O, Caamal JP, Reza M, Herrera-Silveira JA (2013) Carbon stocks of tropical coastal wetlands within the karstic landscape of the Mexican Caribbean. PLoS ONE 8(2):e56569. doi:10.1371/journal.pone.0056569

Ajonina G (2011) Rapid assessment of mangrove status and conditions for use to assess potential for marine payment for ecosystem services in Amanzuri and surrounding areas in the western coastal region of Ghana, West Africa. In: The Marine Ecosystem Services (MARES) Programme, Forest Trends Report, 37 p

Ajonina G, Kairo J, Grimsditch G, Sembres T3, Chuyong G, Mibog DE, FitzGerald C (2013) Assessment of carbon pools and multiple benefits of mangroves in Central Africa for REDD+. In: UNEP, WCMC, CWCS, KMFRI Report, 85 p

Ajonina GN (2008) Inventory and modelling mangrove forest stand dynamics following different levels of wood exploitation pressures in the Douala-Edea Atlantic coast of Cameroon, Central Africa. Mitteilungen der AbteilungenfürForstlicheBiometrie, Albert-Ludwigs- Universität Freiburg. 2008- 2, 215 p

Alongi, D.M. (2002). Present state and future of the world's mangrove forests. Environ Conserv 29: 331–349

Brown S (1997) Estimating biomass and biomass change of tropical forest. FAO Forestry paper 134. FAO, Rome, 76 p

Burford MA, Longmore AR (2001) High ammonium production from sediments in hypereutrophic shrimp ponds. Mar Ecol Prog Ser 224:187–195

Cahoon DR, Hensel P, Rybczyk J, McKee KL, Proffitt CE, Perez BC (2003) Mass tree mortality leads to mangrove peat collapse at Bay Islands, Honduras after Hurricane Mitch. J Ecology 91:1093–1105

Cintron G, Schaeffer-Novelli Y (1984) Methods of studying mangrove structure.In: Snedaker S Snedaker JG (eds) The Mangrove Ecosystem: Research Methods. UNESCO Publication pp 251

Condit NC (1998) Tropical forest census plot: Methods and results from Barro Colorado Island, Panama and a comparison with other plots. Springer, Germany, p 211

Daïnou K, Vermeulen C, Doucet J-L (2008) Consommation de bois dans les zones humides du complexe Ouest du Bénin: Besoins et gestion locale des formations ligneuses. Bois et forêts des tropiques 298(4):13–24

Donato DC (2012) Perspective and parsimony in forest carbon management. Carbon Manag 3(3):227–230

Donato DC, Kauffman JB, Murdiyarso D, Kurnianto S, Stidham M, Kanninen M (2011) Mangroves among the most carbon-rich forests in the tropics. Nat Geosci 4(5):293–297: doi:10.1038/NGEO1123

Giri CE, Ochieng LL, Tieszen Z, Zhu A, Singh T, Loveland J, Masekand N, Duke (2010) Status and distribution of mangrove forests of the world using earth observation satellite data. Glob Ecol Biogeogr 20:154–159

Hanan NP, Kabat P, Dolman AJ, Elbers JA (1998) Photosynthesis and carbon balance of a Sahalian fallow savanna. Glob Change Biol 4:523–538

Husch B, Beers TW, Keuhaw JA Jr (2003) Forest mensuration, 4th edn. Wiley, Hoboken, p 443

Kauffman JB, Donato DC (2012) Protocols for the measurement, monitoring and reporting of structure, biomass and carbon stocks in mangrove forests. Working Paper 86. CIFOR, Bongor, Indonesia

Kutsch WL, Hanan N, Scholes B, McHugh I, Kubheka W, Eckhardt H, Williams C (2008) Response of carbon fluxes to water relations in a savanna ecosystem in South Africa. Biogeosciences 5:1797–1808

Lovelock CE, Ruess RW, Feller IC (2011) CO2 efflux from cleared mangrove Peat. PLoS ONE 6(6):e21279. doi:10.1371/journal.pone.0021279

Merlbold L, Ardö J, Arneth A et al (2009) Precipitation as driver of carbon fluxes in 11 Africain ecosystems. Biogeosciences 6:1027–1041

Pool DG, Snedaker SC, Lugo AE (1977) Structure of mangrove forests in Florida, Puerto Rico, Mexico and Costa Rica.Biotropica 9:195–212

Tomlinson PB (1986) The botany of mangroves. Cambridge University Press, Cambridge, p 419

UNEP (2007) Mangroves of Western and Central Africa. UNEP—Regional Seas Programme/UNEP–WCMC, 88 p

Williams CA, Hanan NP, Baker I, Collatz J, Berry J, Denning AS (2008) Interannual variability of photosynthesis across Africa and its attribution. J Geophys Res 113:G04015. doi:10.1029/2008JG000718

Mangrove Conditions as Indicator for Potential Payment for Ecosystem Services in Some Estuaries of Western Region of Ghana, West Africa

Gordon N. Ajonina, Tundi Agardy, Winnie Lau, Kofi Agbogah, and Balertey Gormey

Abstract

A rapid assessment was carried out on the Greater Amanzule wetlands in Ghana to assess the types and conditions of mangroves and associated ecosystem services (e.g. carbon sequestration, wood collection and tourism) for determining the potential for payment for ecosystem services (PES). A combination of stakeholder discussions and on-the-ground surveys was used to gather information on 18 mangrove sites. The survey showed that over 1,000 ha of mangrove forests exists in scattered pockets of less than 10 ha (in 50 % of the sites), representing nearly 10 % of the known national mangrove coverage of 14,000 ha. The mangroves are estuarine type, generally healthy, and reach canopy height of 30 m. They support livelihoods and ecological securities of the surrounding fishing or farming communities. There was a general community perception that mangrove forests have decreased in area over the last 20 years. Identified threats include harvesting for fuelwood and construction, pollution from domestic and mining waste disruption in the tidal regime at some estuaries. Carbon sequestration remains a valuable service: total aboveground carbon stored in intact mangrove areas ranged from 65 to 422 tC/ha (mean of 185 tC/ha) with estimated aboveground roots (aerial roots) making up 78 % of the carbon stock in some degraded areas. The economic value of mangroves as a source of fuelwood was approximated at US$2,765/ha. A number of factors were identified as contributing to suboptimal governance of mangroves and wetlands. Appropriate PES schemes with improved legal and institutional arrangements are expected to help surmount management challenges.

Keywords

Rapid assessment • Payment for ecosystem services • Mangroves • Ghana

G. N. Ajonina (✉)
CWCS Coastal Forests and Mangrove Programme, BP 54, Mouanko, Littoral Region, Cameroon
e-mail: gnajonina@hotmail.com

T. Agardy · W. Lau
The Marine Ecosystem Service (MARES) Programme, Forest Trends, 1050 Potomac Street N.W, Washington, DC 20007, USA

K. Agbogah · B. Gormey
Coastal Resources Center, P.O. Box MC 11, Takoradi, Ghana

G. N. Ajonina
Institute of Fisheries and Aquatic Sciences, University of Douala (Yabassi), Douala, Cameroon

Introduction

Mangroves cover about 140 km^2 in Ghana. They are limited to very narrow, noncontinuous coastal areas around lagoons in the west of the country; to the east, they are found on the fringes of the lower reaches of the Volta River delta (UNEP 2007). In western Ghana, the most extensive stretches are between Cape Three Points and the border with La Côte d'Ivoire. Six genera of mangroves, typical of west–central African mangroves, are found in Ghana: *Acrostichum aurerum*, *Avicennia germinans*, *Conocarpus erectus*,

S. Diop et al. (eds.), *The Land/Ocean Interactions in the Coastal Zone of West and Central Africa*, Estuaries of the World, DOI: 10.1007/978-3-319-06388-1_13,
© Springer International Publishing Switzerland 2014

Laguncularia racemosa, Rhizophora harrisonii and *Rhizophora racemosa* The open lagoons tend to be dominated by *R. racemosa*, while closed lagoons with an elevated salinity harbour *A. germinans, C. erectus, L. racemosa* and *A. aurerum* (UNEP 2007).

Ghana's mangrove ecosystems are tremendously valuable, providing ecosystem services such as carbon sequestration, protection from storms, floods, erosion, provision of timber and nontimber forest products, processing of waste and nutrient pollution, aquaculture and agriculture support and habitat for aquatic and terrestrial species. In the lower Volta area, the total estimated value, including mangrove-related harvesting for fish smoking or house roofing (US$340/ha) and contribution to marine fisheries (US$165/ha), is over US$500/ha, with total country estimates of well over US$6,000,000/year. These values do not include all the other mangrove ecosystem services (e.g. erosion control, trapping of pollutants and provision of biomass for the detritivore food chain) (Gordon et al. 2009). Yet, as in many other parts of the world, short-term development needs are undermining long-term mangrove health and survival. Between 1980 and 2006, the mangrove area in Ghana fell from 181 to 137 km^2 representing a loss of 24 % (UNEP 2007; Ajonina et al. 2008).

Economic mechanisms have the potential to tip the balance towards restoration, maintenance and protection of mangrove forests. The need for proper valuation of mangrove ecosystem services underpins such mechanisms, including the establishment of any realistic payment for ecosystem services (PES) schemes in any given mangrove or wetland habitat. The objective of this study was to assess the types and conditions of mangroves and associated ecosystem services (e.g. carbon sequestration, wood collection and tourism) in the field for use in feasibility assessment for PES schemes in the mangroves and wetlands in Greater Amanzule and surrounding areas in coastal Western Region of Ghana. This study provides an analysis of the types of mangroves and wetlands, their status and distribution in the coastal Western Region; gathers information about the uses and governance of these wetlands; and provides insights into the use of information gathered in feasibility assessments of the potential for PES in the mangroves and wetlands in the region.

Methods

Site Description

Location of the Survey Area

The survey area lies within the Amanzule wetlands, which is a community-owned wetland with no official conservation status in the eastern and western Nzema traditional areas of the Western Region (Fig. 1). The catchment of the wetland lies within the Western Evergreen Forest zone of Ghana. It covers an area of approximately 100 square km and lies within latitudes 4°53′ and 4°46′ north and longitude 2°00′ and 2°05′ west, 360 km west of Accra. The mangrove areas lie within the Greater Amanzule region in the Jomoro district, Ellembelle, Nzema east and Ahanta west districts of the Western Region.

Biophysical Environment

The region lies in the equatorial climate zone characterized by moderate temperatures. About 75 % of the land area falls within Ghana's high forest zone. The area is characterized by high rainfall with a double maximum and peaks in May–June and October–November each year. The average annual rainfall is 1,600 mm with the relative humidity being 87.5 %, at a mean annual temperature of 26 °C. There is a short dry season (December–March) during which a southwesterly directional wind is experienced with slight Harmattan conditions.

The soil is predominantly Forest Oxysols and Forest Ochrosols-Oxysols intergrades. Its hydrology is driven by six streams, one lake, the Amanzule freshwater lake and one outlet to the sea. The area is subject to seasonal flooding. The Amanzule wetlands is a relatively pristine complex consisting of a freshwater lake, forest, grasslands and rivers. It has the most extensive remaining stand of intact swamp forest in Ghana. These extensive swamps are outside the mangrove areas. Swamps cover about 70 % of the area and associated with a dense hydrological network. Within the area, there is still some logging of an important swamp forest species—*Lophira alata* (ironwood)—for charcoal. The Amanzule wetlands is the only known swamp forest in Ghana and the best example of swamp forest characterized with black humic water. Some 70 % of the site is covered by swamp forest making accessibility difficult, hence contributing to its unspoiled nature. The area has a rich biodiversity and in terms of flora: 33 % of the 237 species of plants identified are endemic to the wetland. Notable species include raffia palm and ironwood (*L. alata*). Faunal reports (GWS 2003) estimate 27 species of mammals, including black-and-white colobus, mona and spot-nosed monkeys; forest squirrels; and Red River hog. Twenty-six species of reptiles and amphibians, including the slender-snouted crocodile and the dwarf crocodile, green mambas and black forest cobra, and 26 species of fish have been recorded. The wetland is relatively rich in indigenous avifauna and also hosts various migrant species. Over 105 species of birds were recorded in an inventory, 65 of which are of global and national conservation interest. The area is classified as an Important Bird Area (IBA) (Birdlife International 2007), and it meets the criteria for designation as a wetland of international importance under the Ramsar

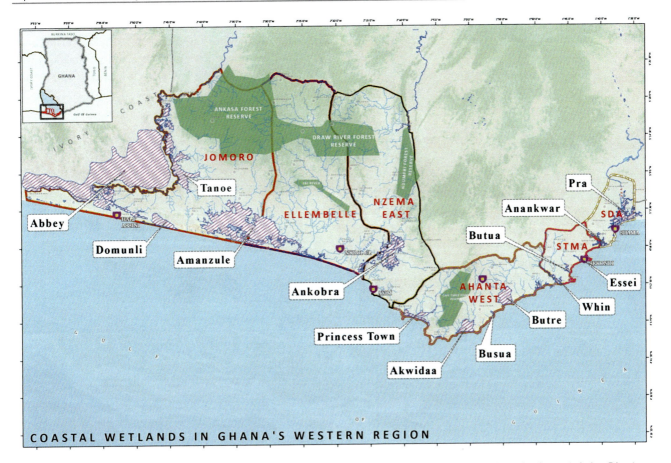

Fig. 1 Map of wetlands distribution in Western Region (Greater Amanzule covers wetlands from Ivory Coast border to Ankobra River)

Convention. Another important biodiversity characteristic of the site is that the coastal sandy beaches are important turtle nesting sites. Species recorded include the leatherback turtle, *Dermochelys coreacea* (common); green turtle, *Cheloniamydas* (common); olive ridley turtle, *Lepidochelys olivacea* (most common); and hawksbill, *Eretmochelys imbricate* (rare) (NCRC/Beyin Beach Resort 2000). Increasing development trends place these wetlands and its biodiversity at further risk. A particular concern is that the wetlands, which are important nursery grounds for many demersal fish species and habitat for birds, mammals and reptiles, are becoming increasingly vulnerable to degradation from development.

Socio-Economic Environment and Human Activities

Generally, the foundation of the economy in Western Region of Ghana and particularly the goods and services traded can be summarized as fish, port services, cocoa, gold, bauxite, manganese, tourism, timber and oil and gas (a new development trend). There is a strong agricultural base, with the agricultural sector employing nearly 60 % of the economically active population in the region. The main crops harvested are rubber, coconut and cocoa, accounting for the largest share of national production. Recently, coconut production has suffered due to Cape Saint Paul wilt disease, which is destroying coconut plantations. Food crop production, on the other hand, is quite low, contributing less than 8 % to national production—which may be due to the lower revenues when compared to other crops (cocoa, coconut, etc.). The six major food crops cultivated in the area are maize, rice, cassava, yam, cocoyam and plantain. Activities within the mangrove areas include fishing, local gin processing, charcoal production, small-scale farming of crops such as coconut, oil palm, cassava, cocoa and plantain, and small-scale trading.

Data and Information Collection and Analysis

Both primary and secondary data collection approaches were used for the study. A combination of stakeholder discussions and on-the-ground surveys was used. A participatory approach was adopted by interactive discussion of basic issues during various consultative meetings and interviews organized with local institutions to gather information about the governance of these wetlands and to discuss the potential for developing a blue carbon project in the region. The institutions included local authorities (district assemblies), private sector operators (especially tourism), community groups, and NGOs, especially Nature Conservation and Research Center (NCRC) and Coastal Resources Center (CRC). Discussions with NCRC and CRC focused on their planned larger-scale carbon mapping work.

Survey of Mangrove Sites

Eighteen mangrove sites were visited during the field trip spanning five days from 11 to 15 September 2011, guided by information on existence of mangroves from many local informants. Data and information collected from each site visited included the following:
- GPS fixes and estimation of mangrove area from maximum width and length for mapping purposes using GIS software (ARC/Info and ArcView),
- mangrove type (estuarine, lagoonal, etc.),
- tidal exchange status,
- health status of the mangrove, based on foliage conditions and live–crown ratios,
- main adjoining vegetation,
- population, institutional and stakeholders issues around mangroves,
- community perception of trends of mangrove area change during the last 10–20 years,
- main activities within mangroves,
- current uses and potential threats,
- options for future management practices and policies.

Assessing Mangrove Ecosystem Services

Carbon storage and sequestration services

The carbon storage and sequestration potential of the mangroves were measured through the establishment of 0.01 ha (10 m × 10 m) temporal sample plots (TSPs) systematically laid at 10-m intervals along a random 30-m transect and compass direction from a fixed point. This is a simplified modification of 6 (10 m × 10 m) plots laid along a 100-m transect used in the central African coastal region of Cameroon (Ajonina 2008). A total of four TSPs were assessed in four mangrove sites (1 TSP per site) and one plot in a degraded mangrove site. The diameter at breast height (at 1.3 m) (dbh) or at 30 cm above the stilt root was measured for all trees within the plot with a diameter tape and stand height determined by means of a Suunto clinometer. Roots and seedlings not measured in the plots were counted in five 1-m^2 plots placed systematically at 1-m intervals along the 10 m × 10 m plot (Fig. 2). Middle diameter and height of the roots and seedlings were also measured.

The tree and stand data obtained were used to obtain estimates for aboveground mangrove forest carbon densities from biomass assessment, following various standard volume estimation procedures (Loetsch et al. 1973; Cailliez 1980; Husch et al. 2003):

Basal area per hectare

$$BA = \pi/4 \sum \left(f_i d_i^2\right)/A \qquad (1)$$

Volume per hectare

$$V_s = BA \cdot H \cdot F \qquad (2)$$

where

BA basal area per hectare (m^2 ha^{-1})
V_s standing volume per hectare (m^3 ha^{-1})
d_i diameter (in metres)
F tree form factor (0.6) (from Ajonina and Usongo 2001)
H canopy height above stilt root (m) or height of roots or seedlings
A plot area in hectares
f_i number of trees in the ith DSR class.

Carbon density was estimated as half the biomass obtained from product of volume and mangrove's mean wood density value of 0.890 t/m^3 (Feamside 1997) and biomass expansion factor of 1.18 (Ajonina 2008).

Mangrove Wood Use and Impacts on Mangrove Forest

A detailed mangrove wood-use survey was carried out in the Effasu fishing community area known for intensive use of mangroves for fish smoking. Mangrove wood stocks owned by households[1] in the area were estimated as well as estimates of turnover rates by members of the household for cooking and fish smoking activities. The information and data from mangrove forest stock assessment were then used to estimate the rate of deforestation (Ajonina and Usongo 2001; Ajonina et al. 2005):

Per household

$$D_h = 52V_h/t \qquad (3)$$

Total rate of deforestation (volume estimate)

$$D_t = nD_h \qquad (4)$$

---

[1] A household was defined in this case as people irrespective of families, sleeping under one roof or living in same house.

Fig. 2 Schematic layout of mangrove stands temporal sample plots (**a**) for measurement of all trees and stems diameter at breast height (1.3 cm) dbh ≥1 cm and (**b**) for counting seedlings and samplings ≤1 cm and stilt roots, with measurement of middle diameter and height of stilt root in 1 m² quadrats systematically laid along transect axis

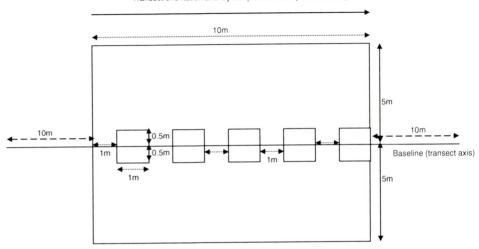

Total rate of deforestation (per hectare estimate)

$$D_r = D_t/V_s \qquad (5)$$

where

- D_h household deforestation rate (m³ household^{-1} year^{-1})
- V_h mean household wood stock (m³)
- t turnover rate (in weeks)
- n total number of households
- D_t total rate of deforestation (m³ year^{-1})
- D_r total rate of deforestation (ha year^{-1})
- V_s mangrove stocking rate (m³ ha^{-1}).

Results and Discussions

Status and Threats of Mangrove Ecosystem Services

General Status of Mangroves

The current general status of the mangroves in the Greater Amanzule wetlands based on the 18 sites surveyed is presented in Table 1 and the mangrove distribution map in Fig. 3.

The survey showed that over 1,000 ha of mangrove forests exists in scattered pockets of less than 10 ha (in 50 % of the sites) in the Amanzule area, representing about 10 % of national mangrove coverage of 14,000 ha following UNEP (2007) subregional survey (Table 1). Most of the mangroves that do not appear in the current vegetation map of the area are of the estuarine mangrove type (61 %) (the area is known to cover about 70 % of swamp forests, including mangroves, but it has been difficult to measure how much has been lost); generally healthy (44 %); reaching a canopy height of 30 m in areas surrounded by coconut plantations (61 %); supporting the livelihoods (from wood demands and use as estimated from this survey); and securities of a surrounding fishing or farming human population, generally of less than 500 people (56 %).

There was a general perception by mangrove communities in 50 % of the sites surveyed that these mangroves have decreased in area over the last 20 years. Identified threats include wood collection for fuelwood, charcoal, construction, etc.; pollution from waste disposal from urban centres and mining activities; and river mouth closure due to sedimentation that impedes tidal flows that maintain mangrove ecological processes. However, there has been no cutting of mangroves in over 60 % of the

Table 1 Appraisal of mangrove status and condition in the Greater Amanzule wetlands (based on 18 mangrove sites), Ghana

Assessment criteria	Ranked category	Number of sites	Percentage
Mangrove type	Estuarine	11	61.1
	Lagoonal	7	38.9
Mangrove areal extent (ha)	0–10	9	50.0
	10–50	1	5.6
	50–100	3	16.7
	100 ha and above	5	27.8
Mangrove canopy height (m)	Less than 10	1	5.6
	10–20	1	5.6
	20–30	11	61.1
	30 and above	5	27.8
Tidal dynamics	Very good (very free tidal flow throughout the year)	6	33.3
	Good (partially blocked in some periods of the year)	8	44.4
	Limited (blocked in some periods of the year)	3	16.7
	Very limited (blocked throughout the year)	1	5.6
Biodiversity status (presence of large mammals especially monkeys within the day)	Many usually seen (more than 10 animals)	3	16.7
	Few usually or occasionally seen (less than 10 animals)	1	5.6
	Not/rarely seen	14	77.8
Health status (visual average proportion of yellow foliage or dead leaves, roots and stems in a given mangrove site)	Very healthy(less than 10 %)	5	27.8
	Healthy (10–30 %)	8	44.4
	Partly degraded (30–50 %)	3	16.7
	Degraded/polluted from waste disposal/mining (more than 50 %)	2	11.1
Main adjoining vegetation	Raphia swamps and other marshes	3	16.7
	Cropland	1	5.6
	Coconut plantations	11	61.1
	Others	3	16.7
Population around mangroves (number of people)	Less than 500	10	55.6
	500–1,000	4	22.2
	1000–5,000	3	16.7
	5000 and above	1	5.6
Community perception of trends in mangrove area change (over the last 20 years)	Increasing	1	5.6
	Stable	8	44.4
	Decreasing	9	50.0
Main activity within mangrove area	Farming	2	11.1
	Fishing	9	50.0
	Urbanization	7	38.9
Current threats	Noncutting (deliberate decision not to cut) or prohibited cutting(by customary laws with fines to defaulters or by conservation organizations)	11	61.1
	Wood collection (for fuelwood, charcoal, construction, etc.)	4	22.2
	Mouth closure due to sedimentation	1	5.6
	Pollution (in situ pollution, waste disposal/mining activities)	2	11.1
Past or current conservation or any intervention efforts by NGOs or partners	Present	7	38.9
	Not present	11	61.1

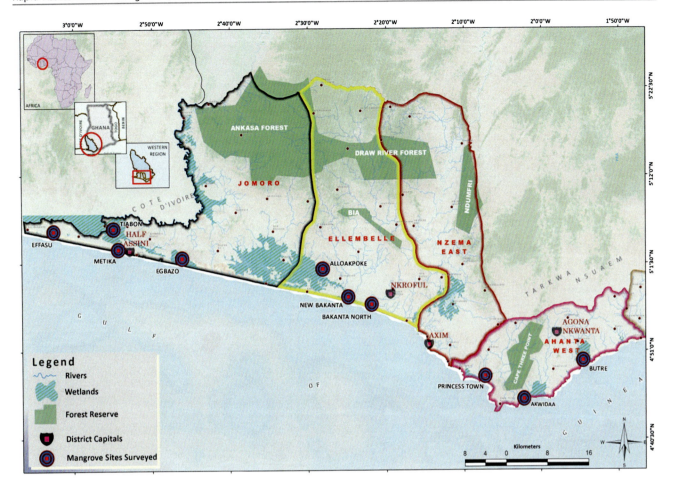

Fig. 3 Map of Greater Amanzule wetlands showing distribution of surveyed mangrove sites

sites due to effective prohibitions. These include community-based by-laws as a result of public education campaigns by NGOs which have been operating in the areas concerned and also the enforcement of traditional norms in some limited cases.

Carbon Storage Potential

The total aboveground carbon stored in intact mangroves in the area (see Table 2 and stem-size class distribution in Fig. 4) ranges from 65 to 422 tC/ha per ha with mean of 185 tC/ha with mangrove aboveground roots (aerial roots), holding in some cases 78 % of the aboveground biomass in degraded mangrove areas. It should be noted that belowground carbon storage accounts for up to 98 % of total whole-ecosystem storage in mangrove ecosystems, based on a recent quantification of whole-ecosystem carbon storage by Donato et al. (2011), so in actual fact, carbon sequestration by mangroves is much larger than measured in this study.

Mangrove Wood Consumption and Impacts on Mangrove Forests

The mangrove is currently valued for fuelwood use at 4,146 Ghanaian cedis (US$2,765) per ha within the Effasu community (Table 3)—well above estimates of US$340/ha in the lower Volta (Gordon et al. 2009). Wood consumption is lower than estimates from a similar typical Cameroon mangrove fishing community (Table 4). Mangrove wood consumption for cooking and fish smoking from the Effasu community is estimated at 16 and 97 m^3/household/year, respectively, and per capita consumption is 1.8 and 10.8 m^3/person/year, respectively. Current deforestation rate for cooking and fish smoking is estimated at 0.06 and 0.34 ha/household/year, respectively, with a per-hectare harvest rate of 0.006 and 0.038 ha/person/year, respectively.

Wood consumption for fish smoking was generally 1.1–1.3 times lower than that for a Cameroonian fishing community owing to the use of more efficient fish smoking ovens.

Table 2 Stand structural characteristics and aboveground carbon stocks of intact and degraded Rhizophora-dominated mangroves in the Greater Amanzule wetlands, Ghana

Community	Type	Status	Stand height (m)	Stand Height above stilt root (m)	Above ground component	Density (Nr/ha)	Basal area /ha (m²/ha)	Volume (m³/ha)	Carbon content Tonnes/ha	As percentage of total	Aboveground root/shoot ratio
Egbazu	Estuarine	Intact	20	17.82	Shoot/stems	2,500	23.43	250.55	111.49	72.71	0.37
					Roots	82,000	46.50	93.88	41.78	27.24	
					Regeneration	4,000	0.31	0.16	0.07	0.05	
					Total				153.34	100.00	
Metika	Lagoonal	Intact	20	16	Shoot/stems	3,000	15.28	146.70	65.28	66.76	0.50
					Roots	32,000	15.91	72.87	32.43	33.16	
					Regeneration	4,000	0.31	0.16	0.07	0.07	
					Total				97.78	100.00	
Bankanta Nor	Estuarine	Intact	20	16.76	Shoot/stems	2,250	69.13	695.14	309.34	73.39	0.36
					Roots	94,000	71.64	251.86	112.08	26.59	
					Regeneration	4,000	0.31	0.16	0.07	0.02	
					Total				421.48	100.00	
Nzelenoanu	Estuarine	Intact	10	7.74	Shoot/stems	6,000	8.74	40.61	18.07	27.84	2.59
					Roots	64,000	44.42	105.00	46.72	72.00	
					Regeneration	6,000	0.47	0.24	0.10	0.16	
					Total				64.90	100.00	
Mean values for intact mangroves					Shoot/stems	3,438	29.15	283.25	126.05	60.2	0.41
					Roots	68,000	44.62	130.90	58.25	39.7	
					Regeneration	4,500	0.35	0.18	0.08	0.1	
					Total				184.37	100.0	
Bakanta (New and Old)	Estuarine	Degraded	15	11.98	Shoot/stems	3,250	17.51	125.84	56.00	19.71	3.96
					Roots	178,000	179.24	498.02	221.62	78.02	
					Regeneration	368,000	28.91	14.45	6.43	2.26	
					Total				284.05	100.00	

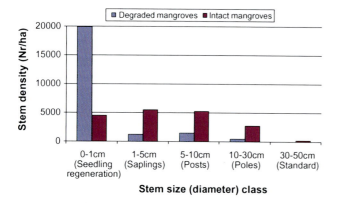

Fig. 4 Stem diameter distribution in mangrove forests of Greater Amanzule wetlands, Ghana

Importance and Opportunities for Local Tourism Industry Development

The Ghana Wildlife Society (GWS) has been playing a critical conservation role in the area, especially as it has introduced small-scale development projects that protect biodiversity while enhancing the economy. In addition, they promote activities that have minimal impacts on the wetlands. To promote tourism, they have built walkways and provided boats to view plant and animal species in the reserve. As a result, the number of tourists has been on the increase since 2006 to more than 10,000 annually (Fig. 5), generating annual revenues of nearly 24,000 Ghanaian cedis (US$16,000).

GWS reaches out to the community and the schools in an effort to educate the population about conservation. In east Nzulezo, they have a task force office that distributes information on protecting the turtle populations in 5 km stretches along the length of the major nesting beaches. The task force works with the local law enforcement to discourage people from eating turtles and turtle eggs. They monitor the turtle nesting sites and collect information on the number of different species, the number of eggs and the distance of the nest from the high tideline. To protect the turtle populations, they will sometimes move the eggs to a hatchery until the turtles are ready for release to the sea. In the last two years, they have recorded over 200 turtle nesting sites. And villagers who once hunted the turtles now volunteer on the task force to protect them. Six of the local communities have donated land to the reserve and work with GWS on various

Table 3 Characteristics of mangrove wood use and per ha value of mangroves for fuelwood within a typical fishing community (Effasu, Greater Amanzule wetlands, Ghana)

Characteristics	Sub-characteristics	Estimates
Population		260
Number of households		29
Household size		9
Fish species smoked		Tele, *Sardenella* spp
Provenance of mangrove wood		Bought from wood cutters
Number of wood stacks		23
Wood stack dimensions (maximum)	Length (m)	2.9
	Width (m)	2.0
	Height (m)	1.4
	Volume (m^3)	8.1
Wood cost	Cost/stack	60 Ghana cedis (US$40)
	Cost/stere	7.41 Ghana cedis (US$4.94)
	Cost/m^3	14.64 Ghana cedis (US$9.76)
	Cost/tonne	16.44 Ghana cedis (US$10.96)
	Cost/ha[a]	4,147 Ghana cedis (US$2,765)

[a] Based on mangrove wood stocking rate (excluding roots) of 283.25 m^3/ha (Table 2)

Table 4 Community fuelwood consumption and impacts on surrounding mangrove forests within the Effasu fishing community (Greater Amanzule wetlands, Ghana)

Region	Activity	Turnover rates (m^3/year)		Deforestation rates (ha/year)		Source
		Per household	Per capita	Per household	Per capita	
West Africa (Effasu mangrove community Ghana)	Cooking	15.83	1.76	0.06	0.006	This survey
	Fish smoking	97.44	10.83	0.34	0.038	
Central Africa (Douala-Edea mangrove villages: Mbiako, Yoyo I & II)	Cooking	19.74	2.50		0.004	Ajonina and Usongo (2001)
	Fish smoking	105.58	13.39		0.024	

projects. In return, the profits are shared among the communities. The people take pride in their communities and the reserve, and the success of the project has provided electricity and better roads in the villages. The people now harvest and store fish instead of turtles and profit from tourist activities including homestays. The efforts of GWS have provided a means of sustainable development for the lagoon and reserve that have caught the attention of the government of Ghana and private tourism operators, especially the Eco-Lodge at Beyin (Fig. 7). Recently, they recognized the success of the Amanzule Conservation and Integrated Development Project during a presentation at a government conference (Fig. 6).

Prospects of Oil and Gas Industry

A number of interesting concerns are posed and questions asked in a recent coastal publication Hen Mpoano—Our Coast, Our Future, Western Region of Ghana (Anon. 2010) concerning booming prospects of oil and gas and the impacts of offshore and onshore construction and operations on both the marine and the terrestrial environments. The publication contends that today in the Western Region, the multiple causes and consequences of ecosystem change are overshadowed by the prospect of oil and gas development. How the exploitation of this new source of wealth, employment and development pressures is managed is said to determine whether the result will be a new and generous source of national income with employment and business opportunities that benefit both the Western Region and the country as a whole, or yet another example of the curse of oil. Questions posed include the following: how will the jobs and the wealth produced be distributed? Will local Ghanaian companies be formed to provide the services required by new industries? Will the local labour force be trained to fill new employment opportunities or will skilled labour be imported and earnings flow overseas? Who will benefit and who will lose? The discovery of oil and gas and associated onshore infrastructure development will undoubtedly impact heavily on natural resources and especially local livelihoods. In addition to the above concerns, the mangrove environments and the delivery

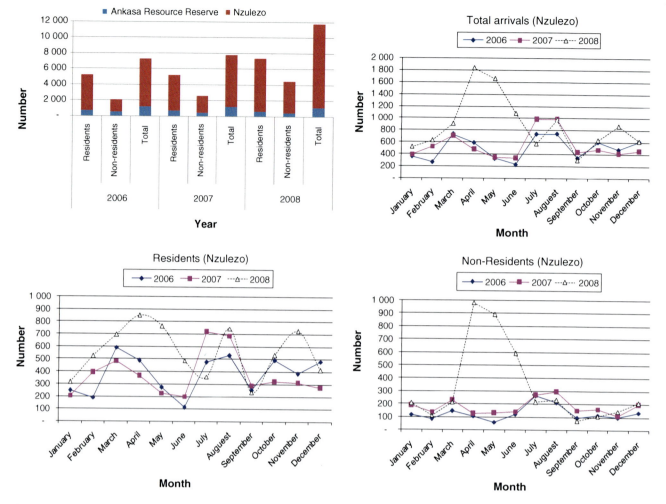

Fig. 5 Monthly and yearly distribution of tourist visits from two touristic sites located around Amanzule wetlands, Ghana (*Source* Data from Ghanaian tourism review)

of services from them will be affected. How to get oil companies to pay for offsets and get involved in mangrove restoration and conservation remains a challenge. They need to be guided within a certain platform on mangrove ecosystem valuation information on what society will lose if the ecosystem services provided by the mangroves are lost, and they must pay costs for their replacement/restoration of these degraded ecosystems.

Institutional Issues and Governances of Wetlands and Potentials for a Coastal and Marine PES Scheme

Institutions and Stakeholders for Mangrove Management

The institutions and stakeholders in the management of mangrove forests in the Western Region of Ghana reflect a broad range of interests at the national, regional and local community level. The role played by each of these stakeholders has a direct or indirect impact on the resource.

At the apex of the stakeholder spectrum is the national government whose policies and programmes for management of mangroves and associated wetlands are implemented through the Wildlife Division of the Forestry Commission and other related agencies such as the Lands Commission and the Department of Town and Country Planning. At the regional level, there is the Regional Coordinating Council. The local level is represented by a number of players or interest groups whose actions or inactions determine the outcome of all mangrove management efforts now and in the foreseeable future. These include, among others, the district assemblies with their substructures, civil society groups, producer groups and trade associations, traditional authorities, commercial organizations and local communities around the mangrove forests. Stakeholders and institutional analysis from the policy to field levels are presented in Tables 5 and 6. In each tier, the major actors are identified, together with their roles and responsibilities, and

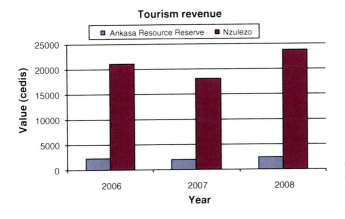

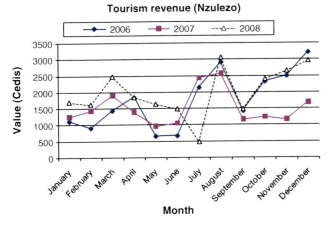

Fig. 6 Monthly and yearly trends in tourism revenue from two touristic sites located around Amanzule wetlands, Ghana (*Source* Data from Ghanaian tourism review, Ghana Tourist Board, Ministry of Tourism, Ghana)

an assessment of the current level of implementation is made with suggested strategies for implementation towards sustainable mangrove management.

The following general observations can be made:
- First-level institutional and stakeholder analysis (central government, department agencies, regional administration, academic research institutions, local government and donor agencies).
  - Weak enforcement of existing legislation.
  - Good structures in place but severely limited by logistical capacity of government departments.
  - Lack of communication and information sharing to intended users.
  - Poor attitude towards mangrove and wetlands which is not taken into account in any development planning processes.
  - Donor funds that address short-term solutions to long-term problems with poor level of private sector participation.
- Second-level institutional and stakeholder analysis (district assemblies, traditional authorities, local communities, NGOs and other civil society groups).
  - Ignorance of roles of wetlands that are taken for granted.
  - Land tenure conflicts.
  - Lack of long-term financing leading to periods of neglect in working with communities as less than 40 % of the mangrove sites have received any intervention (Table 1).
  - Lack of adequate knowledge and capacity for wetland management.

Conclusions and Recommendations

Conclusions

From this rapid assessment exercise, limited in time and by certain information gaps, the following conclusions can be drawn:

- *Types, distribution, values and status of mangroves and associated wetlands in the Amanzule region.*

More than 1,000 ha of mostly estuarine mangrove forests exists in scattered pockets in the Amanzule area, representing about 10 % of national mangrove coverage with most of the mangroves not captured in the current vegetation maps of the area. These mangroves are generally healthy, supporting the livelihood and ecological securities of surrounding fishing or farming human population. There was a general perception by mangrove communities that these mangroves have decreased in area over the last 20 years with identified threats that include wood collection for fuelwood, charcoal and construction; pollution from wastes disposal, especially in around urban centres and mining activities; and mouth closure due to sedimentation that impedes the tidal flows that maintain mangrove ecological processes.

The total aboveground carbon stored in intact mangroves in the area is also important and can be comparable to similar mangrove systems.

The mangrove is currently valued for fuelwood use at 4,146 Ghana cedis (US$2,765) per ha with a significant impact on mangrove forest.

- *Institutional issues and governance of wetlands in the area.*
The various institutions and stakeholders at the national, regional and local community levels have varying interests which should be considered in any effort aimed at managing the wetlands in the area *Feasibility of a PES scheme*.

A PES scheme may be feasible when addressed within the context of the institutional and stakeholders in the region. The challenge is how to get the private sector, especially the oil companies, interested in investing in ecosystem services protection. Already, private tourism initiatives hold promise as some have started green investments. Oil and gas companies need to be guided within a certain platform on mangrove ecosystem valuation information on what society

Fig. 7 Private tourism operators' case of Eco-Lodge at Beyin (the proprietor in red shorts) (Photograph by Nicolas Jengere)

will lose if the ecosystem services provided by the mangroves are lost, and they must pay for their replacement/restoration.

Recommendations

Short-Term Recommendations

- Addressing Stakeholders and Instructional Issues
  - *Development of stakeholders contact database.* All the groups working on mangroves need to be identified through the establishment of a database initiated by either themselves or a state institution like the Wildlife Division of the Forestry Commission.
  - *Organization of an expanded stakeholder forum.* In order to resolve conflict issues on mangrove and coastal resource management issues, an expanded forum of key stakeholders, preferably convened by the government with NGOs playing a facilitating role, is needed to identify and chart out a common vision and workable strategies for mangrove conservation within the general integrated coastal management discussion. The ongoing organization of the proposed coastal forum by Coastal Resources Center (CRC) could be a window opportunity.
  - *Cross-institutional capacity building.* Need to strengthen the operational capacities of governmental organizations by providing them basic logistics, especially transport, etc.
  - *Sharing of research finding.* Mechanisms should be developed to promote communication and sharing of research results either through forums or through any communication media.
  - *Sensitization campaigns.* Need for adequate sensitization campaigns to strengthen the awareness on the importance of mangroves, the threats affecting them and the need to conserve wetlands.
- Completion of Ghana carbon map and mangrove carbon monitoring.

There is need to complete the current Ghana carbon map with the mangrove component using appropriate methodologies, especially those currently implemented within the central African mangrove forests with some establishment of long-term permanent sample plot system for measuring mangrove carbon sequestration and ecosystem dynamics under various human extraction regimes.

Medium- and Long-Term Recommendations

- Addressing the direct on-site threats to mangrove ecosystem services is critically important in order to develop effective management/conservation policies and to lay the groundwork for innovative financing, including PES schemes.
- A focus on areas for potential wet carbon benefits via avoided deforestation or the peat/swamp forests and restoration of mangroves (Table 7) could provide the necessary foundation for PES.

Rapid Assessment of Mangrove Conditions for Potential Payment

Table 5 First-level institutional and stakeholder analysis for mangrove management

Institution/ stakeholder	Roles/responsibilities	Assessment/remarks	Strategies for improvement towards sustainable mangrove management
Central government	Policies, treaties and laws (forestry and wildlife policy 1994, Ramsar treaty 1971, etc.)	Laws in place, but enforcement is weak	There is need for legal reforms that harmonize existing laws/acts to reduce overlap and promote their enforcement
Departments/ agencies	Implementation, regulation, enforcement, compliance, monitoring and evaluation. Wildlife division is present in the region but is largely focused on game reserves, e.g. Ankasa	Structures in place, but are not functioning well due to severe limitations in logistics (vehicles, communication for field monitoring operations)	More training needed (on efficient use and maintenance of logistic equipment) and more resources to be provided by central government. Mangrove conservation objectives need to be emphasized and integrated with wildlife reserve objectives as an integrated ecological framework. This means that protected areas containing mangroves should specifically state in one of their objectives that it will ensure protection and sustainable use of mangrove forests as fisheries nurseries grounds and protection against storms
Regional administration	Complementary and supervisory	Does have capacity needed for mangrove management. Recently efforts are being made by some NGOs to sensitize them	Its role is important and needs to be involved at the policy and implementation levels
Academic and research institutions	They are to provide research information on state of wetlands and conditions of mangroves	There have been research efforts by some academic researchers from University of Ghana and some NGOs such as Wildlife Society and others. However, the research findings are difficult to come by except on the Internet sometimes and some foreign journals	There is the need to coordinate the process of sharing research findings for wider dissemination of information and in order to avoid duplication of efforts
Local government	Regulation, enforcement and implementation	With respect to mangrove and wetland management, the six coastal districts have not done much. The laws have not been enforced due to the general poor attitude towards wetlands. The weak presence of the Wildlife Division in the region has also contributed to this poor state of affairs	Local government structures need to be equipped with adequate knowledge on the importance of wetlands in addition to resources to be able to support the division and other NGOs to protect and conserve the mangroves
Donor/funding agencies	Provision of funds and technical advice	They have been providing funds for long time now, e.g. WB, EU countries and other bilateral donors. Some of these funds addressed the problems in the short-to-medium terms. In most instances, the projects stalled as funding dried up and this has created more problems for resource management	Project planning and funding should be oriented towards long-term goals to ensure sustainability. All partners need to work together to achieve this. The Ghanaian government has a role to play by factoring this into annual national budgets. It should not be the sole responsibility of outside donors. Commercial bodies whose actions impact on the wetlands should also be brought on board, e.g. hard and soft mineral producing companies. Possibilities of a carbon tax/green tax may be explored

- Conservation should be the focus in sites with significant peat/swamp forests, especially Ekumkpole and Amanzule, and significant healthy mangroves, especially Alengezule, Domunli, Kikam, Ankobra, Awhuele, Ezele, Butre, Ndabelah and Ampain.
- Potential areas for restoration include the following:

- *Hydrological restoration with replanting.* For sites with mouth closure due to sedimentation through channel opening to allow normal tidal exchange with mangrove areas or badly degraded sites especially Amanzule and Effasu.
- *Waste removal.* Particularly for urban and peri-urban polluted sites that include Yan and Awine-anloa areas.

Table 6 Second-level local institutional and stakeholder analysis for mangrove management

Institution/ stakeholder	Roles/responsibilities	Assessment/remarks	Strategies for improvement towards sustainable mangrove management
District assemblies	Regulation, permitting enforcement	Laws are not enforced due in part to ignorance about importance of wetlands. Limited resources to visit sites for inspection and enforcement	Workshops should be conducted to educate assemblies. They should be involved in the design and implementation of future projects concerning wetlands backed by some logistics (basic field operation equipment and transport vehicles, fuelling, etc.) for the task
Traditional authorities	Are custodians of the land and are supposed to regulate its usage through the granting of tenure rights, etc.	They have not been playing this role effectively. In most cases, they lose control over the management of land in their areas. Most, if not all of them, are totally ignorant about the importance of wetlands in their areas	Their involvement through education and sensitization is very important in this regard that can help bring about local ownership of the proposed interventions
Local communities	In most cases, depend directly on the resource and are to help manage it	Their direct activities in most cases have been impacting negatively on mangrove management in local communities aggravated by ignorance, tenure right controversies and concerns of settler fisher folks. Limited livelihood opportunities and severe poverty associated with downturn in fishing and associated activities are forcing most of them to exploit aqua forests at a commercial scale. Local perceptions about mangrove forests are not healthy who see them as ordinary trees which can grow again after harvesting	Sensitization of local communities, redefining of tenure rights and resolution of concerns of settler fisher folks are very necessary in addressing mangrove management problems in their areas of jurisdiction. Alternative livelihood schemes need to be identified and promoted in local communities
NGOs and other civil society groups	Inform and educate local communities on sustainable management of mangroves	Problem with securing long-term finance has led to long periods of neglect (in activity with the local communities) in working with communities, e.g. CRC from Italy did a lot of work in the past but has reduced its presence in local communities due to limited funds. Wildlife Society is present in some local communities, but their activities have been limited. The majority of them are concentrated in the regional capitals. Capacities are also limited in the area of scientific and technical appreciation of wetland ecology	All the groups working on mangroves need to be identified through the establishment of a database initiated by either themselves or a state institution like the Wildlife Division of the Forestry Commission. This should not be seen as a means to control their activities. Their capacities should also be built, and long-term financing mechanisms secured

To capture the full range for such wet carbon initiatives, a specific targeted research agenda may be needed including getting reliable estimates of belowground carbon stocks, growth rates and carbon sequestration potentials of mangroves through laying out and periodically measuring permanent sample mangrove plots as ongoing in central African mangroves.

- Policy review leading to reforms may be needed to revise existing laws with more research to enhance policies to encourage stakeholder communication, actual management and enforcement.
- Sustainable funding through private sector participation should be encouraged and should constitute a main focus of mangrove conservation efforts. Donor partners should

Table 7 Site-targeted specific actions for wet carbon initiatives

Site name	Community	Species composition	Extent (ha)	Other characteristics
(a) *For conservation*				
Allengenzule	Allengenzule	*Rhizophora*, *Avicennia*	0.1	Used to be crocodiles, little threats, nonhuman use, etc.
Domunli	Egbazo	*Rhizophora*	75.0	Low wood collection for cooking, construction and minimal threats
Ekumkpole	Metika (1 &2)	Extensive *Rhizophora* stands with adjoining Raphia swamps	250.0	Monkeys usually seen, low wood collection for cooking, construction
Amanzule North	Aloekpoke	Very few *Rhizophora* stands (less than 1/ha), large swamp mixed of ironwood (*L. alata* with tree density of 50/ha with diameter more than 1 m)	200.0	Birds, fishes, wood for charcoal, fishing especially black tilapia with charcoal production
Amanzule	Nzelenoanu	*Rhizophora*	180.0	Abundance of crabs which are collected by the population including fuelwood collection
Amanzule	Kikam City	*Rhizophora* dominated with adjoining	12.0	Nonidentified threats
Ankobra	Tiaboe	*Rhizophora* dominated with adjoining coconut plantations	8.0	Nonidentified threats
Awhuele	Princess town/Anleo village	Mixed mangrove species with adjoining grassy hills with mixed tree crop farms and fallows	75.0	Monkeys present with nonidentifiable threat especially wood harvest prohibited by traditional authorities
Ezele	Akyenum	Mixed mangrove species	150.0	Nonidentified threats
Butre	Butre	Mixed mangrove species	7.5	Nonidentified threats
Ndabelah	Nzimitianu	Large swamp communities with some spotted *Rhizophora* and *Avicennia* stands	5.0	Nonidentified threats
Ampain	Ampain	More or less pure stands of *Avicennia* (900 trees/ha), Conocarpus, *Rhizophora*	2.0	Nonidentified threats
(b) *For restoration through replanting and/or hydrological restoration*				
Amanzule	Bakanta (New and Old) and north	Vast stands of *Rhizophora* with adjoining coconut plantation	475.0	Charcoal production, wood for cooking, wood for construction
Amanzule	Chiko (Half Assini Anowa)	*Rhizophora*	0.5	Threats of mouth closure
Effasu	Effasu	*Rhizophora*	2.6	Heavy wood collection for fish smoking with threats of mouth closure
(c) *For waste removal with hydrological restoration*				
Yan	Princess town/Anleo	Mixed mangrove species	2.0	Area with dumped wastes
Awine-anloa	Fante (Half Assini Fante)	*Rhizophora*	1.5	Though few monkeys occasionally visit the area, there is heavy dumping of wastes in the area

work with all partners to put in place sustainable funding mechanisms based on existing funding sources, especially to stimulate private sector investment.

Acknowledgments The assessment was supported by USAID through the Coastal Resources Center, Ghana, and Forest Trends, USA. The authors are gratefully indebted to persons, projects, programmes and institutions contacted during the study. We appreciate greatly the field assistance and contribution of Nicolas Jengere of Nature Conservation and Research Center (NCRC) and Felix Nani of wildlife division for their invaluable field assistance without which the work could not have been possible. Great thanks are also due Patrick Sarpong proprietor of Beyin Beach Resort for hospitality and responding to the interview. John Mason of NCRC, Frank Hicks (Forest Trends) and Joerg Seifert-Granzin (Forest Trends) provided initial contacts that led to this work. We also thank Esinam Attipoe (Hanns Seidel Foundation, Ghana) for sending complementary information about the area.

References

Ajonina GN (2008) Inventory and modelling mangrove forest stand dynamics following different levels of wood exploitation pressures in the Douala-Edea Atlantic coast of Cameroon, Central Africa. Mitteilungen der Abteilungenfür Fors tlicheBiometrie, Albert-Ludwigs- Universität Freiburg. 2. 215 p

Ajonina GN, Usongo L (2001) Preliminary quantitative impact assessment of wood extraction on the mangroves of Douala-Edea Forest Reserve. Cameroon Trop Biodivers 7(2–3):137–149

Ajonina G, Abdoulaye D, Kairo J (2008) Current status and conservation of mangroves in Africa: An overview. WRM Bull 133, August 2008. http://wrmbulletin.wordpress.com/2008/08/25/current-status-and-conservation-ofmangroves-in-africa-an-overview/

Ajonina PU, Ajonina GN, Jin E, Mekongo F, Ayissi I, Usongo L (2005) Gender roles and economics of exploitation, processing and marketing of bivalves and impacts on forest resources in the Douala-Edea Wildlife Reserve, Cameroon. Int J Sustain Dev World Ecol 12:161–172

Anon. 2010. Hen Mpoano.Our coast, our future, Western Region of Ghana.Building capacity for adapting to a rapidly changing coastal zone, p 64

Barbier EB, Acreman MC, Knowler D (1997) Economic valuation of wetlands. A guide for policy makers and planners. Ramsar convention bureau gland, Switzerland, p 143

Birdlife International (2007) Birdlife's online World Bird Database: the site for biodiversity conservation. Version 2.1. Cambridge, UK: Birdlife International. Available: http://www.birdlife.orgn

Cailliez F (1980) Forest volume estimation and yield prediction, vol 1. Volume estimation, FAO Forestry paper no. 22 FAO, Rome

Donato DC, Kauffman JB, Murdiyarso D, SofyanKurnianto S, Melanie Stidham M. MarkkuKanninen M (2011) Mangroves among the most carbon-rich forests in the tropics. Nat Geosci 4, doi: 10.1038/NGEO1123

Feamside PM (1997) Wood density for estimating forest biomass in Brazilian, Amazonia. For Ecol Manage 90:59–87

Feka NZ, Chuyong GB, Ajonina GN (2009) Sustainable utilization of mangroves using improved fish smoking systems: a management perspective from the Douala-Edea Wildlife Reserve, Cameroon. Trop Conserv Sci 4:450–468

Gordon C, Tweneboah E, Mensah AM, Ayivor JS (2009) The application of the ecosystem approach to mangrove management: lessons for Ghana. Nat Faune 24:30–41

GWS (2003) Amanzule wetlands ecosystem management plan 2003. Ghana Wildlife Society

Husch B, Beers TW, Keuhaw JA (2003).Forest mensuration, 4th edn. Wiley, 443 p

Loetsch I, Zohier F, Haller K (1973) Forest inventory, vol 2. Second Edition BIV, Germany, 469 p

NCRC/Beyin Beach Resort (2000) Sea turtle conservation on the west coast of Ghana: a background report

UNEP (2007) Mangroves of Western and Central Africa. UNEP-regional seas programme/UNEPWCMC. 92 p

Plantation Agriculture as a Driver of Deforestation and Degradation of Central African Coastal Estuarine Forest Landscape of South-Western Cameroon

Patience U. Ajonina, Francis A. Adesina, and Oluwagbenga O. I. Orimoogunje

Abstract

Plantation agriculture has a long history of establishment in Cameroon and is increasing at an unprecedented rate with detrimental impacts on coastal estuarine forest landscape. Remote sensing data from Landsat imageries and geographic information system (GIS) techniques were used to analyse changes in the areal extent of plantations within the coastal Atlantic estuarine forest complex area of Cameroon between the periods 1986, 2000 and 2011 to ascertain the extent of deforestation due to plantation agriculture. Given the base year of 1986 (67,792 ha of plantation), the results showed a 67 and 47 % decrease in the dense coastal estuarine forest coverage in 2000 (14,032 ha) and 2011 (24,564 ha), respectively, in the area and an increase in the area occupied by plantations (51,295 ha in 2000 and 68,340 ha in 2011) giving an annual loss of 3.4 % estuarine forest complex and an increase in plantation area of 0.03 % from the periods 1986 to 2011. There is need for better plantation management practices and policies to curb further loss in estuarine forest cover with consequent implications on the Wouri estuary.

Keywords

Plantation agriculture • Coastal estuarine forests • Impact • Deforestation • Degradation • Forest conversion

Introduction

The tropical rainforest including coastal estuarine forests constitutes the most valuable ecosystem of the planet containing 50–90 % of animal and vegetal species on earth (CIDA 2001). They represent over 57 % of global forest cover and are home to over 500 million people (CIDA 2001). Central and West Africa's 200 million ha of tropical rainforest are the second largest remaining humid tropical forest in the world after the Amazon basin (FAO 1995). They are also known not only for their rich biodiversity but also for the essential ecosystem services such as nutrient and water cycling as well as climate moderation necessary for the proper functioning of the biosphere and the welfare and well-being of humans (Daily 1997; Millenium Ecosystem Assessment 2005; DeFries et al. 2004). Whenever the forest is opened up for agricultural purposes including plantations, the subsisting natural equilibrium in the forest ecosystem is disrupted and the services rendered impaired. Plantation agriculture in particular is established by clearing large areas of natural plant cover for the cultivation of a crop such as oil palm (*Elaeis guineensis* Jacq.), rubber (*Hevea brasiliensis*), cocoa (*Theobroma cacao*), coffee (*Coffea* sp.) and other perennial crops. There are many perspectives of what

P. U. Ajonina (✉)
Department of Geography, University of Buea, Buea, Cameroon
e-mail: pusongo@yahoo.com

F. A. Adesina · O. O. I. Orimoogunje
Department of Geography, Obafemi Awolowo University, Ile-Ife, Nigeria

S. Diop et al. (eds.), *The Land/Ocean Interactions in the Coastal Zone of West and Central Africa*, Estuaries of the World, DOI: 10.1007/978-3-319-06388-1_14,
© Springer International Publishing Switzerland 2014

plantation agriculture is, but it is most frequently referred to as a large scale, usually foreign-owned and specialised high-input/high-output farming system that is export oriented (Courtenay 1980). The plantations owned both on large-scale investors and by smallholders are key sources of income for many farmers and countries in the tropics.

Several studies carried out have shown that deforestation due to agricultural activities often leads to severe transformations and degradation of the coastal aquatic ecosystems, especially through water quality changes. Metals tend to be assimilated in sediment with organic matter, Fe/Mn oxides, sulphides and clay thus forming several reactive components which are harmful to the environment (Wang and Chen 2000; Praveena et al. 2010). Hence, sediment is always regarded as the potential reservoir for metals and plays an important role in adsorption of dissolved heavy metals (Wang and Chen 2000; Praveena et al. 2010). Under different physical and chemical conditions, metals in sediment may leach out into the water column as free ions. In turn, contaminated sediments also act as sources of heavy metals when released into the river water. Metals concentration in river water can be regarded as a good indicator of the river contamination. Metal ions can be either an essential nutrient or toxic to living organisms (Salomons and Förstner 1984). When the metals concentration exceeds standard permissible limits, it would have toxic effects on living organisms and cause negative impact on lower life forms. When these metals enter into the food chain through phytoplankton and are biomagnified in aquatic organisms, they would pose potential risk to human health (Alkarkhi et al. 2009).

Moreover, deforestation and degradation following plantation establishment are also important factors of climate change contributing up to 30 % of greenhouse gases especially carbon dioxide (CO_2) that is known to cause global warming with consequent implications on the rise in sea level (Mitchell et al. 2007) especially as forests are carbon dioxide 'sinks' informing many of the current international efforts at mitigating climate change with current initiatives such as reducing emissions from deforestation and degradation (REDD+). The UN- REDD+ works to benefit the climate by preserving the tropical forest ecosystems to enhance the continuous delivery of environmental goods and services to ameliorate climate change impacts. Therefore, effective management of coastal agroforests as well as plantations will enhance the prospect of deriving environmental benefits of forests from them. This study aims at analysing changes in the areal extent of plantations within the coastal Atlantic estuarine forest complex area of Cameroon between the periods 1986, 2000 and 2011 to ascertain the extent of deforestation due to plantation agriculture as a basis of searching for ways to enhancing sustainability of the plantation ecosystem.

Methods

Site Description

The south-west region of Cameroon is located approximately between latitude 4° and 6° and longitude 8° and 9°45′ east of the equator, covering an area of approximately 24,571 km². It is bordered to the north by the Federal Republic of Nigeria, to the east by the north-west region, to the west by the Atlantic Ocean and to the south by the Littoral Region (See map, Fig. 1). The area is characterised by a humid equatorial climatic type with high rainfall. Due to the orientation of the slopes of Mount Cameroon, it receives the monsoon winds at right angle. As a result of this, the moisture transported by the monsoon from the ocean is deposited in the form of heavy rainfall. The Limbe area has an annual rainfall average of 4,050 mm (Neba 1987). Rainfall reaches a peak of about 7,000 mm in July and August. Debuncha that is the wettest place in West and Central Africa has an annual rainfall of 10,000 mm which is the highest in the region. The relative humidity of the area is strongly influenced by its maritime location and is at 75–80 %. The maximum annual temperature of the area is 25 °C, while the minimum is 24.7 °C suggesting a relatively high constant temperature with a low range. Neba (1987) puts the annual temperature at 24 °C. The vegetation of the area is characterised by the equatorial rainforest type. It forms the discontinuity of the swampy vegetation which is dominant along the Cameroon coastline (Neba 1987). The luxuriant evergreen forest is comprised of a variety of tree species such as *Lophira alata, Sacoglottis gabonensis and Erythrophleum ivorensis*. The largest National Park—Korup National Park—with many rare and endangered species and the Limbe Botanic garden among others are found in this region. As a result of man's intervention, the climax vegetation has been cut down for human activities such as agriculture, communication, transportation and urbanisation. The secondary vegetation is now widespread with the luxuriant vegetation occupying peripheral positions mostly at high altitudes (Neba 1987). One conspicuous feature of human intervention is the domination of the vegetation by Cameroon Development Cooperation (CDC) plantations at the foot of Mount Cameroon (Fig. 2).

Mount Cameroon (4,100 m), the highest physical feature in West Africa and an active volcanic mountain which lastly erupted in May 2000, is found in this region (Neba 1987). The area is well drained and is an important watershed as most of the streams take their sources from Mount Cameroon and flow across the west slopes cutting V-shaped and U-shaped valleys before draining into the sea. The Wouri estuary lies to the east of Mount Cameroon and empties into the Bight of Biafra (Fig. 3). It is fed by the Mungo, Wouri

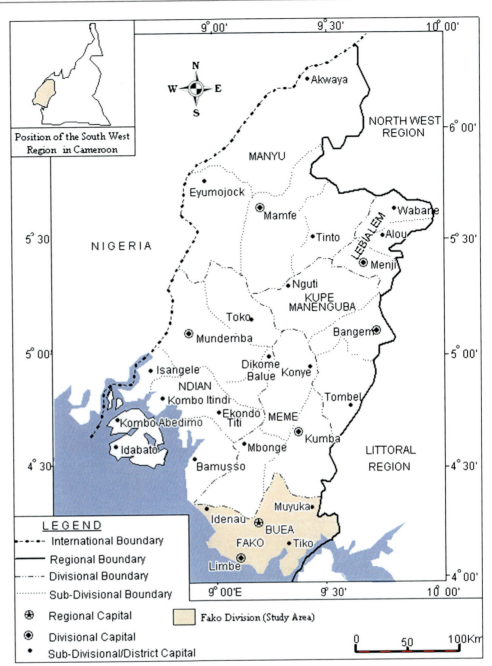

Fig. 1 Map of the south-west region showing the study area

and Dibamba rivers. To the west of the estuary are the slopes of Mount Cameroon covered with banana plantations.

Agriculture is the mainstay of the Cameroon economy. Pesticides are not regulated and also contribute to pollution. Pesticides that have long been banned elsewhere are still in use. The growing population is increasing production of export crops such as coffee, cocoa, bananas and palm oil, using imported pesticides and fertilizers. Typically fertilizers contain urea, ammonia and phosphorus. Pesticides applied are mostly DDT and other derivatives of organohalogens (Sama 1996).

Overview of Plantation Agriculture in Cameroon

Plantation agriculture managed by the CDC is an important economic activity in south-western Cameroon, employing over 15,856 workers and contributing significantly to the national economy. The 2002 statistics of the Consultative Group for International Agricultural Research (CGIAR) estimates that agriculture accounts for 43 % of the GDP of the area. Rubber and banana have been grown in the forest zone of south-western Cameroon for close to a century (CDC Reports). Plantation agriculture began in the south-western region of Cameroon in 1885, when the two leading German

Fig. 2 A three-dimensional image of the area showing the location of plantations

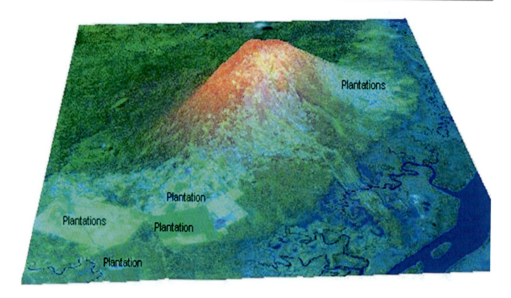

Fig. 3 Map showing the Wouri estuary and the rivers

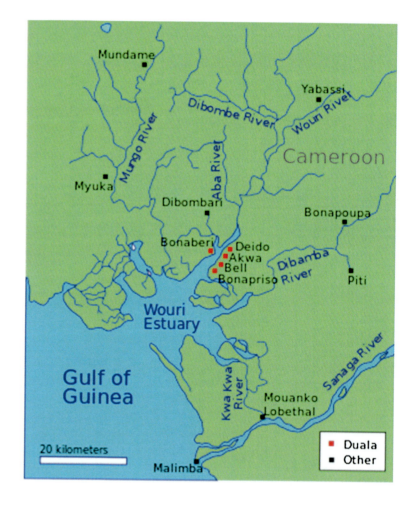

firms in German Cameroon, Woerman and Jantzen und Thormalen, founded two plantation companies (Courtenay 1980). The massive expropriation by these two firms altered the sociology and history of the various Bakweri clans. German companies exploited the traditional land ownership of the native people to their advantage. The vast level land was used for plantation agriculture and the people were forcefully displaced from their homes and herded off onto strange and unfriendly patches of land around the plantations. In a nutshell, the Germans alienated about 400 square miles of the most fertile land around the Mount Cameroon alone and stripped the people of over 200,000 acres of their most fertile lands. The size of the plantations kept on growing under British supervision following the defeat of the Germans after the First World War (Cameroon Development Corporation Report 1970). The spread of plantations slowed during the depression years between 1920 and 1940. However, following the return of peace and favourable conditions for international trade and investments after the end of World War II in 1945, the plantation system started expanding once again. In January 1947, CDC was created to take over and manage the plantations. The expansion continued through the following two decades. It accelerated during the immediate post-Independence era, about 1960–1965, which saw direct involvement of the national governments of the newly independent states in the establishment and running of plantations. Thereafter, the expansion slackened, mainly because of a decline in external investments following the growing political instability and state control of the national economies and the attendant erosion of foreign investors' confidence in them (Halfani and Barker 1984). Since about 1975, the plantation system has witnessed rejuvenation as a strategy for stimulating agricultural production. In the south-western region of Cameroon, the area under plantations increased from about 20,000 in 1960 to over 73,788 ha in 2006 (CDC Report 2006).

Data Collection

Relevant data for the study were collected from both primary and secondary sources. Primary data set was collected from the field with the use of the global positioning system (GPS). Secondary data included data collection from documented sources especially plantation records of the Cameroon Development Corporation on plantation ages, areal extent, cultural practices, yields, economics, among others. Information on the flora and fauna was obtained from organisations involved in conservation such as the Mount Cameroon Project, The German Technical Cooperation (GTZ), Delegations of Forestry and Wildlife in the area.

Landsat satellite imageries for the year 1986, 2000 and 2011 were also obtained from the global land cover facility (GLCF) website. Other sources include journals, internet and books. Due to the fact that Mount Cameroon is most of the time clouded, the factors that governed the choice of the satellite imageries were the quality of the imageries available and time interval.

Data Analysis

GIS and remote sensing were used to determine land cover change following the establishment of plantations in the area between 1986 and 2011. The activities that were carried out include satellite image processing and classification for land cover change detection. Changes in the areal extent of the plantations were analysed using Landsat imageries of the site for 1986, 2000 and 2011 to be able to monitor the extent of deforestation for plantation agriculture that has taken place in the area. The landsat imageries were georeferenced to the coordinate system of the study area (WGS84, projection: UTM zone 32). Erdas Imagine 9.2 software was used in processing and analysing the imageries. Visual interpretation of satellite imageries was enhanced through the use of linear stretching. Clouds and clouds shadows present on images used for the study were reduced through masking techniques in Erdas Imagine Software.

Two main steps were followed in land cover mapping. First, unsupervised image classification was carried out prior to field visits, in order to determine strata for ground truthing. Supervised classification based on maximum likelihood classifier algorithm was then used in the classification of the 1986, 2000 and 2011 images. This was based on 130 training sets or ground control points collected. Expert knowledge was used in selecting the 130 points on spot at good distances away. The sample points collected were used to validate classification results.

Some 25 % of the collected ground control points were used to train the data and the remaining 75 % were used for the analysis. The image classification accuracy was assessed by calculating the Kappa coefficient. The Kappa statistics is an estimate of the measure of overall agreement between image data and the reference (ground truth) data. Its coefficient falls typically on a scale of 0 and 1. It is often multiplied by 100 to give a percentage measure of classification accuracy. Kappa values are classified into 3 groupings: a value greater than 0.80 (80 %) represents strong agreement, a value between 0.4 and 0.8 (40–80 %) represents moderate agreement and a value below 0.4 (40 %) represents poor agreement (Congalton 1996). Postclassification method (Lu et al. 2004) of change detection was used in analysing the result from the land cover maps.

Results and Discussion

Figures 3, 4 and 5 show the land cover change between 1986 and 2011. They indicate that the secondary forest, farmlands and the plantations have considerable overlap in their spectral reflectance. It was however, difficult to separate the dense forest and the plantations based on their

Fig. 4 Land cover changes from 1986 to 2011 around the Mount Cameroun estuarine forest complex

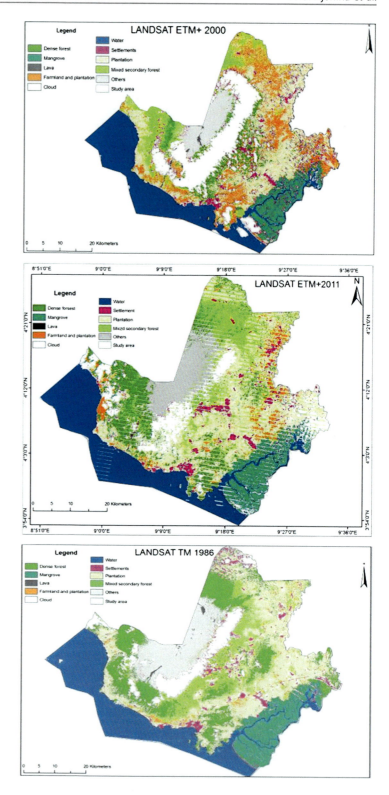

spectral characteristics. The difficulty is probably related to the similarity in the ground cover of both land cover classes. The similarity in the crown cover offers a similar reflectance surface for both land cover types, leading to an overlap in their spectral characteristics. However, textural differences in the land cover classes associated with the regular spacing of trees in the plantations and the fact that plantations are laid out in consecutive regular blocks made easy the demarcation between the plantation and the forest. Hence, the regular spatial arrangement of trees in the plantations possibly contributes to the textural differences between plantations and the forest. The secondary forest and the

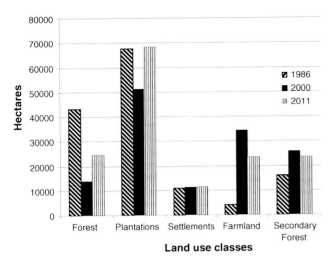

Fig. 5 Land cover change between 1986 and 2011

farmlands were also separated based on the textural differences observed. The Mount Cameroon area is permanently clouded throughout the year. Though the clouds somehow acted as a natural impediment to classification, it was still possible for various classes to be identified. Nine land cover classes were discriminated based on the spectral, textural and structural characteristics observed on the satellite images. These include the following: water, settlements, lava, and the dense forest, plantations, mixed secondary forest, mangrove, farmlands and the clouds. Clouds and clouds shadows present on images were reduced through masking techniques. Figures 3, 4 and 5 show the land cover map of the study area in 1986, 2000 and 2011, respectively.

Accuracy Assessment

The Kappa statistics gave an overall accuracy of 88 %. This value falls within the range described by Congalton (1996) as a strong agreement. These high-accuracy results demonstrate that the combined use of spectral and textural characteristics increased the number of classes in the final classification and also with a good accuracy (Table 1).

Land Cover Change Detection Qualification and Quantification

Land cover change analysis over the period 1986 to 2011 in the area revealed four important land cover changes as can be seen in Fig. 5. The results revealed that the dense forest coverage was lower in 2000 than in 2011. The total surface areas occupied by the dense forest were 43,280.9, 14,032 and 24,564 ha, respectively, for the year 1986, 2000 and 2011. The results revealed a decrease in the dense forest coverage in 2000 and 2011. There were a 67 and 47 % reduction in forest area in the year 2000 and 2011, respectively. More than half of the forest area existing in 1986 was gone in the year 2000 pointing out clearly the impact of deforestation due to plantation agriculture in the area.

The results show that settlements had a steady increase throughout the study period (11,022, 11,312 and 13,512 ha for the year 1986, 2000 and 2011, respectively). The area classified as farmlands increased between 1986 and 2000 and decreased in 2011 (4,149, 34,453 and 23,515 ha). The plantations increased in 2011 and decreased in the year 2000 (Table 2). The year 2000 witnessed a 24 % (16,497 ha) reduction in the area occupied by plantations compared with 1986. The year 2011 witnessed a 33 % (17,045 ha) increase in the surface area occupied by plantations. The decrease observed in the year 2000 was because of CDC policy to surrender land to various communities.

There have been conflicts between CDC and the various Bakweri clans whose lands were taken by the colonial masters for plantation agriculture. In the early 90s, the matter was taken to court by the various Bakweri chiefs. The court ordered CDC to release part of the land to the natives. As from the year 1998, the corporation started allocating land to over 55 village communities, schools, municipal councils, churches, companies and non-governmental organisations (CDC Report 1999). This was in conformity with the procedure for the surrender of CDC land thereby accounting for the decreased observed in the year 2000.

Land Cover Dynamics

The results of land cover change in the area between 1986 and 2011 revealed some important changes which were consistent over the years. From these results, it is observed that the rate of conversion of dense forest to plantations, secondary forest and other land uses dropped from 2,089 ha a year between 1986 and 2008 to 957 ha a year between 2008 and 2011. The reason for this drop can be attributed to an increase in awareness on the dangers of deforestation demonstrated by some conservation NGOs in the area. Plantations expanded during the two periods with the rate of expansion in the second period increasing to almost a third the rate in the first period from 1,178 ha per year to 1,549 ha per year. The difference in the land cover change rates between the two periods is due to some plantation management policies. The CDC policy of expansion may account for this increase.

The observed decrease in forest cover and increase in plantation coverage in the area is a pointer to the fact that plantation encroachment is one of the causes of deforestation in south-western Cameroon. The increase in the surface area occupied by plantations is due to its positive economic impact in the country as a whole which is linked to the high

Table 1 Accuracy assessment classification

Class name	Reference totals	Classified totals	Number correct	Producers accuracy (%)	Conditional kappa for each category (K^)
Unclassified	1	1	1	–	0.8407
Forest	6	7	6	100.00	0.7852
Mixed secondary	4	5	4	100.00	1.000
Water	4	4	4	100.00	0.6588
Others	0	0	0	–	1.000
Farm and plantation	7	4	4	57.14	1.000
Plantation	6	5	5	83.33	1.000
Lava	2	2	2	100.00	1.000
Settlements	13	11	11	84.62	1.000
Cloud	2	2	2	100.00	1.000
Mangrove	3	4	3	100.00	0.7364
Totals	58	58	52		
Overall classification accuracy	= 89.66 %				
Overall kappa statistics	= 0.8834				

Table 2 Surface area occupied by various land uses in the study area

Landuse (ha)	1986	2000	2011
Dense forest	43,280.9	14,032	24,564
Mixed secondary forest	16,253	25,916	23,515
Plantation	67,792	51,295	68,340
Farmland and plantation	4,149	34,453	23,515
Mangrove	21,557	16,286	17,855
Settlements	11,022	11,312	13,512
Lava	835	667	278
Cloud	20,608	40,270	23,238

demand of banana and other plantation produces in the Western world. This has caused many national and international companies to be increasingly involved in this activity. Also, various assistance schemes such as the supply of free seedlings and financial assistance to individuals and community-based organizations such as the common initiative groups through the South West Development Authority (SOWEDA) have made the process of starting a plantation fairly easy.

The role of plantation agriculture in modifying the natural forest cover has been documented by other researchers. Hartemink et al. (2008) attributed the high rate of forest conversion to intensive agricultural plantations in the tropical regions as largely due to the increasing population in the region. Serneels (2001) discovered that changes in the land cover of the Kenyan Embu Highlands were due to the introduction of commercial plantations in the area which resulted to a decrease in the natural vegetation cover.

The struggle to satisfy the basic human livelihood demands is generally accepted as the main trigger behind forest conversion in tropical areas (CIDA 2001). The rate of current deforestation is alarming and remains a critical issue regarding its implication for the global carbon cycle and biodiversity. The annual deforestation rate in the area due to plantation agriculture is 1.3 % which is greater than the FAO (1995) annual deforestation rate of 0.6 % in Cameroon.

Several studies carried out around the world also revealed the negative impacts of plantation agriculture on biodiversity loss. Koh and Wilcove (2008) looking at the impact of oil palm agriculture on tropical biodiversity in Malaysia and Indonesia acknowledged that oil palm plantations in Malaysia and Indonesia have replaced forests and, to a lesser extent, pre-existing cropland. More importantly, their study revealed that the conversion of primary forests and logged forests to oil palm plantations decreases the species richness of forest birds by 77 and 73 %, respectively. A similar analysis conducted for Indonesia between 1990 and 2005 revealed that oil palm-cultivated area in Indonesia increased by 3,017,000 ha (FAO 2007). The effect of deforestation on estuaries is much more rapid in the tropics than in temperate zones because of intense rainfall. Deforestation has been shown to increased soil erosion and sediment loads in rivers typically by a factor of 10 (Wolanski and Spagnol 2000).

Conclusion

Estuaries form a transition zone between river environments and maritime environments and are subject to both marine influences, such as tides, waves and the influx of saline water; and riverine influences, such as flows of fresh water and sediment largely influenced by deforestation. The inflows of both sea water and fresh water provide high levels of nutrients in both the water column and sediment, making estuaries among the most productive natural habitats in the world. The sea water entering the estuary is diluted by the fresh water flowing from rivers and streams. The pattern of dilution varies between different estuaries and depends on the volume of fresh water, the tidal range and the extent of evaporation of the water in the estuary.

Plantation agriculture is a major driver of deforestation and biodiversity loss in south-western Cameroon. Based on the changes observed, better management practices can be put in place to ensure the sustainability of resources and the proper functioning of the estuarine ecosystem for the well-being of man and the environment.

Acknowledgment This was part of the PhD thesis in the Department of Geography, Obafemi Awolowo University, Ile-Ife, Nigeria. We are indebted to the Acting Head of Department, Dr. O. A. Ajala for constant support and others in the department including Professors O. Ekanade, L. K. Jeje, Aloba and Dr. A. Adediji. I also acknowledge the technical support and field assistance of the staff of the Cameroon Wildlife Conservation Society (CWCS) Coastal Forests and Mangrove Programme especially Eugene Diyouke, Laisin Bruno, Kalieu Robert, Ni Thaddeus and Ndze Emmanuel. I am also thankful to the Administration of the Cameroon Development Cooperation for access to the plantations.

References

Alkarkhi A, Ahmad A, Easa A (2009) Assessment of surface water quality of selected estuaries of Malaysia: multivariate statistical techniques. Environmentalist 29(3):255–262

Cameroon Development Corporation (CDC) (1970) Report

Cameroon Development Corporation (CDC) (1999) Report

Cameroon Development Corporation (CDC) (2006) Report

CIDA (2001) Deforestation: tropical rainforest in decline

Congalton RG (1996) Accuracy assessment. A critical component of land cover mapping. Gap analysis. Am Soc. of Photogrammetry and Remote Sensing, pp 119–131

Courtenay P (1980) Plantation agriculture, 2nd edn. Bell and Hyman, London

Daily GC (1997) Nature's services: societal dependence on natural ecosystems. Island Press, Washington DC

DeFries R, Foley J, Asner GP (2004) Land use choices: balancing human needs and Ecosystem Function. Front Ecol Environ 2(5):249–257

FAO (1995) FAO production yearbook: FAO statistics series, Rome, vol 48

FAO (2007) FAOSTAT online statistical service. Available from: http://faostat.fao.org. United Nations Food and Agriculture Organization (FAO), Rome. Accessed Oct 2007

Halfani MS, Barker J (1984) Agribusiness and agrarian change. In: Barker J (ed) The politics of agriculture in tropical Africa. Sage Publications, Beverly Hills, pp 35–63

Hartemink AE, Veldkamp T, Bai Z (2008) Land cover change and soil fertility decline in tropical regions. Turk J Agric For 32:195–213

Koh LP, Wilcove DS (2008) Is oil palm agriculture really destroying tropical biodiversity? Conserv Lett 1(2):60–64

Lu D, Mausel P, Brondizio E, Moran E (2004) Change detection techniques. Int J Remote Sens 25(12):2365–2401

Millenium Ecosystem Assessment (2005) Ecosystems and human well-being: synthesis. World Resources Institute Island Press, Washington DC, p 155

Mitchell AW, Secoy K, Mardag N (2007) Forests first in the fight against climate change. Global Canopy Programme

Neba A (1987) Modern geography of Cameroon. Neba Publishers, Camden

Praveena SM, Aris AZ, Radojevic M (2010) Heavy metals dynamics and source in intertidal mangrove sediment of Sabah, Borneo Island. Environ Asia 3(special issue):79–83

Salomons W, Förstner U (1984) Metals in the hydro cycle. Springer, Berlin

Sama DA (1996) The constraints in managing the pathways of persistent organic pollutants into the large marine ecosystem of the Gulf of Guinea—the case of Cameroon. Intergovernmental Forum on Chemical Safety. Retrieved 27 Feb 2011

Serneels S (2001) Drivers and impact of landuse/land cover change in the Serengeti-Mara ecosystem: a spatial modelling approach based on remote sensing data. University Catholique de Louvain, Belgium

Wang F, Chen J (2000) Relation of sediment characteristics to trace metal concentrations: a statistical study. Water Res 34(2):694–698

Wolanski E, Spagnol S (2000) Environmental degradation by mud in tropical estuaries. Reg Environ Change 1:152–162

Assessment of Mangrove Carbon Stocks in Cameroon, Gabon, the Republic of Congo (RoC) and the Democratic Republic of Congo (DRC) Including their Potential for Reducing Emissions from Deforestation and Forest Degradation (REDD+)

Gordon N. Ajonina, James Kairo, Gabriel Grimsditch, Thomas Sembres, George Chuyong, and Eugene Diyouke

Abstract

We present results of the field assessment using a total of fifteen 0.1 ha mangrove permanent sample plots (PSPs) in four selected countries in Central Africa, including: Cameroon, Gabon, Republic of Congo and Democratic of Republic, which together account for 90 % of mangroves in Central Africa. Above- and belowground carbon stocks were computed using data from the PSPs in all four countries. Long-term monitoring data in Cameroon were used to estimate carbon sequestration rates. Four major carbon pools were considered: aboveground carbon, belowground root carbon, deadwood and the soil organic carbon. All the eight mangrove species described in Central Africa were encountered in the study. The dominant species in Central Africa is *Rhizophora racemosa*, and it occupies more than 70 % of the forest formation. The average stand density ranged from a low of 450 tree/ha in degraded forest of RoC to a high of 3,256 tree/ha in undisturbed stands of Cameroon. Standing volume ranged from a low of 213 m^3/ha in RoC to a high of 428 m^3/ha in Cameroon; corresponding to aboveground biomass values of 251 and 505 Mg/ha, respectively. Together with the deadwoods, the total vegetation biomass in the study area ranged from a low of 394 Mg/ha in RoC to a high of 825 Mg/ha in Cameroon. Mean diameter increment for primary and secondary stems was 0.15 cm/year. This translates to above- and belowground annual biomass increments of 12.7 and 3.1 Mg/ha/year, respectively. Total ecosystem carbon in undisturbed system was estimated at 1520 ± 164 Mg/ha with 982 Mg/ha (or 65 %) in belowground component (soils and roots) and 538 Mg/ha (35 %) in the aboveground components. Carbon density differed significantly ($p < 0.05$) with forest

G. N. Ajonina (✉) · E. Diyouke
CWCS Coastal Forests and Mangrove Programme, BP 54
Mouanko, Cameroon
e-mail: gnajonina@hotmail.com

J. Kairo
Kenya Marine and Fisheries Research Institute, PO Box 81651
Mombasa, Kenya

G. Grimsditch · T. Sembres
UNEP, PO Box 30552-00100 Nairobi, Kenya

G. Chuyong
University of Buea, BP 63 Buea, Cameroon

G. N. Ajonina
Institute of Fisheries and Aquatic Sciences, University of Douala
(Yabassi), Douala, Cameroon

S. Diop et al. (eds.), *The Land/Ocean Interactions in the Coastal Zone of West and Central Africa*, Estuaries of the World, DOI: 10.1007/978-3-319-06388-1_15,
© Springer International Publishing Switzerland 2014

conditions. The least total ecosystem carbon of 808 ± 236 Mg/ha was recorded in heavily exploited forests, translating to CO_2 equivalent of 2,962 Mg/ha. Undisturbed mangrove forests sequester annually 16.5 MgC/ha against 6.9 MgC/ha for degraded systems. Certain recommendations are made to improve and consolidate these estimates especially through validation of cover change, continuous monitoring PSP as well the development of site specific allometric equations for mangroves in Central Africa.

Keywords

Carbon accounting • Mangroves • REDD+ • Central Africa

Abbreviations

dbh	Diameter at breast height
DRC	Democratic Republic of Congo
FAO	Food and Agriculture Organization
PSP	Permanent sample plot
REDD	Reducing emissions from deforestation and forest degradation
RoC	Republic of Congo
UN-REDD	United Nations Reducing Emissions from Deforestation and Forest Degradation Programme
UNEP	United Nations Environment Programme
UNEP-WCMC	United Nations Environment Programme World Conservation Monitoring Centre

Introduction

Mangroves of West and Central Africa extend over 20,144 km^2, representing 59 % of the African mangroves or 11 % of the total mangroves area in the World (UNEP-WCMC 2007). These forests are particularly important for subsistence economies, providing harvestable wood and non-wood products, as well as ecosystem services such as shoreline protection, fish habitat and climate change mitigation through carbon sequestration. However, over-exploitation, conversion pressure and pollution effects have degraded or reduced mangroves in the region by about 20–30 % over the last 2 decades. Climate change effects now threaten the remaining mangroves in the region through increased precipitation and sedimentation. The consequences of current rates of mangrove deforestation and degradation in Central Africa are potentially enormous as these seriously threaten the livelihood security of coastal people and reduce the resiliency of mangroves to mitigate climate change effects.

'Reducing Emissions from Deforestation and forest Degradation' (REDD+) is an emerging international financial mechanism enabling tropical countries to get rewarded for their efforts in reducing CO_2 emissions from deforestation and forest degradation, and a number of Central African countries have embarked on ambitious national reforms and investments to improve forest landscapes management in order to benefit from REDD+. There are opportunities for mangroves to be included in national REDD+ strategies, especially in the light of recent findings that indicate that mangroves can store several times more carbon per unit area than productive terrestrial forests (Donato et al. 2011). Although mangroves cover only around 0.7 % (around 137,760 km^2) of global tropical forests (Giri et al. 2011), degradation of mangrove ecosystems potentially contributes 0.02–0.12 Pg carbon emissions per year, equivalent of up to 10 % of total emissions from deforestation globally (Donato et al. 2011). These make mangroves suitable candidate for REDD+ projects.

Fig. 1 Map showing the location of selected mangrove countries

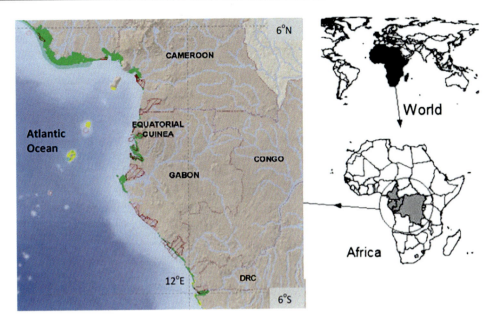

Previously, no study existed in the Central Africa region quantifying mangrove carbon stocks, sequestration rates and possible emissions in response to their degradation. A key challenge for successfully implementing any REDD+ Project is the reliable estimation of biomass carbon stocks in forests. A reliable estimation of forest biomass has to take account of spatial variability, forest allometry, wood density and management regime. Many studies have been published on aboveground carbon stocks in tropical forests around the world (Komiyama et al. 2005), but limited studies exist on belowground root biomass and soil carbon. Knowledge is even more limited for mangroves, where localized allometric equations for different mangrove species are limited.

This chapter presents results of field assessment in the four selected countries in Central Africa, including: Cameroon, Gabon, RoC and DRC, which together account for 90 % of mangroves in Central Africa. The information can serve as a contribution to further improve our global understanding of the climate change mitigation potential of mangroves and a basis to establish initial baselines in future mangrove projects and REDD+ strategies in Central Africa.

Study Approach and Methodology

Descriptions of Project Area

Four pilot areas in Central Africa were selected for the study, including Cameroon, Gabon, DRC and RoC (Fig. 1; Table 1). Collectively, these pilot countries contain 90 % of mangroves in Central Africa. Further, DRC and the RoC are part of the UN-REDD programme countries; while Cameroon and Gabon take part in the World Bank Forest Carbon Partnership and have the highest mangrove covers in Central Africa. The following general criteria were used in selecting study sites:

- the forest structure and composition appear to be typical of other sites in the region,
- waterways and canals are reasonably navigable even during low tides to allow for access and transportation of equipment and materials,
- different forest conditions are represented,
- the area is not so readily accessible that sample plots may be illegally felled.

Biophysical Characteristics

A variety of habitat types (coastal lagoons, rocky shores, sandy beaches, mudflats, etc.) characterize the Central African coastline with a vast array of rivers flowing from the hinterlands into the Atlantic Ocean. The confluences of these rivers with marine waters form suitable conditions for the development of outstanding giant mangrove vegetation in the region that also harbours the world's second largest tropical rainforest. The climate in Central Africa is mainly equatorial characterized by abundant rains (3,000–4,000 mm in Cameroon, 2,500–3,000 mm in Gabon and RoC and 772 mm in DRC) and generally high temperatures with monthly average of 24–29 °C, with a dry season spanning November to March in Cameroon and June to October in DRC. A typical climate diagram in Central Africa (Cameroon) is given in Fig. 2. September is normally the month with the highest rainfall, while December has the least.

Table 1 Selected sites within the central African mangroves for ecosystem service assessment

Country	Number of mangrove sites	Study site	Forest conditions
Cameroon	5	South West region, Bamasso mangroves	Undisturbed transboundary mangroves near Nigeria border
		Littoral region, Moukouke	Undisturbed mangroves at the Cameroon estuary
		Littoral region, Yoyo mangroves	Heavily exploited mangroves of Cameroon estuary
		Littoral region, Youme mangroves	Moderately exploited mangroves of Cameroon estuary
		South region, Campo mangroves	Undisturbed mangroves transboundary mangroves at the Ntem estuary
Gabon	4	Province de l'Estuaire, Commune de Libreville	Undisturbed mangroves of Akanda National Park
		Province de l'Estuaire, Commune de Libreville	Heavily exploited peri-urban mangroves
		Province de l'Estuaire, Commune de Coco-Beach	Moderately exploited transboundary mangrove near Equatorial Guinea
		Province de l'Estuaire, Commune de Coco-Beach	Undisturbed mangrove Emone-Mekak estuarine
RoC		Département de Pointe Noire	Heavily exploited peri-urban of Louaya
		Département de Pointe Noire	Moderately disturbed mangroves located within the touristic centre of Songolo town
		Département du Kouilou	Undisturbed mangroves transboundary in Gabon–Angola border
DRC		Province du Bas-Congo, district de Boma the only mangrove zone in DRC entirely in Muanda Mangrove Park and	Heavily exploited mangrove within Marana Line
			Moderately exploited mangroves
			Undisturbed mangrove of Île Rosa Tompo

(continued)

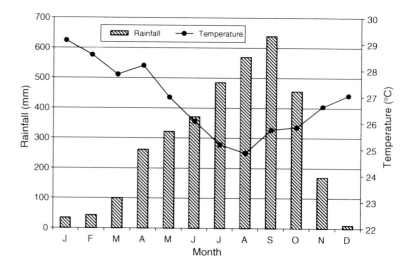

Fig. 2 Typical climate diagram in Central Africa. This particular diagram is for Doula-Edea Reserve, Cameroon

Table 2 Mangrove woody species found in the pilot areas

Mangrove species	Country			
	Cameroon	Gabon	RoC	DRC
Avicennia germinans	x	x	x	x
Conocarpus erectus	x	x		
Laguncularia racemosa	x	x		
Rhizophora harissonii		x		
Rhizophora mangle		x		
Rhizophora racemosa	x	x	x	x
Associated species				
Hibiscus sp	x	x		
Phoenix sp		x		
Total	5	8	2	2

Composition and Distribution of Mangroves in Central Africa

Mangrove formation in Western and Central Africa is characterized by low species composition common with new world mangroves (Tomlinson 1994). In Central Africa, there are 8 mangrove species of economic importance (UNEP-WCMC 2007). The largest blocks of mangroves in the region are found in deltas and large rivers estuaries in Cameroon and Gabon (UNEP-WCMC 2007). The dominant species is *Rhizophoraceae racemosa* which accounts for more than 70 % of the forest formation. The species fringes most shorelines and river banks, attaining up to 50 m in height with tree diameter of over 100 cm around the Sanaga and Wouri estuaries marking one of the tallest mangroves in the world (Blasco et al. 1996, p. 168). Other important mangrove species in the region are *Rhizophora mangle*, *Rhizophora harissonii*, *Avicennia germinans* (Avicenniaceae), *Laguncularia racemosa* and *Conocarpus erectus* (both Combretaceae) (Table 2). Undergrowth in upper zones can include the pantropical *Acrostichum aureum* (Pteridaceae) where the canopy is disturbed. *Nypa fruticans* (Arecaceae) is an invasive mangrove palm introduced in Nigeria from Asia in 1910 and has spread to Cameroon.

Common mangrove associates in Central Africa include Annonaceae, *Cocos nucufera* (Areaceae), *Guiboruti demensei* (Caesalpiniaceae), *Achornea cordifolia* (Euphorbiaceous), *Dalbergia ecastaphylum* and *Drepenocarpus lunatus* (both Fabaceae), *Pandanus candelabrum* (Pandanaceae), *Hibiscus tilaeceus* (Malvaceae), *Bambus avulgaus* (Poaceae) and *Paspalum vaginatum* (Poaceae), among others (Ajonina 2008).

Socioeconomic Importance

Fishing is a major economic activity along the West-Central African coastline especially in Central Africa with a population of about 4 million living in or around mangrove ecosystems (UNEP-WCMC 2007). About 60 % of fish harvested in these rural areas is of artisanal origin. Open drying, salting, icing, refrigerating and smoking are the common methods used to preserve fish in the region (Feka and Ajonina 2011 citing others). Mangrove wood is widely preferred for fish smoking within coastal areas of this region because of its availability, high calorific value, ability to burn under wet conditions and the quality it imparts to the smoked fish (Oladosu et al. 1996). Fish-smoking and fish-processing activities are largely responsible for more than 40 % degradation and loss of mangroves in the West-Central African coastal region (UNEP-WCMC 2007).

Quantification of Carbon Pools

Carbon density was estimated with data from existing and newly established rectangular 0.1 ha (100 m × 10 m) permanent sample plots (PSP). Existing PSPs in Cameroon provided an excellent opportunity to model stand dynamics and carbon sequestration potential of the mangroves in the region. Based on mangrove area coverage in each country 5 PSPs in Cameroon, 4 in Gabon, 3 in RoC and 3 in DRC were selected for the study (Table 1). Measurement protocol consisted of species identification, mapping, tagging and measurements of all trees inside the plot using modified forestry techniques for mangroves (Pool et al. 1977; Cintron and Novelli 1984; Kauffman and Donato 2012). Transect and plots boundaries were carefully marked and GPS points taken. Detailed procedures for establishment of PSP are given in Ajonina (2008). Four carbon pools were considered in the present study, including: vegetation carbon pools (both above and belowground), litter, coarse deadwood and soil.

Measurement of Vegetation Carbon

An important carbon stock in forestry is the aboveground component. Trees dominate the aboveground carbon pools and serves as indicator of ecological conditions of most forests. In each PSP, three plots of 20 m × 10 m were established along transect at 10-m intervals (Fig. 3a). Inside the plots, all trees with diameter of the stem at breast height (dbh_{130}) ≥ 1.0 cm were identified and marked. Data on species, dbh, live/dead and height were recorded for all individuals. In *Rhizophora*, dbh was taken 30 cm above highest stilt root. Aboveground roots and saplings $(dbh \leq 1$ cm) were sampled inside five 1-m^2 plots placed systematically at 1 m intervals along the 10 m × 10 m plot (Fig. 3b). Newly recruited saplings were enumerated, while missing tags were replaced by reference to initial plot maps.

Fig. 3 **a** Schematic layouts of mangrove forest stands permanent sample plots. **b** Roots and sapling inventories (after Ajonina 2008)

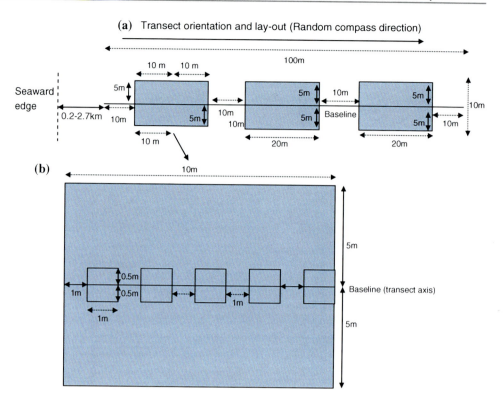

Dead and Downed Wood

Dead wood was estimated using the transect method whose application is given in Kauffman and Donato (2012). The line intersect technique involves counting intersections of woody pieces along a vertical sampling transect. The diameter of deadwood (usually more than 0.5 cm in diameter) lying within 2 m of the ground surface were measured at their points of intersection with the main transect axis. Each deadwood measured was given a decomposition ranking: rotten, intermediate or sound.

Soil Samples

Mangrove soils have been found to be a major reservoir of organic carbon (Donato et al. 2011), and given the importance of this carbon pool, we describe the methodologies used to calculate soil carbon in detail. Soil carbon is mostly concentrated in the upper 1.0 m of the soil profile. This layer is also the most vulnerable to land-use change, thus contributing most to emissions when mangroves are degraded. Soil cores were extracted from each of the 20 m × 10 m plots using a corer of 5.0 cm diameter and systematically divided into different depth intervals (0–15, 15–30, 30–50, and 50–100 cm); following the protocol by Kauffman and Donato (2012). A sample of 5 cm length was extracted from the central portion of each depth interval to obtain a standard volume for all sub-samples. A total of 180 soil samples were collected and placed in pre-labelled plastic bags—Cameroon (60 soil samples), Gabon (48), RoC (36) and DRC (36). In the laboratory, samples were weighed and oven-dried to constant mass at 70 °C for 48 h to obtain wet:dry ratios (Kauffman and Donato 2012). Bulk density was calculated as follows:

$$\text{Soil bulk density} \, (\text{g m}^{-3}) = \frac{\text{Oven} - \text{dry sample mass (g)}}{\text{Sample volume (m}^3)} \quad (1)$$

where, Volume = cross-sectional area of the corer × the height of the sample sub-section.

Of the dried soil samples, 5–10-g sub-samples were weighed out into crucibles and set in a muffle furnace for combustion at 550 °C for 8 h through the process of loss on ignition (LOI), and cooled in desiccators before reweighing. The weight of each ashed sample was recorded and used to calculate organic concentration (OC). Total soil carbon was calculated as:

$$\text{Soil C (Mg/ha}^1) = \text{bulk density (g/cm}^3) \\ * \text{ soil depth interval (cm)} * \% \, C \quad (2)$$

The total soil carbon pool was then determined by summing the carbon mass of each of the sampled soil depth.

Data Analysis and Allometric Computations

General field data were organized into various filing systems for ease of analysis and presentation. Both structural and biophysical data were entered into prepared data sheets.

Later, the data were transferred into separate Excel Work Sheets containing name of the country, zone and other details of the site. Sample data sheets for different data types are given in the Annex 1. Standing volume was determined using locally derived allometric relations from sample data with dbh as the independent variable:

$$v = 0.0000733 * D^{2.7921} (R^2 = 0.986, n = 677) \qquad (3)$$

where

v volume
D diameter of the stem for
the range: $1 \text{ cm} \leq D \geq 102.8 \text{ cm}$

Biomass conversion/expansion factor (BC/EF), which is the ratio of total aboveground biomass to stand volume, and shoot/root ratio (SRR) developed by Ajonina (2008), Ajonina, D. R. R. Pelz, and Chuyong (2012, Tree and stand volume equations for mangrove forests in the Atlantic coastal region of Cameroon, Central Africa, Unpublished manuscript), Ajonina, D. R. R. Pelz, and Chuyong (2012, Tree biomass expansion, partitioning and shoot-root ratio models for above- and below-ground carbon stock estimations for mangrove forests in the Atlantic coastal region of Cameroon, Central Africa, Unpublished manuscript) were used for the estimation of total tree biomass and carbon densities. The BC/EF used in the study was 1.18 (Ajonina 2008) which is comparable to that reported for humid tropical forests by Brown (1997).

Tree, Stand Dynamics and Carbon Sequestration Estimations

Using PSP in Cameroon, we estimated periodic annual increment (PAI) of the forest as a function of mortality and recruitment of seedlings at the beginning and end of each growing period. Development of detailed carbon sequestration estimates will, however, require long-term studies on regeneration, stand dynamics and also the distribution pattern of the seedlings under mother trees.

Deadwood

Deadwood volume was estimated using the protocol by Kauffman and Donato (2012):

$$\text{Volume } (\text{m}^3/\text{ha}) = \Pi^2 * \frac{\sum_{i=1}^{n} d_i^2}{8L} \qquad (4)$$

where $d_i = d_1, d_2 \ldots d_n$ are diameters of intersecting pieces of deadwood (cm) L = the length of the intersecting line (transect axis of the plot) generally $L = 20$ m being the length of each plot or 100 m being the length of transects. Deadwood volumes were converted to carbon density

estimates using the different size-specific gravities provided by Kauffman and Donato (2012).

Results and Discussion

Floristic Composition and Distribution

Structural attributes (tree height, basal area, stand density, species composition, etc.) of the mangroves of Central Africa are provided in (Tables 2, 3). All the mangroves described in Central Africa were encountered during the present study (Table 2). The dominant and prominent species is *R. racemosa* that occur in expansive pure stands across the countries. There were only two species that were found in RoC and DRC. These results are in conformity with earlier surveys (e.g. UNEP-WCMC 2007; Ajonina 2008; Ajonina et al. 2009) and confirm Central African mangroves as being generally species poor as compared to the Indo-west pacific mangroves that may have up to 52 species (Tomlison 1986; Duke 1992; Spalding et al. 2010). Common mangrove associates that were encountered in the pilot areas include *Hibiscus* sp, *Phoenix* sp and *A. aureum*.

There is no obvious zonation that is displayed by the dominant mangrove species in Central Africa. The seaward side as well as creeks is mostly occupied by *R. racemosa*, whereas *R. mangle*, *A. germinams*, and *A. aureum* mosaic covers the middle and outer zones. In a few places in Cameroon, we found the invasive *Nypa* palms growing in association with *R. mangle* and *R. racemosa* on creek margins.

Stand Density, Volume and Biomass

Table 3 provides vegetation inventories for Central Africa mangroves. The average stand density ranged from a low of 450 tree/ha in heavily exploited forest of RoC, to a high of 3,256 tree/ha in undisturbed stands of Cameroon. In most undisturbed plots, the stem density decreased exponentially with increasing diameter. These are typical reversed 'J' curves for stands with a wide range of size classes and by inference also age classes (Fig. 4). This pattern was, however, distorted in heavily exploited mangroves stands in the region, where size classes above 30 cm were literally missing.

Standing volume ranged from a low of 213 m³/ha in RoC to a high of 428 m³/ha in Cameroon; corresponding to aboveground biomass values of 251 and 505 Mg/ha, respectively. Together with the deadwoods, the total vegetation biomass in the study area ranged from a low of 394 Mg/ha in RoC to a high of 825 Mg/ha in Cameroon to (Table 3).

Table 3 Structural characteristics of mangroves in Central African

Country	Tree density (Nr trees/ha)	Max height (m)	Mean diameter (cm)	Basal area (m²/ha)	Stand volume (m³/ha)	Aboveground biomass (Mg/ha)	Belowground biomass (Mg/ha)	Dead woods (Mg/ha)	Total biomass (Mg/ha)
Cameroon	3,256	52.1	4.6	25.1	427.5	505	306	15	825
Gabon	1,467	41	9.5	24.5	288.9	341	151	21	512
RoC	1,667	25.2	7.7	18.8	213	251	122	20	394
DRC	1,267	27	9.1	24.5	346.9	409	185	69	663

All stems with DBH > 1.0 cm inside PSPs plots were measured
Extract of calculation from Ajonina (2008) as follows
AGB = BEF$_{ABG}$ * stand volume
BEF = 1.18, BGB = BEF$_{BGB}$eqn * trunk volume = $(1.385 * Diam^{-0.4331})$ * trunk volume
Where BEF$_{BGB}$Equation = $(1.385 * Diam^{-0.4331})$

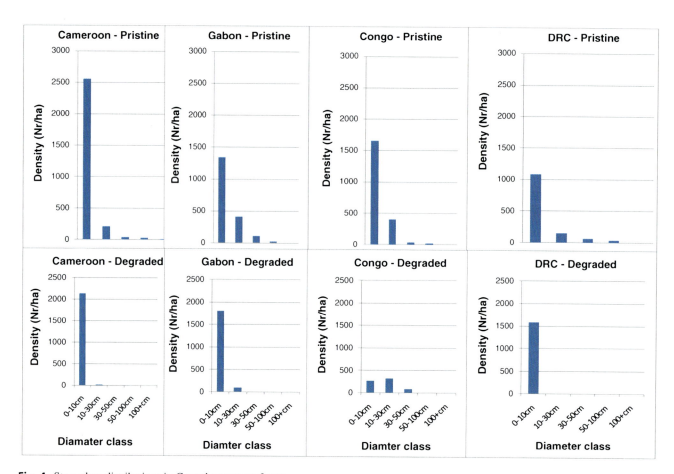

Fig. 4 Stem class distributions in Central mangrove forest

Carbon stocks

Soil Organic Carbon

There was high variability in the concentration of soil organic carbon ($p < 0.05$) with undisturbed sites showing higher carbon concentrations than exploited forests. Across the region, the average quantity of soil organic carbon amounted to 827 ± 170 Mg/ha. The undisturbed stands recorded the highest amount of mean soil organic carbon (967 ± 58 Mg/ha; Table 4). This was followed by heavily and moderately exploited sites that recorded an average SOC of 774 ± 163 and 741 ± 190 Mg/ha, respectively. The results are in conformity with high content of organic carbon that is associated with mangrove sediments (Donato et al. 2011, found an average of 864 Mg/ha in the Indo-Pacific; Adame et al. 2013, found up to 1,166 Mg/ha in the Mexican Caribbean). Alluvial deposition from multiple rivers flowing through the mangroves into the Atlantic

Table 4 Soil organic carbon (*SOC*) along the different forest conditions in Central Africa mangroves

Forest condition	Soil Depth (cm) 0-15	15-30	30-50	50-100	Total (Mg C/ha)
Undisturbed	157.8 ± 22.8	182.4 ± 70.7	230.5 ± 39.9	396.7 ± 108.6	967.4 ± 57.6
Moderately exploited	169.1 ± 34.5	140.0 ± 45.6	167.2 ± 86.3	303.9 ± 198.0	780.2 ± 162.9
Heavily exploited	130.1 ± 18.1	147.0 ± 33.6	156.6 ± 58.4	306.8 ± 195.5	740.6 ± 189.6

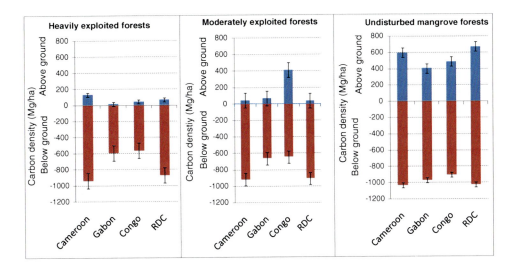

Fig. 5 Partitioning of carbon stocks in mangroves of Central Africa under different conditions

Table 5 Total ecosystem carbon stocks in mangroves of Central Africa under different perturbation regimes

Pools	Heavily Exploited Trees Mg/ha	SE	Moderately Exploited Mg/ha	SE	Undisturbed Mg/ha	SE
Aboveground						
Live component	58.0	50.4	123.3	179.7	467.1	70.0
Dead component	6.1	3.7	16.4	18.1	70.6	85.2
Total Aboveground	64.1	49.9	139.6	181.4	537.7	116.5
As % total	7.2	4.0	14.1	16.6	35.1	4.2
Belowground						
Tree-roots	3.1	1.4	12.1	18.8	15.1	4.2
Total Soil	740.6	189.6	773.6	162.9	967.4	57.6
Total Belowground	743.6	190.9	785.7	149.8	982.5	60.8
As % total	92.8	4.0	85.9	16.6	64.9	4.2
Total ecosystem carbon stock (Mg/ha)	807.8	235.5	925.4	137.2	1520.2	163.9

Carbon pools of trees (aboveground) were calculated as the product of tree stand biomass multiplied 0.5 CO_2 value is derived by multiplying C stocks by 3.67, the molecular weight ratio of CO_2 to C

Ocean could explain high organic carbon content in the soils of even mangroves that are in degraded conditions. There was high variation in SOC in the 50–100-cm depth as compared to the rest of the zones (Table 4, Fig. 5).

Total Ecosystem Carbon

Based on the four major carbon pools accounted in this study, total ecosystem carbon in undisturbed mangrove of Central Africa was estimated at 1,520 ± 164 Mg/ha with 982 Mg/ha (or 65 %) in belowground component (soils and roots) and 538 Mg/ha (35.0 %) in the aboveground biomass (Fig. 5). Total ecosystem carbon stocks differed significantly ($p < 0.05$) with forest conditions. The lowest ecosystem carbon of 808 ± 236 Mg/ha was recorded in moderately degraded forests, translating to CO_2 equivalent of 2,962 Mg/ha (mean: 808 ± 236 Mg/ha) (Table 5). These figures are comparable to other studies around the world, which have shown average values of

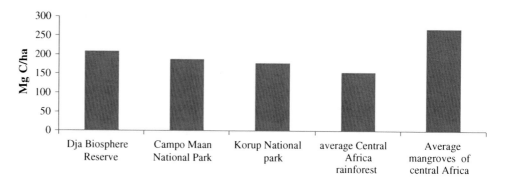

Fig. 6 Aboveground carbon stocks of selected terrestrial rainforest in Congo basin and the mangroves sampled in this study

1,023 ± 88 MgC/ha in the Indo-Pacific (Donato et al. 2012) and 987 ± 338 MgC/ha in Mexico (Adame et al. 2013).

Although it is clear that undisturbed forests contain the largest amounts of carbon, the difference between moderately exploited and highly exploited systems is less clear. The relatively high carbon contents in exploited systems could be explained by the fact that soils in exploited systems could be receiving carbon input from outside the system through flood water, alluvial deposits and tides. High soil carbon figures in highly exploited as well as moderately exploited forests in RoC and DRC were influenced by a peri-urban setting that suffers pollution effects. Furthermore, the relatively high carbon deposits in soils of exploited systems shows that not all soil carbon is oxidized and emitted to the atmosphere when the system becomes degraded; part of this remains captured in the soil column. The significant difference in carbon stocks between nondisturbed and moderately exploited systems points to the possibility that mangroves release carbon stocks relatively quickly after degradation, even if degraded moderately, and that it is important for mangroves to remain in completely undisturbed states if they are to maintain maximum carbon values.

Carbon Dioxide (Greenhouse Gas) Emission Potential

The most vulnerable carbon pools following mangrove deforestation and degradation are the aboveground carbon as well as soil carbon from the top 30 cm. Estimating emissions from land-use change was conducted using the uncertainty-propagation approach detailed in Donato et al. (2011). For the mangroves of Central Africa, a conservative low-end estimate of conversion impact was used, with 50 % aboveground biomass loss, 25 % loss of soil C from the top 30 cm, and no loss from deeper layers. Use of low-end conversion impact in the current study is justified by low-level reclamation of mangroves for aquaculture and agriculture in Central Africa. Using these conservative estimates, we estimate that 1,300 Mg of carbon dioxide would be released per ha of cleared pristine mangrove in Central Africa. A recent report estimates that 771 km^2 of mangrove were cleared in Central Africa between 2000 and 2010 (UNEP-WCMC 2012), equating to estimated emissions of 100,152,000 Mg of carbon dioxide. Of course, not all the carbon dioxide is released immediately, and these emissions occur over decades.

Comparison with Adjacent Central African Rainforests of the Congo Basin

Ecosystem carbon storage reported in the mangroves of Central Africa is among the largest for any tropical forest (IPCC 2007). We made comparisons of mangrove carbon stocks with some of the reported carbon stocks of the terrestrial Congo basin rainforest (Fig. 6). For consistency, we have only utilized aboveground biomass, as most of the studies in terrestrial forests lacked belowground carbon stocks. Aboveground carbon pools were 209 Mg/ha in Dja Biosphere Reserve (Djuikouo et al. 2010), 188 Mg/ha Campo Ma'an National Park (Kanmegne 2004) and 178.5 Mg/ha in Korup National Park (Chuyong, unpublished data), all in Cameroon. The average aboveground carbon pool for undisturbed rainforest in Central Africa was 154 Mg/ha. This is lower than the 538 MgC/ha found in the mangroves of this study, underscoring the value of mangroves as carbon stocks. When soil carbon is added to the equation, the difference between the carbon storage potential of mangroves and terrestrial rainforests could become even greater.

The extremely high carbon content of mangroves compared to terrestrial forests is often explained by the high levels of organic carbon in the soil, which is typical of many coastal ecosystems including seagrasses and salt marsh. The reason for the high organic carbon content in the soil is the accretion rates of these ecosystems as they keep up with sea-level rise, sometimes over thousands of years, and trap detritus and sediments from tidal movement and alluvial deposits. Most terrestrial ecosystems reach maximum carbon content in their soils over decades or even centuries, but coastal ecosystems can keep on accreting over millennia

Table 6 Biomass accumulation in the Central African mangrove forests

Disturbance regimes	Mean periodic annual increment				
	Diam (cm/year)	Basal area (m^2/year)	Volume (m^3/year)	AGB (tonnes/ha/year)	BGB (tonnes/ha/year)
Heavily exploited	0.34	0.05	0.35	0.38	0.40
Moderately exploited	0.42	1.67	9.66	10.43	3.35
Undisturbed	0.06	0.02	25.34	27.36	5.67
All regimes	0.15	0.56	11.78	12.72	3.14

Figures are annual size-specific growths under different exploitation regimes

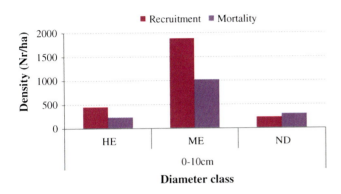

Fig. 7 Recruitment and mortality in mangrove forests

Table 7 Carbon sequestration in mangrove forests under different exploitation regimes

Exploitation regime	Biomass (MgC/ha/year)		
	AGC	BGC	Total
Heavily exploited	0.2	0.2	0.4
Moderately exploited	5.2	1.7	6.9
Undisturbed	13.7	2.8	16.5
Average	6.4	1.6	7.9

and create sediment deposits several metres deep. Emissions of this stored carbon can be avoided by maintaining mangroves in an intact state through REDD + activities, policies and projects.

Carbon Sequestration in Central African Mangrove Forests

Forest Dynamics: Growth and Biomass Accumulation

Net growth was higher in moderately exploited forests (ME) than in heavily exploited (HE) and undisturbed (UND) (Fig. 7, Table 6). This implies that there is a threshold level for exploitation to guarantee stand development. FAO (1994) recommends a minimum of 12 trees/ha parental mangrove trees (standards) be retained during harvesting operations to act as seed bearers for the next generation.

Although it is still early to foretell the nature of future forest in Central Africa mangroves, mortality rate observed in the present study is in conformity with the FAO (1994) values of 50 % loss observed during the 1–10 years growing period.

Apart from Cameroon, growth data were not available for other mangrove areas in the region. Mean annual diameter increment (MAI) for primary and secondary stems under different management regime was 0.15 cm/year. This translates to above- and belowground annual biomass increment of 12.7 and 3.1 Mg/ha/year, respectively. The values are consistence with published productivity data in Malaysia (Ong et al. 1993) and Kenya (Kairo et al. 2008). As expected, heavily degraded forests had the lowest biomass increment, whereas the moderately exploited and undisturbed forests had higher biomass increment (Table 6).

Carbon Sequestration

Carbon sequestration rates were found to vary with forest conditions (Table 7). Aboveground components had higher sequestration rates (6.4 MgC/ha/year) compared to belowground carbon pools (1.6 MgC/ha/year). Undisturbed forests sequestered on average 16.5 MgC/ha/year against 0.4 and 6.0 MgC/ha/year by heavily and moderately degraded systems, respectively. Mean sequestration rate for all forest conditions was 7.9 MgC/ha/year.

Conclusion

Mangrove forests in Central Africa are very carbon rich with carbon stocks in natural undisturbed forests in trees more than 2–3 times that of adjacent tropical rainforest. About 65 % of carbon stocks in natural undisturbed mangroves are stored in the soil layers with higher proportions in some disturbed forests. The large reservoirs of carbon stored by the exceptional and gigantic mangrove systems of Central Africa are thus important for climate change mitigation. We estimate that undisturbed mangroves contain

1,520 ± 164 Mg/ha with 982 Mg/ha (or 65 %) in the belowground component (soils and roots) and 538 Mg/ha (35.0 %) in the aboveground biomass. In moderately exploited mangrove ecosystems, 91.7 % of total ecosystem carbon was found in the soil component. These figures are higher than other studies around the world (Indo-Pacific and Mexico), but given the gigantic nature of these trees (up to 50 m high and 1 m diameter), and the large alluvial deposits in the soils from the River Congo, this is certainly possible. Using conservative estimates, we estimate that 1,300 Mg of carbon dioxide would be released per ha of cleared pristine mangrove in Central Africa. These estimates were made using the carbon values collected in the field in Central Africa. A recent report estimates that 771 km^2 of mangrove were cleared in Central Africa between 2000 and 2010 (UNEP-WCMC 2012), equating to estimated emissions of 100,152,000 Mg of carbon dioxide, although of course this carbon dioxide would be emitted over a time span of decades. Therefore, the mangroves of Central Africa could be among the most carbon-rich ecosystems in the world, and of value for climate change mitigation internationally.

Continuous monitoring through mangrove permanent plot systems would improve the quality of the data. Regular re-measurement of permanent mangrove forest plots would allow the gauging of not only dynamics of carbon but also general mangrove ecosystem dynamics (growth, mortality, recruitment) for carbon and other Payment for Ecosystem Services initiatives, as well as for providing baselines for REDD+ strategies in the region. More allometric studies for African mangroves would further improve the quality of the data and would allow the development of location and species-specific equations. Data collection can also be improved by the strengthening of existing networks and partnerships such as the African Mangrove Network.

REDD+ strategies can incentivize and support conservation, sustainable management of forests and enhancement of forest carbon stocks. Strengthening the existing networks (African Mangrove Network, the East African Mangrove Network, etc.) can generate a large-scale impact of mangrove forest protection and restoration initiatives through reforestation and sustainable management techniques as well as building capacities in various domains of mangrove conservation and sustainable management. Sustainable forest management practices to reduce mangrove deforestation can address some of the main causes of deforestation in the region, notably wood for fish smoking and also growing urbanization. To reduce use of wood for fish smoking, improved technology for fish-smoking stoves could be introduced that would generate more heat and energy from less wood, thus decreasing consumption. Deforestation from urbanization could be reduced by

ensuring that mangrove protection is integrated into coastal and marine protected area networks that are properly enforced and policed. The network of mangrove and marine protected areas could include seaward extensions of existing coastal parks in order to conserve biodiversity and in order for mangroves to fully provide their role as hatcheries and nursery grounds for aquatic fauna, as well as shoreline protection against erosion and storms. The results showing the high value of mangroves in this chapter are not only relevant to planning of networks of marine protected areas, but also to all integrated coastal and marine spatial planning. Information of the high value of ecosystem services provided by mangroves can be integrated into spatial planning exercises; for example so that conservation targets for ecosystem services for local communities can be determined and planned for. This could improve the well-being of communities in the area that benefit from the ecosystem services provided by mangroves.

Overall, this chapter provides a case for the inclusion of mangroves in national REDD+ strategies given their high carbon value, and also the levels of threat to the ecosystem and the associated rates of loss in the region. We hope that it can serve as a baseline study for future carbon market or climate change mitigation strategies, as well as providing evidence for the high value of mangrove ecosystems. Furthermore, it points to the mangroves of Central Africa being an exceptional ecosystem on a global scale, with higher carbon stocks measured here than in other mangroves or even adjacent rainforests.

Acknowledgments This project was implemented by the Cameroon Wildlife Conservation Society (CWCS) and the World Conservation Monitoring Centre (WCMC),with financial and technical support from the United Nations Environment Programme (UNEP) and United Nations programme on Reducing Emissions from Deforestation and Forest Degradation (UN-REDD), and the Kenya Marine and Fisheries Research Institute (KMFRI). The authors are indebted to all those who assisted the project by providing information, support and facilities especially: Constant ALLOGO (CARPE, IUCN Gabon); Bernard Henri VOUBOU, UNDP, Gabon; Léandre M EBOBOLA Ministry of Forests & Water, Gabon; Mme Marie AYITO, Director of Aquatic Ecosystems, Gabon; FélicienJoël BODINGA, Deputy Director of Aquatic Ecosystems, Gabon; Dr Emmanuel ONDO ASSOUMOU, Geography Department, Omar Bongo University, Gabon; Germain KOMBO, Jean Felix ISSANG, Marcel MPOUNZA, UNDP, Congo; MFOUTOU Gaston, Ministry of Sustainable Development, Forests Economy and Environment, Congo; Jerôme MOKOKO, WCS-Congo; Jean Pierre KOMBO, Focal Point of Abidjan Convention, Congo; Akenzenee OGNIMBA, Ministry of Sustainable Development, Forests Economy and Environment, Congo; Pierre Justin MAKOSSO, Mairie de PN; Jean Simplice MADINGOU, Department of Forestry, Congo; Antoine BITA, Department of Environment, Congo; Roland Missilou BOUKAKA, Conservator Conkouati-Douli National Park; Basile NIAMATELE, Conkouati-Douli National Park, Congo; Vincent KASULU SEYA MAKONGA, Ministry of Environment, Conservation of Nature and Tourism, DRC; CosmaB. WILUNGULA, Director of the Congolese Institute for Nature Conservation (ICCN), DRC; COLLET Mangrove Marine Park, DRC;Urbain ASANZI,

Mangrove Marine Park, DRC; Louis NGUELI MPAYI, Mangrove Marine Park, DRC; Peter LUKAMBA LUNDENGO, OCPE, DRC; MBUNGU NDAMBA, ACODES, DRC.

References

Adame M, Kauffman B, Medina I, Gamboa J, Torres O, Caamal J, Reza M, Herrera-Silveira J (2013) Carbon stocks of tropical coastal wetlands within the Karstic landscape of the Mexican Caribbean. PLoS ONE 8:1–13

Ajonina G, Tchikangwa B, Chuyong G, Tchamba M (2009) The challenges and prospects of developing a community based generalizable method to assess mangrove ecosystems vulnerability and adaptation to climate change impacts: experience from Cameroon. FAONature Faune 24(1):16–25

Ajonina GN (2008) Inventory and modelling mangrove forest stand dynamics following different levels of wood exploitation pressures in the Douala-Edea Atlantic coast of Cameroon, Central Africa. Mitteilungen der AbteilungenfürForstlicheBiometrie, Albert-Ludwigs-Universität Freiburg, 215 p

Blasco F, Saenger P, Janodet E (1996) Mangroves as indicators of coastal change. Catena 27:167–178

Brown S (1997) Estimating biomass and biomass change of tropical forest. FAO Forestry paper 134. FAO, Rome, 76 p

Cintron G, Novelli YS (1984) Methods of studying mangrove structure. In: Snedaker S, Snedaker JG (eds) The mangrove ecosystem: research methods. UNESCO Publication, 251 p

Djuikouo MNK, Doucet JL, Nguembou CK, Lewis SL, Sonke B (2010) Diversity and aboveground biomass in three tropical forest types in the Dja Biosphere reserve, Cameroon. Afr J Ecol 48:1053–1063

Donato DC, Kauffman JB, Murdiyarso D, Kurnianto S, Melanie SM, Kanninen M (2011) Mangroves among the most carbon-rich forests in the tropics. Nat Geosci 4:293–297

Feka NZ, Ajonina GN (2011) Drivers causing decline of mangrove in West-Central Africa: a review. Int J Biodivers Sci Ecosyst Serv Manage. doi:10.1080/21513732.2011.634436

Food and Agriculture Organization (FAO) (1994) Utilization of Bonga (Ethmalosa fimbriata) in West Africa. Fisheries Circular No. 870. Food and Agriculture Organization. Rome

Kanmegne J (2004) Slash and burn agriculture in the humid forest zone of Southern Cameroon: soil quality dynamics, improved fallow management and farmer's perception. PhD thesis, Wageningen University, 184 p

Kauffman JB, Donato DC (2012) Protocols for the measurement, monitoring and reporting of structure, biomass and carbon stocks in mangrove forests. Working Paper 86, CIFOR, Bongor, Indonesia

Komiyama A, Poungparn S, Kato S (2005) Common allometric equations for estimating the tree weight of mangroves. J Trop Ecol 21:471–477

Oladosu OK, Adande GR, Tobor JG (1996) Technology needs assessment and technology assessment in the conceptualization and design of Magbon-Alande—Fish smoking—Drying equipment in Nionr. Expert consultations on fish smoking technology in Africa. Report No. 574. FAO, Rome, pp 76–80

Pool DG, Snedaker SC, Lugo AE (1977) Structure of mangrove forests in Florida, Puerto Rico, Mexico and Costa Rica. Biotropica 9:195–212

UNEP-WCMC (2007) The mangroves of West-Central Africa. UNEP-WCMC report, 92 p

UNEP-WCMC (2012) Status and threats to mangrove forests in Cameroon, Gabon, Republic of Congo and Democratic Republic of Congo between 2000 and 2010 and the potential impacts of REDD+. UNEP-WCMC report, 23 p

Governing Through Networks: Working Toward a Sustainable Management of West Africa's Coastal Mangrove Ecosystems

Dominique Duval-Diop, Ahmed Senhoury, and Pierre Campredon

Abstract

The West African coastal environment's extremely productive and biologically diverse estuaries and mangrove ecosystems have suffered increasing stress due to natural and human-induced pressures. Conserving biodiversity in this region is full of complexity as a result of the myriad connections inherent in natural ecosystems and the variety of perspectives and interests arising at multiple scales and out of varying social and cultural contexts. Therefore, a participatory system of interregional governance is necessary in order to develop appropriate solutions to achieve effective conservation. Two case studies are presented that demonstrate the usefulness of the networked governance approach to engage actors as all levels in the preservation of mangrove ecosystems. The Regional coastal and marine conservation partnership in West Africa (PRCM) and the West African network of marine protected areas (RAMPAO) demonstrate the effectiveness of coordinating local actions with the development of national and regional policies. Challenges remain including the impact of competing goals, communication difficulties, uncertain funding, unequal capacity, and political instability.

Keywords

Networked governance • Mangrove conservation • Marine protected area network • Cross-scale dynamics • Participation

D. Duval-Diop (✉)
West African Regional Network of Marine Protected Areas
(RAMPAO), Mamelles Villa F46, Dakar, Senegal
e-mail: ddiop@consultfermina.com

A. Senhoury
Mobilization and Coordination Unit Regional Partnership for the
Conservation of the West African Coastal and Marine Zone
(PRCM), S/C UICN BP 4167, Nouakchott, Mauritania

P. Campredon
International Union for Conservation of Nature (UICN), Bissau,
Republic of Guinea-Bissau

Context

The West African coastal and marine environment is composed of extremely productive upwellings, estuaries and mangrove swamps, rich fishing zones, and ecosystems that are home to biologically diverse habitats and species. Local and national economies depend on assets such as sand and shells, minerals, oil, and tourism. A critical resource for these economies and for the food security of coastal populations remains the fish produced and harbored by these ecosystems.

However, these ecosystems have suffered increasing stress due to both natural and human-induced changes emanating from a number of sources. Intensified environmental degradation due to irresponsible exploitation of

S. Diop et al. (eds.), *The Land/Ocean Interactions in the Coastal Zone of West and Central Africa*, Estuaries of the World, DOI: 10.1007/978-3-319-06388-1_16,
© Springer International Publishing Switzerland 2014

mineral resources, concentrated pressure on fishing stocks, and urbanization of coastal areas presents challenges that call for effective strategies and coordinated action. Furthermore, in implementing solutions to one set of problems, conflicts often arise as regional stakeholders strive to preserve their ecosystems while effectively developing other sectors (agriculture, extractive industries, trade, finance, and fishing) (PRCM 2010).

Such conflicts render the management of West African natural resources complex. The complexity increases when the elements of individual ecosystems are interconnected and interdependent, and more so in *regional* environmental systems. One excellent example is the mangrove estuaries that play a critical ecological and economic role in coastal countries throughout West Africa. Eight true mangrove species are found in West Africa and the mangroves of West and Central Africa represent 13.2 % of global mangrove coverage (Spalding et al. 2010). Characterized by a high level of biodiversity and biological productivity, these ecosystems offer abundant fish and wood resources, which support agriculture and fishing and other economic activities. Mangroves play an important role in sustaining coastal fisheries and in acting as breeding and nursery grounds for many commercial species (Ong and Gong 2013). They also provide refuge for numerous endangered species, filter sediment runoff from human and natural activities, and serve as natural buffers against the erosive power of waves and rising seas (Wolf 2012).

Over the years, mangrove ecosystems in the West African region spanning Mauritania to Sierra Leone have experienced an accelerated rate of degradation. In spite of their significance, a poor understanding of the value of the services they provide has led to intensified human efforts to convert them for agricultural use, to clear them for residential and infrastructure developments, and to extract wood for salt production, fish smoking, and unsustainable timber harvesting (Rönnbäck 1999; Dayton et al. 2005; Mangrove Charter 2009; PRCM 2007). Sea-level rise and drought caused by climatic variation are also accelerating the degradation. The subsequent decline in revenue coming from resources extracted from this ecosystem has a detrimental effect on the people who depend on mangrove ecosystem services. Resulting increases in the poverty of indigenous coastal populations can further destabilize the ecosystem as they exert even more pressure on the natural resources to alleviate income losses (Dayton et al. 2005).

Understanding the nature of the interactions between human activities and ecological systems is the main focus of those who study 'coupled human and natural systems or systems in which human and natural components interact' (Lui et al. 2007). These systems can be characterized as entities that have layered hierarchies where people and nature form complex webs of interactions across organizational levels, and spatial and temporal scales. Positive or negative feedback from both human and natural actions, direct and indirect effects, and the emergence of new behaviors and properties serve to accelerate change and complicate our ability to understand these processes in order to reduce the vulnerability and degradation of mangrove ecosystems. The globalization of modern world social and economic systems has increased the need to take into account the spatial coupling of natural ecosystems since 'local couplings are influenced by broad-scale processes that in turn act in the context of still larger-scale processes and ultimately global-scale processes' (Lui et al. 2007, p. 642).

When an endangered natural resource or ecosystem such as mangroves physically extend beyond the artificial lines of national borders, and the impacts of local actions and localized natural events are felt at broader scales, the development of national and regional policies to better manage and protect them requires a high level of coordination at the regional scale (Van Lavieren et al. 2012). Because many of the activities that threaten the survival of mangroves occur in local communities in close proximity to these ecosystems, local conservation efforts must also accompany sub-regional approaches. The challenge is that policy making to establish legal protection mechanisms most often occurs at the national level and favors local and national priorities. Therefore, solutions demand a substantial and meaningful engagement of a myriad of actors at multiple scales. 'Tackling complex policy problems requires multi-level governance systems that work at multiple, interlinked levels, promoting learning and cooperation' (Jones 2011, p. 22).

Applying an understanding of the network of interactions arising between human social systems and the environment is a first and necessary step to developing a regional approach to mangrove conservation. Just as the systems in each human body work side by side to make it function, human networks, if they work together, can increase the resiliency of mangrove ecosystems (Quill 2012). By harnessing the knowledge that human-driven networks deeply influence and are affected by natural cycles, these arrangements can surmount the challenges of insufficient technical and financial resources that hinder the effective implementation of public policies governing coastal planning and management. Furthermore, networks and a greater understanding of their role in human–natural interactions can mitigate conflicts of interest occurring across scale, and among institutions and sector-based policies, thus increasing the coherence mangrove conservation efforts.

This chapter demonstrates how the system of interregional governance implemented in the West African coastal region, harnesses networks, and applies complex resolution processes that implicate a variety of actors. Two networks are presented in this chapter as illustrations of governance through collaboration that have produced major successes in the West African marine and coastal region over the past several years. Thematic networks that target a specific functional unit, such as the marine protected area, are shown to achieve success in developing strong relationships among actors who share knowledge about how to best preserve and manage critical habitats and thus take coherent and effective action. But they have also encountered many of the challenges that typically arise when networked governance approaches are applied to complex policy problems. These networks—the Regional Coastal and Marine Conservation Partnership (PRCM), the West African network for marine protected areas (RAMPAO)—have revealed the utility and challenges of using networked governance models to address mangrove conservation.

Complexity and Governance

Species losses are often the 'result of interactions between a number of highly context-dependent causal factors' (Blaustein and Kiesecker 2002, p. 597). Therefore, instead of focusing on single factors that may endanger a species or critical habitat, Blaustein and Kiesecker (2002) assert the need to understand the complex interactions among multiple factors affecting ecosystems in order to fully understand the causes of biodiversity loss. Such an approach allows the examination of how human actions such as habitat destruction, overexploitation of natural resources, and the release of contaminants interact with environmental factors to exacerbate species or habitat losses.

Human–environment interactions and policy solutions occur within the intricate structures of ecosystems (Fig. 1). These systems are themselves naturally in a constant state of flux and change due to environmental forces such as climate, gravitational pull, and the amount of precipitation or carbon dioxide in the air (Blaustein and Kiesecker 2002; Lui et al. 2007; Dayton et al. 2005). This state of constant action and reaction creates an atmosphere of seeming chaos where the causes and effects of various changes are difficult to distinguish. However, the dynamic nature of ecosystems actually indicates the capacity of such systems to engage in continual adaptation. For example, the Science for the Protection of Indonesian Coastal Ecosystems (SPICE III) program discovered the presence of a new faunal species in the Segara Anakan Lagoon, representing a habitat adaptation in response to a high concentration of organic pollutants in the sediment (ZMT 2012). Social systems such as policy governance[1] systems are in a similar state of constant

transition, meaning that opportunities to influence change are always available to be exploited by actors within the system (Waldrop 1994; Huitema et al. 2009). The combination of human activities, such as the clearance of mangroves for aquaculture, and environmental pressures, such as sea-level rise, necessitate the development of a different policy framework for mangrove conservation.

This chapter employs the concept of networked governance as a useful framework for managing the often overlapping or conflicting conservation goals that simultaneously attempt to enhance human well-being and ecological resilience (Hirsch et al. 2010; McShane et al. 2011). We present two case studies of how networks can be used to mitigate the degradation of mangrove ecosystems and the negative effects emanating from conservation policy implementation. Enhanced collaboration among various government bodies, and between those bodies and non-governmental stakeholders, is a key ingredient to the success of networked governance. If we acknowledge that both natural and human agents contribute to environmental degradation through complicated interactions at multiple scales, then conservation efforts will be more coherent.

Networks offer a powerful tool for enhancing coordination and synergy, and for making sense of the trade-offs. They represent an important mechanism that works to capitalize on the effectiveness of local communities in addressing complex problems that have regional and even global consequences. Lessons learned on the ground from experience and experimentation can be effectively transmitted to the national and regional levels through networked governance to effect policy change. This model of governance permits each participating institution and entity to focus on its core mission while multiplying the impact of its individual actions. This multiplier effect is transmitted through the interconnections or relationships that connect the various parts of these human networks. These connections are sustained through information flows, or, in some cases, through contractual arrangements (Meadows 2008). The network brings together the combined might of individual organizations whose missions complement each other and address different aspects and scales of complex policy and conservation problems.

While Huitema et al. (2009) conclude that there is little empirical evidence proving that polycentric governance systems (systems characterized by multiple centers of authority or control) are more flexible and resilient than centralized hierarchical systems, many case studies (including this one) have demonstrated the advantages of systems that distribute ecosystem management responsibilities across scales and actors (Imperial 2005; PRCM 2012). Indeed, effectiveness increases as opportunities to collaborate are multiplied, allowing the development of trust.

---

[1] Following Huitema et al. (2009), in this study, governance is defined to include the gamut of formal and informal structures and relationships that are implicated in governing.

Fig. 1 Human–coastal and marine ecosystem interactions

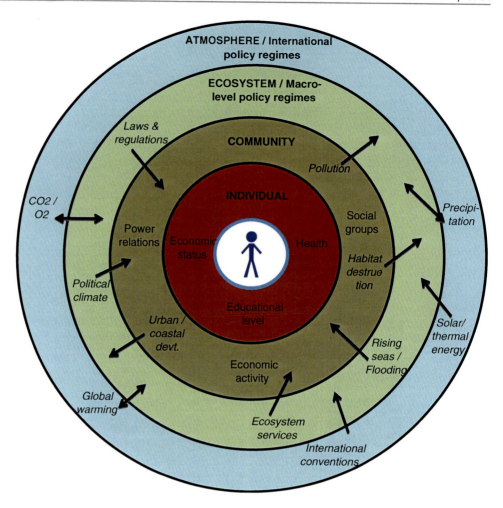

The Importance of Networks in Governance for Conservation

Recent years have borne witness to the proliferation of various types of networked governance models and their application to a multitude of policy areas and economic activities (Borgatti and Foster 2003; Peter 2007). Traditional forms of governance organized around bureaucratic command-and-control hierarchies have ceded more ground to new forms of network governance (Lukas 2013). This is particularly true in policy arenas distinguished by a significant amount of fragmentation, where political authority over the policy problem is shared across many leaders.

These newer forms of governance gather both public and private stakeholders who generally share similar visions and values and agree to work in more or less formal arrangements to achieve their shared goals. At their most basic level, networks represent and are shaped by social ties and interactions, and at a broader level encompass group processes and systems involving a range of actors (Peter 2007). 'Such autonomous, self-organized systems, conceptualized as "polycentric governance", have been shown to enhance innovation, learning, adaptation, cooperation and trustworthiness, and can help achieve more effective, equitable and sustainable outcomes at multiple levels' (Jones 2011, p. 21).

Whether initiated from the top-down or emerging from the ground up, networks exhibit a variety of structures that span formal arrangements to the more informal where individual members possess a higher degree of information and autonomy. Regardless of the diversity of their arrangements, the horizontal nature of networks differs significantly from traditional governance hierarchies, which are characterized as vertical structures that hinder timely communication and decision making. The 'new multi-organizational forms arising from collaborative endeavour' create policy opportunities (Skelcher 2005, p. 5).

Decentralizing the formulation and implementation of policies is one way of empowering lower levels and smaller scales. And because '…there is too often a mismatch between the scale of what is known about the world and the level at which decisions are made and actions taken' (ibid. p. 22), the implementation of sustainable solutions must be coordinated across scales. Networked governance allows this to happen by taking advantage of the fact that '…the decentralized, fluid form of a network and the autonomy of each member allows for decision making at the most appropriate level for the citizen' (Goldsmith and Eggers 2004, p. 38).

The move toward networked governance acknowledges the need to promote learning and cooperation at multiple and linked scales. Such loosely structured arrangements depend on collaboration and connect smaller governance systems and capable actors operating at the local level with actors who are capable of addressing macro-level, regional issues.

If the network functions well, individuals and local institutions are empowered to effectively join their efforts with those of national networks that interact with regional ones. Furthermore, knowledge sharing must be promoted between entities in flexible, non-hierarchical ways for the capacity of organizations to be enhanced. Dedeeurwaedere (2005) states that the function of networked governance 'is to create a synergy between different competences and sources of knowledge in order to deal with complex and interlinked problems' (p. 2). In all effective networks, the empowerment of local actors to effectively contribute to shared agendas is a major strength of governance. Thus, a commitment to capacity building and mutual learning is a necessary component of successful networked governance in the domain of conservation and natural resource management. Lastly, collaboration is ensured when all members buy-into and accept the polycentric institutional arrangements that are characteristic of networks, thus accepting shared power between many different decision-making units and scales. The networks that are highlighted in this chapter are generally organized to allow mutual learning and joint problem solving at local scales and experimentation with possible solutions at the regional level.

Case Study of Networked Governance in Practice in West African Mangrove Conservation Efforts

The networked governance approach is exemplified in the PRCM (in West Africa), which has for the last ten years acted as a network of networks that has empowered each participating organization to focus on its core conservation mission while multiplying the impact of individual actions (PRCM 2008, 2012a). Effective governance networks such as the PRCM bring together the combined might of organizations whose missions complement each other and simultaneously address different aspects of complex problems. This has allowed all stakeholders to work collectively toward the shared goal of the conservation and sustainable management of the West African region's natural resources and ecosystems.

This section demonstrates the pertinence of harnessing networked governance models, in particular the PRCM to confront the threats to West Africa's ecosystems. Figure 2 below summarizes how PRCM interventions in policy, capacity building, and investment in local areas are targeted at solving the complex multi-faceted problem of conservation in West Africa.

The PRCM: A Brief History

Before the PRCM was founded, coastal zone conservation efforts in West Africa were limited to a few marine protected areas (MPAs) and a small number of scattered projects (PRCM 2012a). In response to the need for a coordinated approach, a Regional Coastal Planning Network (RESOCOTAO) was set up in 1997. Designed as a network of expertise, the RESOCOTAO set out a number of guidelines, which foreshadowed the advent of the PRCM, most notably with respect to the need to address ecoregional issues.

It was with this in mind that a workshop on 'Priorities for coastal conservation in West Africa' was held in St. Louis, Senegal in 2000. The workshop's participants were struck by the strong similarities in the priorities stated by the representatives of the countries involved, with special emphasis being placed on the establishment of MPAs, sustainable management of fisheries resources, and mangrove biodiversity conservation. It was at this workshop that the principle of collaboration among the international organizations gained acceptance, a principle that was to become official soon thereafter when a Memorandum of Understanding and Partnership was signed by the International Union for Conservation of Nature (IUCN), the Worldwide Fund for Nature (WWF), the International Foundation for the Banc d'Arguin (FIBA), and Wetlands International.

The strategy emerging from the workshop embraced a shared vision and regional approach to conservation based on: an understanding of the central role played by local communities, a belief in the effectiveness of shared governance, an understanding that the cultural dimension is inextricably linked to the environment, the direct linkage to the issue of fisheries, and the need for strong institutions.

The project portfolio was subsequently presented at a regional workshop held in Dakar in 2003, which was the venue of the first meeting of the program's technical and financial partners and of the official launch of the PRCM. It was there that a Memorandum of Understanding was signed with the Permanent Secretariat of the Sub-regional Fisheries Commission (SRFC), whose geographical scope covers the same countries as the PRCM. This MOU affirmed the

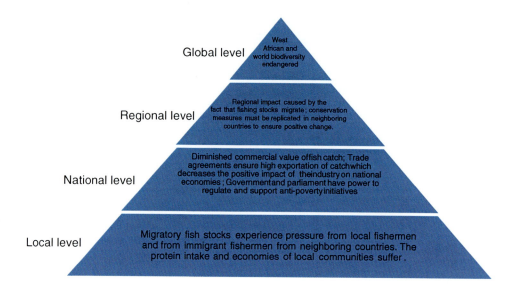

Fig. 2 PRCM impact on complex problems of conservation in West Africa

States' commitment to marine conservation, and for the PRCM, it represented a crucial linkage with government policies and their harmonization at the regional scale.

The coalition emerged as a result of the natural evolution of a process. Even its funding model evolved naturally within this historical process since the PRCM's two largest donors, the MAVA Foundation and the Embassy of the Netherlands in Dakar, had already been funding coastal zone conservation initiatives for many years and were involved in the planning process. And as a result of this evolution, the entire process was already deeply rooted and the prominent members of the coalition had long-standing relationships based on trust and friendship, thus conferring tremendous strength to the overall architecture of the initiative.

The initial grouping of some forty institutions from six countries in a regional program with one shared vision required considerable effort by all. Not only did new paradigms and changes of scale in approaching problems have to be adopted, but also new models for operating and managing relationships. This called for major effort and investment long before project-based work actually commenced.

The objective of the PRCM during its first two phases (2004–2012) was to promote, with other actors in the region, a shared vision of regional conservation priorities and to divide up the responsibilities for reaching this vision according to the specific competencies of each organization. The articulation and coordination of these activities were meant to create a coherent program of interventions that not only built synergy but also decisively influenced public policies relative to the development of the coastal zone and the exploitation of its resources.

The PRCM West African Mangrove Initiative

As a coalition, the PRCM is equipped with coordination and communication competences. It maintains a broad reach, which touches a large number of preservation and management problems occurring in the West African coastal and marine zone. Through this structure, the PRCM was able to combine programmatic activities with networking activities for maximum impact in its first two phases (PRCM 2012a). For example, projects falling under the conservation component helped to improve the management of MPAs. Project leaders then coordinated with efforts to protect specific species such as manatee or marine turtles (Duval-Diop 2012). Those initiatives were then supported by broader participative governance and management projects. All of these efforts then informed the advocacy and networking efforts undertaken in projects falling under the integrated governance component.

This combination of on-the-ground pilot experience and regional policy advocacy is evident in the West African Mangrove Initiative (WAMI) funded and coordinated by the PRCM from 2007 to 2010 (Duval-Diop 2012; PRCM 2012a, b). Using a participative approach grounded in local communities, the project helped to conserve and restore mangrove ecosystems in six countries (Fig. 3) and to improve the well-being of the local communities. With a budget of approximately 480,000 euros, the project was implemented in several phases including: the establishment of baseline data and reference studies; the transfer of knowledge on mangrove restoration and management to local communities; the implementation of pilot activities relating to restoration and alternative livelihoods best practices; and the identification of gaps and inconsistencies

Fig. 3 Mangrove coverage (*Source* RAMPAO 2014)

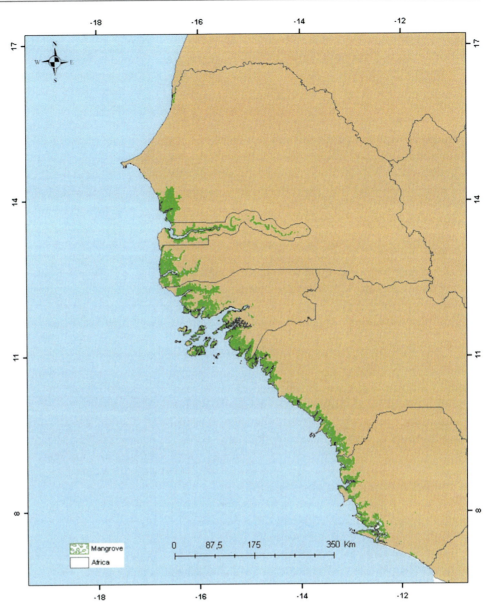

in national policies, laws and management mechanisms for the conservation of mangrove habitats. These actions subsequently led to the signing of a regional charter on the conservation and sustainable reuse of mangroves by six countries in the region (Mauritania, Senegal, Gambia, Guinea, Guinea Bissau, and Sierra Leone). The common vision expressed in this charter focuses on a respect of joint principles while implementing nationally defined action plans that reflect to local realities (Mangrove Charter 2009).

The process by which this consensus was reached emphasized a respect of each country's national priorities and context (Duval-Diop 2012). For example, in Mauritania, mangroves are completely protected and cannot be exploited by local populations. However, in other countries such as Sierra Leone or Guinea Bissau, the ecosystem services provided by mangroves represent a significant amount of income for populations. Attempting to forbid the exploitation of mangroves would have failed. Instead in these countries, the pursuit of sustainable management practices that follow the general guidelines described in the regional charter is more practical and feasible.

The high level of engagement and motivation of coastal communities who now understand the need to conserve their resources and who have the ability to do so will ensure the sustainability of the successes described above. Without building the awareness of local population, this engagement and the sustainable adoption of conservation practices would remain insurmountable.

Huitema et al. (2009) assert that the effectiveness of collaboration in networked governance can be influenced by how activities are ordered. Therefore, tasking various collaborating partners with the simultaneous implementation of actions was more effective than the sequential ordering of activities (Sproule-Jones 2002). An ambitious endeavor, implementation of this model yielded many positive results and changed the landscape of conservation and natural resource management in the region. The WAMI project, in particular, was relevant and strategic for the PRCM and contributed directly to better management and protection of the mangrove zone at the local and national levels (Borner and Guissé 2010). It also had an impact at the international level when a coalition of PRCM partners successfully lobbied parties to the Abidjan Convention in 2012 to adopt a motion to develop an additional protocol to the convention on the mangrove (PRCM 2012b).

Furthermore, the PRCM was able to effectively create other networks that impact specific thematic areas. The partnership then worked to build synergies between those networks and the actors within them to maintain its global, systematic approach.

From the relationships created over years of working shoulder to shoulder on the preservation of endangered sharks and rays, to the ties reinforced by mutual struggles to create MPAs, the PRCM has stimulated innovation and promoted learning, adaptation, and cooperation. All of these elements are essential to attaining more effective, equitable, and sustainable results across all scales and are appreciated by various actors throughout the region (Fig. 4).

The PRCM: A Force for Capacity Building

A critical role of governance networks is the promotion of learning and cooperation at multiple and linked scales. A collaborative system, that connects smaller governance systems and people operating at the local level with actors who are capable of addressing broader-scale, regional issues, depends on building the human capacity of local actors. They are most effective when generally organized to allow the knowledge acquired at local scales that emerges from experimentation with possible solutions to percolate to the national and regional levels. Decentralizing the formulation and implementation of policies is one way of empowering lower levels and smaller scales. This is particularly important when different aspects of a complex problem may be experienced at different scales, *and* where the potential for implementation of sustainable solutions should be coordinated across scales. But this cannot be accomplished without the meaningful engagement of local communities, which in turn depends on their ability to fully engage with the process.

Capacity building as a term is not easy to define precisely, because of the breadth of areas that it touches. However, at its most basic level, capacity building deals with people, organizations, communities, and the process of improving the effectiveness of what everyone does. Because people are one of the main culprits responsible for biodiversity and habitat loss and are more affected by the conservation of environmental resources, they must be the prime targets for ensuring its protection. Expanding and then channeling human capacity is therefore fundamental to preserving our environment and its diverse ecosystems. In the West African region and beyond, conservation that is effective in the long-term hinges on linking dedicated individuals and institutions that possess the ability and assets to confront the pressures facing our natural world (Duval-Diop and Meriaux 2012; PRCM 2012; FIBA 2012).

Strengthening capacity is also a way of levelling an often lopsided playing field and ensuring equity in the face of external actors who hold a wealth of resources and knowledge. The old adage holds true—'knowledge is power.' When local populations are empowered with knowledge that enables action, they can then take the lead in conservation efforts. While external interventions can be useful in the short term, particularly in helping to raise awareness of external pressures that are difficult to perceive at the local level, lasting conservation that is grounded in a new way of regarding the environment and that leads to changed behavior must come from local communities and institutions.

Therefore, a key function of governance networks is to connect different competencies and capacity gaps with sources of knowledge in order to truly enable the engagement of local stakeholders (Dedeeurwaedere 2005). In all effective networks, the empowerment of local actors to effectively contribute to shared agendas is a major strength of governance. Thus, a commitment to capacity building is also a necessary component of successful networked governance in the domain of conservation and natural resource management.

The PRCM invested a significant amount of resources (26 % of total resources in 2011 alone) to build the capacity of both local institutions and individuals to understand the nature of the problems affecting local ecosystems, to do data collection and monitoring, to contribute to the formulation of policy solutions as well as to implement local project solutions (Fig. 5). Experimentation on the ground through pilot projects allowed the collection of data and information on best practices, which informed the development of tools used in training and capacity building. Local actors were linked through networks with regional actors, thus enhancing their ability to advocate in the policy arena. This was often accomplished through facilitating dialogue and exchange among regional experts to share best practices. Through workshops and training, the PRCM also built the capacity of organizations by connecting regional technical expertise with people who needed that expertise on the ground.

This strategy truly levels the playing field between on-the-ground work and high-level policy making and balances

Fig. 4 Testimonies from PRCM actors

'Of all the environmental coalitions I have known, the PRCM is without a doubt one of the most innovative and the most effective and it can serve as a model for other marine and coastal zones in the world... The PRCM is mobilizing civil society to wield greater influence and has the ear of governments.'
Luc Hoffmann, Honorary President of the MAVA Foundation

'We are a global village... Whatever goes wrong in Senegal or Guinea-Bissau or Mauritania, it will surely affect Gambia. But if we are talking, we are networking, we are working together as unit, then we are sure that we are addressing the problems of common interest, as a country and as a subregion.'
Kebba N. Sonko, Permanent Secretary for the Ministry of Forestry and the Environment, Gambia

'The PRCM is probably one of the most powerful coordinating and partnership mechanisms to promote conservation anywhere in Africa, because it brings together governments, nongovernmental organizations, local institutions, international institutions all together in a partnership which doesn't have to have thick structures.'
Paul Siegel, Scientific Advisor, WWF Marine Program in Africa

the power between local and national/regional agents, something which local actors value. In the words of Augusta Henriques, Secretary General of the national NGO Tiniguena and winner of the Ramsar Award:

> The challenge is to invest in several stakeholders and institutions at every level. But above all, the importance of strengthening the grassroots level must not be overlooked, since it is the communities that anchor the process in the field. Indeed, the sum of the PRCM's field experience represents enough potential influence to propose solutions at the ecoregional level. In other words, solutions must spring from local and national situations and experiences pooled together in an ecoregional perspective.

The WAMI project invested in building the capacity of local communities in mangrove restoration and in alternative livelihoods to facilitate the conversion away from the economic activity of harvesting mangrove timber. A total of 30 individuals including 14 women received training in mangrove restoration, solar salt production and the use of improved fish smoking ovens (Borner and Guissé 2010). While the project planned to facilitate peer learning exchanges at the regional level, budget constraints and differences in language and climate allowed only exchanges between neighboring countries (Borner and Guissé 2010; WAMI 2010).

For example, reciprocal visits allowed the Gambian communities of Buram and Bali Mandinka to benefit from the expertise of the Senegalese community of Dasselamé Serer, where awareness for the conservation of mangroves is well established and where reforestation activities are part of a well-established annual program. Exchanges were also facilitated between Guinea and Senegal to demonstrate and share the experience of producing solar salt.

Challenges in the PRCM Networked Governance Model

However, networked governance is not without challenges. One of the greatest challenges to governing in networks includes the presence of differing and competing goals, since networks often gather stakeholders whose interests simultaneously overlap and clash. Because the Atlantic coast of Africa has a high concentration of population and industries, the need for mangrove habitat conservation is often in conflict with the need to encourage rapid growth. Competition is further exacerbated by the fact that short-term subsistence needs are best addressed by high-revenue generating activities such as mangrove cutting while conservation priorities and actions generate benefits in the long term (UNEP 2007). The fact that the PRCM partnership includes both non-governmental and governmental organizations means that even conservation goals do not always align. Unequal capacity and poor national and local

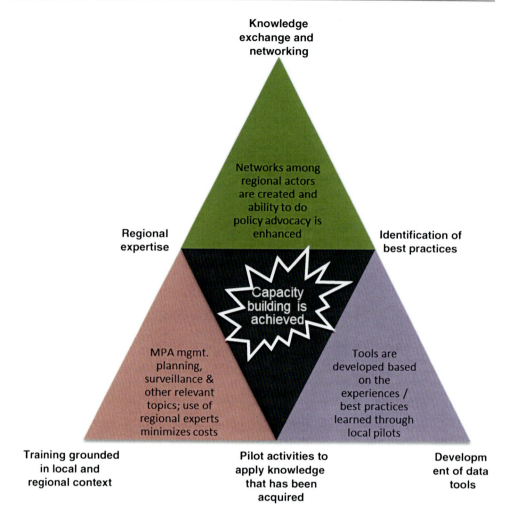

Fig. 5 PRCM regional capacity building model

governance mechanisms can cause local communities to cede to the agendas of well-funded international organizations or NGOs. Achieving a common vision is particularly difficult when organizations at all scales compete for funding and projects that reflect their priorities.

The tension between competition and collaboration can become high, when a group creates a network and then ends up competing with members of the same network. In this situation, distrust and a refusal to share information can result. Furthermore, in networked governance, the hierarchy of responsibility is replaced by a hierarchy of interactions, which sometimes allows certain institutions to avoid completing tasks and makes ensuring accountability difficult. The difficulty of coordinating activities can increase the transaction costs related to the extensive consultations needed to develop and enforce agreements. All of these problems are exacerbated when the network is particularly extensive and comprises a host of diverse actors and interests like the PRCM.

Additional obstacles to effective networked governance comprise communications failures, and data and capacity shortages. In networks, roles and responsibilities are often dispersed, which contributes to communication difficulties. Creating accessible meeting spaces, harnessing technology, and funding opportunities for exchanges are helpful but can sometimes be costly. In a region where there are multitudes of local languages, where the Internet connection can be unreliable in many areas, and where electricity outages are frequent, even the simplest communication efforts can be challenging. For example, language differences, cultural barriers, and sheer distance worked against greater communication among various regional actors in the WAMI project (Borner and Guissé 2010; WAMI 2010). Furthermore, diffuse and ill-defined authority makes democratic decision-making processes difficult particularly when power and capacity is unequally distributed spatially and organizationally (Skelker 2005).

Furthermore, ensuring the durability and financial sustainability of such partnerships is often a major obstacle to the long-lasting changes they hope to achieve. Like any program that depends on uncertain funding, the PRCM is not a permanent structure, particularly in times of economic crisis. However, the motivation for working together will only increase over time and will remain a necessity in the long run. It is therefore necessary to design structures for consultation and collaboration for which the operating and transaction costs are kept low, while producing a set of services compelling enough to sustain the interest of their users.

Harnessing Thematic Networks: The Case of the West African Network of Marine Protected Areas

The sustainable collaboration of actors and the coordination of efforts within the regional area are enriched by the support of thematic networks, working on a voluntary basis and at reduced costs. Thematic networks can ensure a greater convergence of interests and goals can facilitate communication and permit an enhanced focus on specific issues. With this in mind, the PRCM has been responsible for the creation of several formal and informal thematic networks such as the Alliance of Parliamentarians and Local Officials for the Environment (APPEL) and the Regional Network of Marine Protected Areas in West Africa (RAMPAO). "The existence of these networks meets the need to structure consultation and collaboration in a way to promote the consistency of interventions and to strengthen capacities to lobby" (PRCM 2012b, p. 11).

The thematic network described below combines elements of self-organization and active steering to arrive at major successes in the West African marine and coastal region over the past several years. Essentially, its ultimate goal is to shape broad policies governing the conservation and management of natural resources and ecosystem services in the coastal and marine areas of the PRCM ecoregion. The RAMPAO network of MPAs presents a learning opportunity for all those engaged in conservation of ecosystems such as mangroves in the region and in the world.

RAMPAO: An Effective Tool for Mangrove Conservation

MPAs are dedicated to protecting sensitive areas such as seagrass beds, mudflats, mangroves, and coral formations that play a specific role in natural resource regeneration and biodiversity conservation. MPAs in West Africa are typified by the presence of human communities who are their traditional inhabitants. Having lived in close contact with their environment for generations, these communities possess valuable knowledge, which is a tremendous asset for environmental management. Far from being sealed-off units, MPAs are in fact areas that produce resources and knowledge that, in turn, maintain the vitality of other areas far beyond their boundaries. And 'if designed correctly and managed well, MPAs have an important role to play in protection of ecosystems and, in some cases, enhancing or restoring the productive potential of coastal and marine fisheries' (IUCN-WCPA 2008, p. 3).

In order to ensure careful management, MPAs have devised novel solutions to the problems they face in the fields of development, natural resource management, research, and surveillance. They have also proven to be testing grounds for sustainable development practices, which engender lessons learned and best practices that can benefit other processes outside their borders.

The networking of individual MPAs connects these areas based on ecological and/or sociopolitical factors. As defined in UNEP-WCMC (2008), an MPA network is a 'collection of individual MPAs or reserves operating cooperatively and synergistically, at various spatial scales and with a range of protection levels that are designed to meet objectives that a single reserve cannot achieve (p. 20).' Indeed, the connectedness of mangrove habitats, the flow of oceanographic currents that carry developing fish larvae from one estuary to another, or the migration routes of various turtle species provide the rationale for regional action to preserve these habitats. The replication of the number of MPAs that protect a particular habitat such as mangroves is an important criterion in the design of resilient ecological MPA networks.

Given that MPAs reflect well-defined thematic, spatial, and institutional realities and that they are an ideal terrain of action for international conservation organizations, it was only logical that they were granted a central place within the PRCM. The first manifestation of this was the development of a regional strategy based on a vision which is still relevant 12 years later: 'A coherent network of marine protected areas, managed by strong institutions using the participatory model, which value natural and cultural diversity to contribute to the sustainable development of the region.' With clearly stated political support for the strategy, its implementation has resulted in the expansion of MPAs in the region.

Between 2004 and 2011, nine MPAs were established within the PRCM ecoregion. The 15 MPAs already in existence in four countries at the time the strategy was adopted swelled to approximately thirty MPAs in six countries today (seven counting Sierra Leone, where the process to establish the Yawri Bay MPA is well underway). Most of these protected areas are members of a network created in 2007 by the PRCM and officially recognized by the States, the Regional Network of Marine Protected Areas of West Africa—RAMPAO (Fig. 6).

RAMPAO's mission is to preserve and strengthen the marine ecoregion of West Africa by maintaining and effectively managing an ecologically coherent set of critical habitats. Networking ecosystems and critical habitats is an

Fig. 6 Map of the RAMPAO MPA members

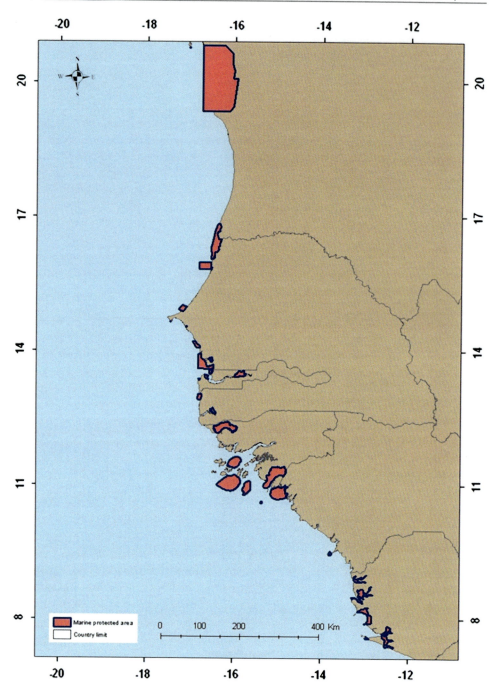

important strategy for tipping the balance in the degradation of our natural resources. An international review of regional and national MPA networks found that regional (multi-country) networks tend to progress best when operating under a coherent and robust coordinating framework and when national parties demonstrate commitment through treaties or other agreements (UNEP-WCMC 2008). In 2010, the RAMPAO gained the support of 15 national fisheries and environmental ministers, thus improving its legitimacy in negotiating and facilitating regional change. Because the RAMPAO encompasses both human and ecological concerns and touches the temporal and spatial scales of ecological systems, the network is better equipped to guarantee long-term sustainability than would a single MPA, which is particularly important for the contiguous mangrove habitat that spans the countries of Mauritania, Senegal, Gambia, Guinea, Guinea Bissau, and Sierra Leone.

According to the RAMPAO Secretariat, 24 MPAs in the network distributed in 5 countries (Mauritania, Senegal, Gambia, Guinea, and Guinea Bissau) identify mangroves as key habitats in their management plans. In a survey conducted by the Secretariat in 2009, 12 of these MPAs reported habitat degradation rates ranging between 10 and 30 % (RAMPAO 2010). Table 1 shows the proportion of

Table 1 Change in mangrove cover and protection levels in PRCM countries

Country	Number of mangrove species	% change in mangrove cover (1980–2006)	% mangroves located in protected areas
Mauritania	3	39, 3	62, 5
Senegal	7	−23, 8	42, 5
Gambia	7	−17, 5	3, 5
Guinea Bissau	6	8, 7	35, 5
Guinea Conakry	7	−31, 9	0, 26
Sierra Leone	6	−37, 3	14, 5

(*Sources* UNEP 2007; Tendeng et al. 2012)

mangroves under protection versus the rate of change in their coverage over approximately 25 years.

In the case of Mauritania and Guinea Bissau, the high amount of protection afforded these ecosystems corresponds to an increase in the size of this ecosystem prior to the establishment of the RAMPAO. In the case of Senegal, however, mangrove coverage decreased between 1980 and 2006 in spite of the high level of protection. To enhance the effectiveness of MPAs in general and to enhance their specific capacity to protect critical mangrove habitats, the PRCM funded 2 projects at over $2 million euros between 2008 and 2012. Key non-governmental PRCM partners, including the FIBA and the International Union, implemented these projects for IUCN that made concerted efforts to improve the management, participative governance mechanisms, and networking in several MPAs in the region.

End of project reports revealed an improvement in the management of MPAs and their natural resources through the implementation of activities such as the development and implementation of updated management plans, monitoring and evaluation of the effectiveness of management, ecological and species monitoring, capacity building in conservation and mangrove restoration, and the establishment of functional community governance structures and a pool of expertise (FIBA 2012a, b). The involvement of more stakeholders (communities, professionals especially fishermen, authorities, etc.) was a key factor in the success of these projects. The evaluation of the project to support the RAMPAO network showed that significant progress was made in the management in 6 MPAs in 4 countries (Bamboung, Niumi, Urok, Orango, Joao Vieira et Tristao) between 2008 and 2012 (FIBA 2012c).

While these effects cannot be entirely attributed to RAMPAO network, the evaluation found that the impact of the network in ensuring a sustainable management of critical and endangered habitats in these areas to be positive. Furthermore, surveys conducted during the evaluation revealed that 57 % of network members believed that the RAMPAO had a positive impact on building the management capacity of MPA staff, which directly impacts the ability to implement effective mangrove conservation measures.

Project leads also found that the inclusion of MPAs in a formal network was extremely beneficial for the MPA managers, especially at the beginning of the MPA management planning process. The RAMPAO facilitated technical exchanges between site managers and specialists, and networking at the human and ecological levels using sound science to develop coherent and effective action. At the same time, the network has nurtured the development of strong relationships between the human actors. It has organized events and activities that promote the exchange of information, leading to mutual learning that helps create synergies between MPAs. The network will continue to enhance the capacity of MPA actors through mutual learning and exchange to further the progress made by the two PRCM projects. It will also support additional efforts to ensure that all mangrove habitats benefit from consistent and coherent protection, effective management, and quality monitoring regardless of their home countries.

However, while the RAMPAO has proven that networked governance is an effective strategy for improving coastal resources management, it confronts many obstacles. Financial sustainability to support the costs of convening members and other network activities is a major concern. Attracting greater financial resources to support the continued improvement of individual member MPAs continues to be difficult. Furthermore, although the network strives to ensure ecological coherence, gaps exist in the representation of key habitats such as corals and seamounts, and the connectivity between MPAs is little understood and requires more scientific study (Tendeng et al. 2012). Political instability and shifting stakeholders makes continued training necessary in spite of the lack of resources. Yet, despite of these challenges, the commitment of myriad stakeholders at all scales ensures the network will continue to grow and increase its effectiveness.

Conclusion

The complexity of interconnected human and natural systems tests traditional natural resource management assumptions and practices. Moreover, the achievement of management objectives depends on the extent to which these policies and practices account for the complexities inherent in these systems. This is especially true given the fact that management of mangrove ecosystems is complicated by the prevalence of indirect interactions between people and nature. All types of actors—government, civil society, non-governmental, youth and others—have important roles to play in the design and implementation of management and conservation solutions. The case studies presented in this chapter have revealed the successful application of the networked governance model to the conservation and natural resource management of the West African ecoregion. Acknowledging the scale of interactions and the tight connection between local economic decisions and the global decisions and actions that influence them has facilitated simultaneous action at multiple scales regarding mangrove conservation. The PRCM's multi-faceted approach combines the practice of funding projects on the ground with national policy-making and advocacy and regional collaboration. Incorporating various styles of networks into one structure and creating diverse contractual arrangements that join various levels of government agencies and other civic institutions have resulted in the dismantling of barriers such as country borders or hierarchical government lines and have effectively engaged a myriad number of stakeholders.

Lessons learned reveal the need for continued engagement and investment in this model. For example, a focus on both the ecological and human networks that exist between MPAs, as the RAMPAO does, can ensure a sustainable management of critical and endangered habitats such as the mangrove. Thematic networks that target a specific functional unit, such as the marine protected area, have greater success in developing strong relationships among actors who share knowledge about how to best preserve and manage similar habitats and thus take coherent and effective action. However, in other networks, opportunities to extend program impact and create synergy can be squandered. For example, the current structure of the PRCM that is based primarily on voluntary engagement has made ensuring accountability difficult and sometimes limited impact. Additionally, the transaction costs of communication and coordination are often high and have caused attempts to ensure accountability to suffer. The hierarchy of interactions that exists in networked governance has often made assigning tasks and responsibilities extremely challenging in many instances, particularly when the benefits accruing to individual actors are unclear. Furthermore, investing in science and knowledge creation in order to better understand the impact of human actions on natural cycles has sometimes been bypassed in favor of implementing immediate solutions.

Goal incongruence is especially problematic when the initiator of the network ends up competing against parts of the network for scarce resources. Thus, in the case of the PRCM, rules which cover matters such as the duty to share information proved difficult to implement. Moreover, the commitment to consensus and collaboration can mask the fact that 'stronger partners may be able to take advantage of weaker partners' (Agranoff 2003). In the PRCM network, conflicts of interest often arose when one of the large international NGOs dominated decision-making processes as opposed to building the capacity of local actors. The desire to participate and maintain a presence in dialogues and the many meetings that took place warred against the willingness of local stakeholders to participate when NGOs failed to prioritize the deep engagement of local actors. Networks that attempt to influence governance typically involve coordination between multiple layers of government, civil society, community-based organizations, non-profit organizations, and others. The differing constituencies that are served complicate such arrangements. Further, uneven power balance may exist. Because certain MPAs have successfully established bilateral arrangements with external funders as is the case with the Banc d'Arguin National Park, they have stronger management structures than newly established MPAs such as Tristao in Guinea Conakry. This persistent challenge of differing motivations and interests necessitates a constant dialogue and consensus building.

As many researchers have noted (Quill 2012), networks acting within and in connection to other networks have the potential to spread risks or to boost resilience and diminish vulnerability. In spite of the challenges presented in this chapter, networks such as the PRCM and the RAMPAO can effectively confront the challenges facing mangrove habitats in West Africa. If they continue to engage actors at all levels and foster on-going collaboration that is grounded in knowledge, achieving healthy coastal mangrove environments and resilient communities that protect these ecosystems remains an attainable goal.

References

Agranoff R (2003) Leveraging networks: a guide for public managers working across organizations. IBM Endowment for the Business of Government

Alcorn JB, Zarzycki A, de la Cruz LM (2010) Poverty, governance and conservation in the Gran Chaco of South America. Rights and Resources http://www.rightsandresources.org/documents/files/doc_1688.pdf. Accessed 27 November 2012

Blaustein A, Kiesecker J (2002) Complexity in conservation: lessons from the global decline of amphibian populations. Ecol Lett 5:597–608

Borgatti SP, Foster PC (2003) The network paradigm in organizational research: a review and typology. J Manage 29(6):991–1013

Borner M, Guissé A (2010) WAMI final evaluation report. Wetlands International Africa and IUCN, Dakar

Bouma J, Huitema D (2010) Socio-economic vulnerability: conservation-development trade-offs and agency in multi-level governance processes. Deliverable report supported by funding from the European Community's Seventh Framework Programme [FP7/2007-2013] under grant agreement No. 211392, Amsterdam. www.livediverse.eu. Accessed 11 Oct 2012

Burlat P, Bescombes B, Deslandres V (2003) Constructing a typology for networks of firms. Prod Plann Control 14(5):399–409

Dayton P, Curran S, Kitchingman A, Wilson M, Catenazzi A, Restrepo J, Birkeland C, Blaber S, Saifullah S, Branch G, Boersma D, Nixon S, Dugan P, Davidson N, Vörösmarty C (2005) Coastal Systems. In: Hassan R, Scholes R, Ash N (eds) Ecosystems and human well-being: current state and trends: findings of the condition and trends working group. Island Press, Washington, DC, pp 513–549

Dedeurwaerdere T (2005) The contribution of network governance to sustainable development. Working paper: Les séminaires de l'Iddri, No. 13, Paris

Duval-Diop D (2012) Case studies on best practices for a cherished and protected biodiversity. PRCM, Nouakchott

Duval-Diop D, Meriaux S (2012) Capacity—The cornerstone of effective conservation: capacity building toolbox for conservation in West Africa. PRCM, Nouakchott

FIBA (2012a) Rapport Technique de fin de projet: Appui au renforcement de l'efficacité de gestion des Aires Marines Protégées (AMP). Fondation International du Banc d'Arguin, Dakar

FIBA (2012b) Rapport technique de fin projet: Appui au renforcement du Réseau d'Aires Marines Protégées d'Afrique de l'Ouest (RAMPAO) et à la mise en œuvre de son plan de travail. Fondation International du Banc d'Arguin, Dakar

FIBA (2012c) Evaluation finale du projet PRCM—FIBA—Appui au renforcement du RAMPAO et à la mise en œuvre de son plan de travail. PRCM/Oreade-Breche, Dakar

Goldsmith S, Eggers WG (2004) Governing by network. Brookings Institution Press, Washington, DC

Hirsch PD, Adam WM, Brosius P, Zia A, Bariola N, Dammert JL (2010) Acknowledging conservation trade-offs and embracing complexity. Conserv Biol 25(2):259–264

Huitema D, Meijerink S (eds) (2009) Water policy entrepreneurs. A research companion to water transitions around the globe. Edward Elgar, Cheltenham

Huitema D, Mostert E, Egas W, Moellenkamp S, Pahl-Wostl C, Yalcin R (2009) Adaptive water governance: assessing the institutional prescriptions of adaptive (co-) management from a governance perspective and defining a research agenda. Ecol Soc 14(1):26–45

Imperial MT (2005) Analyzing institutional arrangements for ecosystem-based management: lessons from the Rhode Island Salt Ponds SAM Plan. Coastal Manage 27:31–56

IUCN-WCPA (2008) Establishing marine protected area networks—making it happen. IUCN World Commission on Protected Areas, National Oceanic and Atmospheric Administration, and the Nature Conservancy, Washington, DC

Jones H (2011) Taking responsibility for complexity. ODI working paper 330. http://www.odi.org.uk/sites/odi.org.uk/files/odi-assets/publications-opinion-files/6485.pdf. Accessed 22 Nov 2012

Laegdsgaard P, Johnson C (1995) Mangrove habitats as nurseries: unique assemblages of juvenile fish in subtropical mangroves in eastern Australia. Mar Ecol Prog Ser 126:67–81

Liu J, Dietz T, Carpenter SR, Folke C, Alberti M, Redman CL, Schneider SH, Ostrom E, Pell AN, Lubchenco J, Taylor WW, Ouyang Z, Deadman P, Kratz T, Provencher W (2007) Coupled human and natural systems. AMBIO 36(8):639–649

Lukas, MC (2013) Political transformation and watershed governance in Java: actors and interests. In: Governing the provision of ecosystem services studies in ecological economics, vol 4, pp 111–132

Mangrove Charter (2009) Charter and action plan for sustainable mangrove management in the PRCM Region: Mauritania, Senegal, The Gambia, Guinea Conakry, Guinea Bissau and Sierra Leone. Produced by the West African Mangrove initiative funded by the PRCM

McGinnis M (2005) Costs and challenges of Polycentric Governance. Paper presented at Workshop on Analyzing Problems of Polycentric Governance in the Growing EU, Berlin

McShane T, Hirsch PD, Trung TC, Songorwa AN, Kinzig A, Monteferri B, Mutekanga D, Thang HV, Dammert JL, Pulgar-Vidal M, Welch-Devine M, Brosius JP, Coppolillo P, O'Connor S (2011) Hard choices: making trade-offs between biodiversity conservation and human well-being. Bio Cons 144:966–972

Meadows DH (2008) Thinking in systems. a primer. Chelsea Green Publishing, Vermont

Ong JE, Gong WK (2013) Structure, function and management of mangrove ecosystems. ISME Mangrove educational book series No. 2, international society for mangrove ecosystems (ISME), Okinawa, Japan and International Tropical Timber Organization (ITTO), Yokohama, Japan

Peter R (2007) Networked governance or just networks? Local governance of the knowledge economy in Limerick (Ireland) and Karlskrona (Sweden). Polit Stud 55:113–132

Peterson J (2003) Policy networks. Institute for Advanced Studies, Vienna

Pierre J, Peters BG (2000) Governance politics and the state. Macmillan, Basingstoke

PRCM (2007) Evaluation cartographique sur l'étendue, les valeurs écologiques, économiques et socioculturelles des mangroves des pays du PRCM: Mauritanie—Sénégal—Gambie—Guinée Bissau—Guinée—Sierra Léone—Rapport de synthese. Produced by the West African Mangrove initiative funded by the PRCM

PRCM (2008) Assessment of the activities of phase 1 (2004–2007). PRCM, Nouakchott

PRCM (2010) Annual Report of Activities 2010. PRCM, Nouakchott

PRCM (2012a) Pulling together in the same direction: a coalition to address coastal zone challenges in West Africa: lessons learned through the PRCM (2003–2012). PRCM, Nouakchott

PRCM (2012b) Renewed coalition to overcome challenges along West African Coastline: annual report. PRCM, Nouakchott

Quill E (2012) When networks network. Science news September 22, 2012

RAMPAO (2010) Access database on MPAs. West African network of Marine protected areas, Dakar

Rönnbäck P (1999) The ecological basis for economic value of seafood production supported by mangrove ecosystems. Ecol Econ 29:235–252

Skelcher C (2005) Jurisdictional integrity, polycentrism, and the design of democratic governance. Governance 18(1):89–110

Spalding M, Kainuma M, Collins L (2010) World Atlas of Mangroves. A collaborative project of ITTO, ISME, FAO, UNEP-WCMC, UNESCO-MAB, UNU-INWEH and TNC. Earthscan, London

Sproule-Jones M (2002) Institutional experiments in the restoration of the North American Great Lakes environment. Can J Polit Sci 35(4):835–857

Stoker G (1998) Governance as theory. Int Soc Sci J 155:17–28

Tendeng PS, Ba T, Karibuhoye C (2012) Ecological gap analysis of the Regional Network of Marine Protected Area in West Africa (RAMPAO)—Final Report. RAMPAO, FIBA, and PRCM, Dakar

TNC, WWF, CI, WCS (2008) Marine protected area networks in the Coral Triangle: development and lessons. TNC, WWF, CI, WCS and the United States Agency for International Development, Cebu City, 106 p

Tomlinson PB (1986) The botany of mangroves. Cambridge University Press, Cambridge (Reprinted in 1996)

UNEP (2007) Mangroves of Western and Central Africa. UNEP-Regional Seas Program/UNEP-WCMC

UNEP-WCMC (2008) National and regional networks of Marine protected areas: a review of progress. UNEP-WCMC, Cambridge

USAID (2013) Networked Marine Protected Areas (MPAs) key to conservation, productivity. http://philippines.usaid.gov/programs/energy-environment/success-stories/networked-marine-protected-areas-mpas-key-conservation-productivity. Accessed on 14 Aug 2013

Van Lavieren H, Spalding M, Alongi D, Kainuma M, Clüsener-Godt M, Adeel Z (2012) Securing the future of mangroves—A policy brief. UNU-INWEH, UNESCO-MAB with ISME, ITTO, FAO, UNEP-WCMC and TNC 53 pp

Waldrop M (1994) Complexity: the emerging science at the edge of order and chaos. Penguin Books, London

West African Mangrove Initiative (2010) Project Activity Report. Compiled by Wetlands International and IUCN, Dakar

Wolf B (2012) Ecosystem of the mangroves. NRES 323—International Resource Management. University of Wisconsin-Stevens Point

World Rainforest Movement (2008) Current status and conservation of mangroves in Africa: an overview. WRM Bulletin 133

ZMT (2012) ZMT Report 2011/2012. Leibniz Center for Tropical Marine Ecology, Bremen, Germany

The Importance of Scientific Knowledge as Support to Protection, Conservation and Management of West and Central African Estuaries

S. Diop, J.-P. Barusseau, and C. Descamps

Abstract

The concluding chapter of this volume of the series "Estuaries of the World", focused on Western and Central African coastal zone, highlights the complexity of such coastal environments and ecosystems. This chapter insists as well on the need for an increased knowledge of the structure and functions of these coastal ecosystems, while highlighting the life-supporting ecosystem goods and services they provide to humanity, including both the scientific and management implications. The challenge ahead of us will be to enhance the networking possibilities and to increase the capacities of young scientists working on African coastal zones.

Keywords

Networking possibilities • Capacities development of young African scientists • Ecosystem goods and services • Life-supporting ecosystem goods and services

As highlighted in this volume of the series "Estuaries of the World" (EOTW) focused on Western and Central African coastal zone, estuarine and other coastal ecosystems represent the basic linkages between the land and the sea. They are complex, biologically productive and important, both for human existence and environmental sustainability, whether in Africa or other regions of the world. Completion of this book devoted to the integrated management and

sustainable use of these important ecosystems, therefore, achieves several major goals. One important goal is increased knowledge of their structure and functions of these linking ecosystems, as well as highlighting the life-supporting ecosystem goods and services they provide to humanity, including both the scientific and management implications. Another goal is enhancing the networking possibilities and increasing the capacities of young scientists working on African coastal zones. Considered in this context, this publication represents a model of collaboration for the protection, conservation and management of the estuaries and other coastal systems of West and Central Africa and their resources in a sustainable manner. Because such information and data are fundamental to the goal of monitoring and analysing the status of estuarine and other coastal ecosystems, pursuing such development and exchange of data and information, while also maintaining effective partnerships and networking for easier access to estuarine-related information and data for West and Central Africa, is essential. The challenge ahead of us is to constitute viable centres that include systems and approaches

S. Diop (✉)
Universite Cheikh Anta Diop de Dakar, Dakar-Fann,
BP 5346, Dakar, Senegal
e-mail: sal-fatd@orange.sn; esalifdiop@gmail.com

J.-P. Barusseau
CEFREM, Universite de Perpignan, Via Domitia—52,
avenue Paul-Alduy, 66860, Perpignan, France
e-mail: brs@univ-perp.fr

C. Descamps
Institut Fondamental d'Afrique Noire—Cheikh Anta Diop,
BP 206, Dakar, Senegal
e-mail: cyrdescamps@yahoo.fr

S. Diop et al. (eds.), *The Land/Ocean Interactions in the Coastal Zone of West and Central Africa*, Estuaries of the World, DOI: 10.1007/978-3-319-06388-1_17,
© Springer International Publishing Switzerland 2014

for quality information exchange effective for informed decision-making, including their expansion to other regions of the African continent.

Based on the results of the investigations contained in this book, a number of conclusions have been reached, as follows:

- New perspectives exist to analyse coastal regions in West and Central Africa, especially from the viewpoint of combining functional ecology studies with a pressure/risk assessment approach, including relevant socio-economic aspects;
- Some original results in this book highlight the close links between the management and governance of such sensitive, vulnerable estuarine systems and coastal environments;
- New elements of research methodologies and knowledge of mangrove forests and ecosystems in Western and Central Africa have been recognised;
- The information and data in this book provide an opportunity for young scientists to develop new scientific approaches and research capabilities for analysing these complex coastal environments, while also enhancing their skills and building their capacities;
- On-the-ground networks have been established between West and Central African research communities through extensive partnerships along the western and central coasts of Africa.

This publication offers a good opportunity and platform for African scientists to publish their study results and research findings on coastal and marine ecosystems, with properly peer-reviewed materials that ensure the credibility of this series.

One can ask what the perspectives in the near future are regarding studies and research on coastal and marine ecosystems in Western and Central Africa. There are several answers to this question:

First, to include and reinforce a more systematically integrated ecosystem approach in future studies directed to estuaries and coastal areas in Africa;

Second, to use modelling tools, mapping and monitoring for more in-depth study of estuarine environments and their sensitivity to bio-geo-chemical and human-induced impacts and changes; and

Third, to better identify and facilitate options for solutions, based on an integrated assessment approach that includes socio-economic tools, while also highlighting the importance of key economic sectors, including fisheries and critical habitats within estuaries, damming, urbanisation and tourism, aquaculture, provision for mangrove ecosystems goods and services and other important sources of income for the livelihood of local populations, etc.

With this background, it is important to use the ecosystem approach as a framework for such research and studies in order to deal with the complex and dynamic nature of ecosystems in a comprehensive and holistic manner. This will help tackling the degradation of ecosystems services or mainstreaming development and planning processes into environmental management of coastal systems in Africa.

One particular area requiring greater attention is the assessment of the vulnerability of estuarine ecosystems (particularly mangrove communities and species) to climate change. Such study can assist in determining the level of exposure and sensitivity of populations and resources, as well as their adaptive capacities, to climate change risks and impacts in Western and Central Africa coastal areas.

Other areas meriting further investigations include:

- The estuarine and other coastal ecosystems with a high economic value which fulfil functions and services important both for biodiversity and human well-being;
- The selection of most sensitive estuarine and other coastal ecosystems sites to be monitored and protected, in order to better assess the state of degradation and the biodiversity and ecosystem services upon which coastal populations depend;
- The understanding and evolution of estuaries and other coastal ecosystems, and their trends and anticipated changes, including the services they provide to humans and ecosystem maintenance;
- The ecological services and the values of estuarine and other coastal ecosystems and their biodiversity, for supporting and/or maintaining biodiversity offsets;
- The management of estuaries and coastal ecosystems (including use of spatial planning and assessment tools) in order to contribute to establishment of sustainable networks of coastal and marine conservation and protected areas; and
- The status of national legislation for managing estuaries and coastal ecosystems for sustainable use within the networks of coastal and marine protected areas.

In conclusion, these areas are among the many examples for developing future scientific research priorities for these important and vulnerable aquatic ecosystems. Indeed, estuaries and coastal ecosystems are very important for people living near the littoral areas of West and Central Africa. They provide food, shelter, wood and other substantial goods and services, as well as contributing to coastal and marine biodiversity. They also are critical from a global perspective of the earth's natural environment, in that they were formed through strong feedback relations between biota, human well-being, landforms, water flows and the atmosphere. To this end, the authors hope this book, which is the result of extensive cooperation among many partners, authors and editors, constitutes an important source of documentation and information for researchers, practitioners, scientists, students and other specialists dealing with the scientific, socio-economic and governance issues associated with sustainable use of coastal and marine environments in the West and Central Africa region.

Index

A
Anthropogenic impacts, 6, 24

B
Barrier bar complex, 65
Benin, 140

C
Cameroon estuary, 98, 101, 125, 180
Carbon budget, 139, 146, 147
Central Africa
  coastal environments, 207
Cetaceans, 97, 101, 103
Climate changes
  extreme events, 19
  models, 18
  planning, 17
CO_2 emission, 140, 178
Coast
  conservation, 15
  erosion, 93, 98
  general, 2, 3
  human use, 10, 11
  profiles, 14
  fragility, 9, 15
  physical properties, 4, 28, 89
Congo, 186

D
Deforestation, 162
Development of coastal built-up and urban areas, 10
DRC, 117

E
Ecotourism, 98, 119
  impact, 120
  management, 121
Ecosystem goods and services, 207
Estuaries
  circulation, 79, 80
  classification, 110
  dam, 24, 26

environmental changes, 24
estuarine sentries, 21
faunal diversity, 97
general, 14, 21, 100
hydrodynamics, 23
management, 20, 21
plume, 80, 94
sediment, 90
tidal regime, 89
vulnerability, 24

F
Fauna
  diversity, 99
  vulnerability, 111
  regulation and management, 105
Forest, 15, 139, 148, 151, 184

G
Ghana, 123, 151
Governance, 191

H
Human-coastal interactions, 192

I
ICZM, ix

L
Lagos Lagoon, 65
Lagoon pollution
  bioaccumulation, 72, 75
  heavy metals, 69
  impact, 73
  organochlorine pesticides, 72
  polycyclic aromatic hydrocarbons, 72
Land cover changes, 59
Langue de Barbarie, 23
Lido breaching, 23, 27, 38
  opening, 28
  impacts, 32, 45

S. Diop et al. (eds.), *The Land/Ocean Interactions in the Coastal Zone of West and Central Africa*, Estuaries of the World, DOI: 10.1007/978-3-319-06388-1,
© Springer International Publishing Switzerland 2014

L (*cont.*)
Litterfall, 125
    composition, 133
Littoral drift, 14, 78

M
Management
    ecotourism, 114
    estuary, 23
    fauna, 104
    mangrove, 159
    networks, 191, 204
    plantation, 167
    pollution, 77, 86
    risk, 45
    watershed, 41
Mangrove
    biodiversity, 15
    biophysical environment, 16, 140, 152
    Central Africa, 178
    cross-scale dynamics, 191
    degradation, 59, 141
    distribution of species, 128, 139, 146, 181, 183
    litterfall, 123
    management, 163
    sample plot, 125, 142
    socioeconomic environment, 142, 153, 178, 181
    structure, 126, 128
    tourism, 158
Marine mammals, 98

N
Natural areas, 15
Navigation hazards, 93
Niger Delta, 77, 110
    Bonnie and Cross River estuaries, 87, 88, 93
    Calabar estuary, 117
    environments, 77, 115
    Escravos and Forcados estuaries, 119
    Imo River estuary, 118
    Qua-Iboe River estuary, 118
    morphology analysis, 109
Nutrients, 68

O
Ocean currents, 79
Oil spill, 84

P
Payment for ecosystem services, 151
Plume, 80, 94
Pollution management, 77, 84
PRCM, 192
Protection strategies, 140, 161, 163, 192

R
RAMPAO, 201
Remote sensing technique, 50, 171
    normalized difference vegetation index, 56
Risk
    linked to management, 46
    pollution, 71, 104, 156
    sea-level rise, 46
River discharge, 50

S
Salinization, 49
    saline intrusion, 26
    salinization of land, 34, 49, 60
Saloum inverse estuary, 50
Sand bars, 90, 95
Sand spit, 24, 61
    quaternary evolution, 35
Sea turtles, 98, 100
Senegal river, 24, 41
    morphology, 36
    suspended sediment, 27
Shelf, 14, 78, 84, 100
    sediment, 13
    currents, 79
Sirenians, 97, 101

U
UNEP, vii

W
Water birds, 98, 101, 102
Watershed management, 41
Wouri estuary, 167

Z
Zonal wind field, 27